HEALTH ASSESSMENT OF ENGINEERED STRUCTURES

Bridges, Buildings and Other Infrastructures

HEALTH ASSESSMENT OF ENGINEERED STRUCTURES

Bridges, Buildings and Other Infrastructures

Editor

Achintya Haldar

University of Arizona, USA

NEW JERSEY • LONDON • SINGAPORE • BEIJING • SHANGHAI • HONG KONG • TAIPEI • CHENNAI

Published by

World Scientific Publishing Co. Pte. Ltd.

5 Toh Tuck Link, Singapore 596224

USA office: 27 Warren Street, Suite 401-402, Hackensack, NJ 07601

UK office: 57 Shelton Street, Covent Garden, London WC2H 9HE

Library of Congress Cataloging-in-Publication Data

Health assessment of engineered structures : bridges, buildings, and other infrastructures / edited by Achintya Haldar (University of Arizona, USA).

pages cm

Includes bibliographical references and index.

ISBN 978-9814439015 (alk. paper) -- ISBN 978-9814439022 (ebook) -- ISBN 978-9814439039 (mobile)

1. Structural health monitoring. 2. Structural analysis (Engineering) 3. Engineering inspection. I. Haldar, Achintya, editor of compilation.

TA656.6.H43 2013

624.1'71--dc23

2012046514

British Library Cataloguing-in-Publication Data

A catalogue record for this book is available from the British Library.

In-house Editor: Amanda Yun

Printed in Singapore.

Preface

Health assessment and monitoring of engineered structures has become one of the most active research areas and has attracted multi-disciplinary interest. Since the available financial recourses are very limited, extending the life of existing bridges, buildings and other infrastructures has become a major challenge to the engineering profession worldwide. Structural health just after a natural event (strong earthquakes, high winds, etc.) or a man-made event (explosions, blasts, etc.) have also become a part of the overall health assessment protocol. In the context of nondestructive evaluation (NDE), the main thrust has been to locate defects in structures at the local element level and then decide what to do. Several advanced theoretical concepts required to detect defects have been proposed. At the same time, improved and smart sensing technologies, high-resolution data acquisition systems, digital communications, and procedures related to noise contamination in measured information and high-performance computational technologies have been developed for implementing these concepts. The general area is now commonly known as structural heath assessment (SHA) or structural health monitoring (SHM). Some of the related areas are in the development phase. As the area matures, more new areas are being identified to implement the concept.

In spite of these developments in analytical and sensor technologies, the implementation of these concepts in assessing structural health has been limited for several reasons. An attempt has been made here to identify some of the major works, their merits and demerits, and future challenges. Without this book, the information may not be available in an organized way for interested parties who are not expert in the area. The thoughts and recent works of some scholars who are providing leadership in developing related areas need to be readily available to benefit students (undergraduate and graduate), researchers (university and industrial), and practitioners (government and private). This edited book provides a sampling of some exciting developments. It is designed for readers who wish to the study and advance the state of the art in many related areas.

The book consists of twelve chapters. They could be grouped into several theme topics but this is not done since some of the chapters cover multiple themes. However, they are organized according to their major emphasis. SHM/SHA evolved over the time, from simple visual inspections, to hitting something with a hammer and listening to the sound, to exciting the structures (non-model or model based) by static and dynamic loadings, to the most recent development of wireless sensors. After discussing the general concept in Chapter 1, various currently available methods of structural health assessment are presented. Long term monitoring of structures is presented in a separate chapter. Smart sensors are routinely used to assess structural health. Sensor types, platforms and advanced signal processing algorithms, and uncertainty associated with every aspect of the process for practical applications are presented. Wireless collection of sensor data, sensor power needs and on-site energy harvesting are also discussed. Each chapter is authored by the most active scholar(s) in the area. The chapters are summarized below for ready reference.

In Chapter 1, Cross, Worden and Farrar introduce the SHM concepts and cover many of the issues and challenges faced by those who wish to implement it. The main focus is on the application of SHM to civil infrastructure. Because of the differences in scale between civil structures and aerospace structures and the fact that the former always operate in uncertain uncontrolled environments, SHM for civil infrastructure presents particular challenges which are emphasized in this chapter. Because of the lack of failure data for real structures, developmental research for civil SHM is reliant on large-scale benchmarking exercises and some of the most comprehensive and successful cases are presented. The chapter also discusses some of the major barriers to implementation for automated SHM.

The original Damage Locating Vector (DLV) method is extended for health monitoring of a wider class of structures in Chapter 2 by Quek, Tran, and Lee. The normalized cumulative energy (NCE) is introduced as a damage indicator to facilitate the identification of damage in structural elements with varying internal stresses and capacities along their lengths. The identification process requires the formulation of the flexibility matrix using either static or dynamic responses captured by sensors. If a smaller than required set of sensors is used, a larger set of "damaged" elements is identified. An intersection scheme is introduced to filter out the actual damaged elements. If wireless sensors are used, intermittent loss of data in wireless transmission is unavoidable. Thus, a data reconstruction technique is presented for use with the DLV method.

Qiao and Fan presented Dynamics-based Damage Identification in Chapter 3. They presented four damage identification algorithms: Two-dimensional Gapped

Smoothing Method (GSM), Strain Energy-based Damage Index Method (DIM), Uniform Load Surface (ULS) and Generalized Fractal Dimension (GFD). They compared these algorithms using numerically generated and experimentally obtained data on a two-dimensional plate.

In Chapter 4, Nasrellah, Radhika, Sundar, and Manohar presented Simulation Based Methods for Model Updating in Structural Condition Assessment. They discussed structural model updating methods that combine various tools comprising of state-space modeling based on Markov vector approach, Bayesian and maximum likelihood estimation methods, Markov chain Monte Carlo and sequential Monte Carlo simulation techniques, and finite element method for structural modeling for both statically and dynamically applied single and multiple loadings and nonlinear structural behavior.

A brief overview of the various stochastic filtering strategies in the context of structural health assessment is presented in Chapter 5 by Sarkar, Raveendran, Roy, and Vasu. In particular, the authors focused on dynamic structural system identification by the sequential Monte Carlo sampling based Bayesian filtering, also known as particle filters, relatively new in the field. To include static inverse problems within the dynamic filtering framework, a pseudo dynamic approach is also discussed. A few numerical results from recent research are also provided.

Das and Haldar summarized research activities at the University of Arizona on SHA covering over a decade in Chapter 6. They discussed how they changed their research activities as new challenges arise. Initially, the research team proposed a novel idea on identifying defects at the element level in the finite element context using only noise-contaminated response information without any information on excitation. Later, when they realized that response information may only be available at a small part of the structure, they integrated the previous development with the Extended Kalman Filter concept leading to the development of GILS-EKF-UI. They verified it using analytically generated and experimentally obtained response information.

In Chapter 7, Hou, Hera, and Noori presented their work on wavelet-based techniques for SHM. It also provides an overview on wavelet transform. The application of wavelet and various wavelet-based tools, developed by the authors, are discussed. The advantages and disadvantages of these techniques for the diagnosis of structural health are presented.

Huang, Salvino, Nieh, Wang, and Chen introduced HHT for structural health monitoring in Chapter 8. With the introduction of the time-frequency representation, one could monitor the dynamical behavior in more detailed and quantitative way. Specifically, the time-frequency representation also enables to quantify the nonlinearity, which by its self might not indicate damage, offers an early indicator of structural deficiency. Based on this approach, the HHT based structure health monitoring has been applied to three types of structures: the civil infrastructures, the ship structures and the aircraft all with laudable results. The authors also discussed few challenges to implement the concept for structural health monitoring.

Koh and Zhang presented the use of genetic algorithms for structural identification and damage assessment in Chapter 9. They first discussed the characteristics of structural identification from the optimization perspective. A recently developed uniformly sampled genetic algorithm with gradient search is then presented, as an efficient means to tackle the problem of optimization-based structural identification. The proposed method combines the complementary strengths of global search and local search. The global search strategy involves quasi-random sequence sampling, multi-species exploration, and adaptive search space reduction. The use of a small number of uniform samples enables preliminary exploration in the solution space so as to shorten the learning curve considerably. The local search is then used to carry out fine tuning for better accuracy. The proposed strategy is shown by numerical and experimental studies to achieve substantial improvement in identification efficiency and accuracy for identification of known mass and unknown mass systems.

Feng, Gomez and Zampieri presented Health Assessment of Highway Bridges Using Vibration-Based Response Data in Chapter 10. By using recorded bridge response data, the authors extracted change in the structural condition embedded in the measurements. The response data come in the form of either ambient/traffic or seismic-induced vibration-based. They compared several structural health assessment techniques using response information obtained from large scale shaking table tests. They also studied heath of three existing instrumented bridges in Southern California.

In Chapter 11, Modares and Mohammadi identified the prevalent challenges in designing sensor networks for structural health monitoring applications and presented the recently evolving technologies in embedded computing, energy harvesting and wireless communications. In particular, continuous monitoring with spatially distributed passive acoustic sensors is highlighted. Introduction of smart wireless sensor nodes enables local processing and decision making capability. In addition, the combination of smart power management and energy harvesting techniques results in sustainable sensor operations, critical for broader adoption of structural health monitoring applications. In this chapter, the authors provided an overview of sensor types that can be employed in SHM system applications in measuring a variety of structural parameters, such as displacements, strains, stresses, cracks, fatigue damage, corrosion, moisture, temperature, and applied loads.

In Chapter 12, Oruklu, Saniie, Modares, and Mohammadi presented sensor data wireless communication, sensor power needs, and energy harvesting. Wireless sensor networks have been proposed extensively over the past several years as a means of alleviating instrumentation costs associated with structural health monitoring of civil infrastructure. However, low data throughput, unacceptable packet yield rates, and limited system resources have generally plagued many deployments by limiting the number of sensors and their sampling rate. The sensor networks present challenges in three broad areas: energy consumption, network configuration and interaction with the physical world. Therefore, the development of sensor networks requires technologies from three different research areas: sensing, communication, and computing (including hardware, software, and algorithms). The next generation of the structural health monitoring sensors needs to be low-cost, low-power, self-healing, self-organized, and compact. They observed that many structural health monitoring systems that are available today are only applicable to specific structures and lack the versatility needed to cover a whole host of distress conditions. They identified several implementation strategies for future applications.

This book is unique in terms of its content, which addresses many emerging research areas where the available information is scarce or not yet properly formulated. The recent thoughts and opinions of experts presented in this book are expected to accelerate the development of these areas. The book provides the reader with a wealth of insight and a unique global and multidisciplinary perspective. It is hoped that this book will convey to the reader the excitement, advances, promise, and challenges in overall structural health assessment and monitoring.

Achintya Haldar
Tucson, Arizona

displacements, strains, stresses, cracks, fatigue damage, corrosion, and [illegible] [illegible] and applied loads.

[illegible] in Smart, wireless, and [illegible] sensors and sensing [illegible] sensor [illegible] environments [illegible] Wireless sensor networks have been proposed extensively over the past several years as a means of alleviating existing limitations associated with structural health monitoring of [illegible] structures. However, low data transmission rates, packet losses, and limited system resources have severely prevented many deployments [illegible] [illegible]. The sensor networks present challenges [illegible] [illegible] configuration [illegible] with the physical world. [illegible] development [illegible] from [illegible] hardware, software, and algorithms. [illegible] structural health monitoring [illegible] need to [illegible] low power, self-[illegible] [illegible] [illegible] [illegible] [illegible] [illegible] [illegible] [illegible] implementation strategies for future applications.

This [illegible] information [illegible] [illegible] [illegible] [illegible] [illegible] [illegible] [illegible] [illegible] and [illegible] [illegible] [illegible] [illegible] [illegible] [illegible] [illegible] [illegible] [illegible] [illegible] [illegible] [illegible] the [illegible] [illegible] [illegible] [illegible] [illegible] [illegible] monitoring.

[illegible] H. [illegible]
Tucson, Arizona

Contents

Chapter 1

Structural Health Monitoring for Civil Infrastructure

E.J. Cross[1], K. Worden[1] and C.R. Farrar[2]
[1]*Dynamics Research Group, Department of Mechanical Engineering, University of Sheffield, Mappin Street, Sheffield S1 3JD, UK*
**E-mail: e.j.cross@sheffield.ac.uk*

[2]*Engineering Institute, MS-T001, Los Alamos National Laboratory, Los Alamos, NM 87545, USA*

This article is intended as an introduction to structural health monitoring (SHM) and attempts to cover many of the issues and challenges faced by those who wish to implement it. A main focus will be on the application of SHM to civil infrastructure. Because of the differences in scale between civil structures and aerospace structures and the fact that the former always operate in uncertain uncontrolled environments, SHM for civil infrastructure presents particular challenges which are discussed here. Because of the lack of failure data for real structures, developmental reasearch for civil SHM is reliant on large-scale benchmarking exercises and some of the most comprehensive and successful are described here with some case study material presented for one of them - the Z24 bridge. The paper also discusses some of the major barriers to implementation for automated SHM.

1. Introduction: SHM Ideology

Structural Health Monitoring (SHM) is, in short, any automated monitoring practice that seeks to assess the condition or health of a structure. Its beginnings as an area of interest to engineers can be traced back as far as the time when tap-testing for fault detection became common, although the field didn't really become established in research communities until the 1980s, when much interest was generated in the structural condition of oil rigs, and later in aerospace structures and their health [Doebling *et al.* (1996)]. Nowadays, SHM is a popular and still growing research field, which is more and more becoming a focus of the civil infrastructure community. In fact, as Wenzel observes [Wenzel (2009a)], bridges (which are arguably the paradigm for critical infrastructure) have always been monitored, with this activity varying between visual inspection and continuous monitoring with dense sensor networks.

1.1. *The aims of SHM*

At its heart, SHM has great aspirations that, if achievable, would hugely benefit society at large. The ideal that SHM strives towards is to be able to monitor (in an automated fashion) a structure in such a way that any damage introduced, or any growth of inherent faults, would be immediately detectable. Further to this, the aim is that, after detection, any fault could be located and its severity inferred so that decisions can be easily made as to what actions need to be taken next (e.g. immediate halt to use of the structure, immediate repair, etc.). These global objectives for SHM have been well formalised in Rytter's hierarchy [Rytter (1993)], which classifies these aims into 'levels' of increasing difficulty as follows:

Level 1 (DETECTION): The method gives a qualitative indication that damage might be present in the structure.
Level 2 (LOCALISATION): The method gives information about the probable position of the damage.
Level 3 (ASSESSMENT): The method gives an estimate of the extent of the damage.
Level 4 (PREDICTION): The method offers information about the safety of the structure, e.g. estimates a residual life.

Although a number of changes and additions to this hierarchy have been suggested in the literature [Worden and Dulieu-Barton (2004)], perhaps the most meaningful is to add a level pertaining to classification; at the risk of repetition, this renders the structure:

Level 1 (DETECTION): The method gives a qualitative indication that damage might be present in the structure.
Level 2 (LOCALISATION): The method gives information about the probable position of the damage.
Level 3 (CLASSIFICATION): The method gives information about the type of damage.
Level 4 (ASSESSMENT): The method gives an estimate of the extent of the damage.
Level 5 (PREDICTION): The method offers information about the safety of the structure, e.g. estimates a residual life.

In this slightly doctored hierarchy the order of levels 2 and 3 is debatable; however, the addition of the classification level is highly desirable from a civil infrastructure point of view. In the first case, civil structures are highly heterogeneous at both the global and material levels and classification of damage type is essential if any prognosis is to be possible. Secondly, and perhaps more importantly, the owners and operators of civil infrastructure are arguably as interested in *performance monitoring* as actual health monitoring. Wenzel [Wenzel (2009a)] observes that there is a direct link between performance and operation, citing the example of determining optimal values for traffic speed or frequency on bridge operation. This will be discussed a little more later, for now it should be observed that if SHM is conducted by monitoring for anomalies, it is clearly essential to distinguish between performance and health anomalies as confusion between the two could lead to a structure being needlessly taken from service; the classification level is essential here.

1.2. *Potential benefits of SHM*

With regards to the advantages of any SHM system able to fulfill these global objectives, the first and most obvious benefit is increased human safety, indeed, unsurprisingly much of the research in this field has been motivated by disasters such as bridge collapses and aeroplane crashes, where many lives have been lost (see [Lynch (2007)], for example, for more details). Even at the lowest level of SHM - a detection of damage or degradation of structural condition could be hugely beneficial if used to provide an early warning that a structure may be unsafe. Additionally, with an automated detection system using non-visual assessment, any areas of a structure that are difficult or impossible to access can be assessed that otherwise may have been neglected in a visual inspection routine, again increasing safety.

Other arising benefits will come from the regime change that comprehensive SHM could bring about. Currently civil and aerospace structures undergo routine inspection and maintenance at specific time intervals, for example bridge inspections in the USA are scheduled every two years [Lynch (2007)], and commercial aircraft undergo a thorough inspection after a given number of flight hours/cycles. A time based approach to management of structural assets such as this, firstly, has the implication that any unexpected faults occurring in between scheduled inspections may go un-noted and cause danger to life, or cause unnecessary stress on other structural components. Inversely, the set time scales for inspections may be overly conservative; if a structure continues to be in good health, the costs of thorough inspections could essentially have been saved. In the case of routine maintenance, where structural components may be replaced even if they are in excellent condition, the economic impact may be even greater. SHM has the ability to solve both sides of this issue, as monitoring has the potential to become continual and maintenance and repair could become condition-based. A switch to condition-based maintenance could also reduce the downtime a structure may undergo for routine and emergency maintenance, which, in turn, would be of economic and environmental benefit.

1.3. *Disambiguation: what SHM is not*

There are a number of research fields very closely related to SHM, which can be seen, depending on one's point of view, as either overlapping with, or perhaps even encompassed within SHM. For disambiguation, it seems sensible to discuss them shortly here, a more detailed discussion can be found in Worden and Dulieu-Barton, (2004). Although tap-testing was mentioned above as marking the beginnings of SHM, the example really belongs to the field of *Non-Destructive Testing* (NDT) or *Non-Destructive Evaluation* (NDE). NDE or NDT concerns the assessment of a structure or component's health through (offline) non-damaging procedures and examples of tools commonly used for NDE are x-ray, electron microscopy, measurement of acoustic emissions and full scale vibration tests. Although all of these techniques may be used for SHM purposes, NDEs currently most commonly occur as one-off planned events, often applied to a small area of a structure where damage is suspected to have occurred. This is a different approach to SHM where monitoring aims to be continuous and global. In the future, it is likely that NDE inspection will form the basis for distinguishing between health and performance anomalies for civil infrastructure where this cannot be accomplished automatically. It is therefore true to say that NDE may be incorporated as part of an SHM system, but not vice-versa.

Another field related to SHM is *Condition Monitoring* (CM). CM largely concerns the health of rotating machinery; it has seen many successes and has been, in some areas, accepted as part of every day practice by industry. Its commercial success relative to SHM can be attributed to a number of factors that simplify the monitoring process: the machinery operate in a controlled environment, it is usually easy to access and is typically on a small scale. Importantly, rotating machinery have been found to exhibit well defined dynamic responses for particular fault categories, which makes fault detection and identification a more easily attainable goal than it perhaps is for SHM [Randall (2011)].

It should be noted that one thing SHM is not, is *monitoring*. Simple collection of data does not constitute SHM, however complex and comprehensive the sensor network is. Figure 1 shows how the number of sensors deployed on bridges has increased in the last two decades. If such sensor networks stream data continuously, the amount of data stored is huge; however, this is of no value unless performance or health knowledge is extracted, and this is far from the norm. As an example of showing how available data is used (or not) to inform SHM practice, a simple experiment was performed. A search for research papers on *Web of Science* with the words 'Tsing Ma' (one of the most densely instrumented bridges) yielded 19 papers. Of the 19 papers, six were concerned with aerodynamics, one with sensor design and one was unrelated to engineering. Of the remaining 11 papers, seven were concerned with building FE or modal models of the bridge (admittedly motivated by SHM) using acquired data for validation; three were concerned with using models to identify fatigue 'hot spots'; only one paper proposed a method for real-time performance monitoring using measured data [Ni *et al.* (2008)]. This experiment is clearly not exhaustive and is intended simply as an illustration, a search over abstracts, for example, would have undoubtedly produced more papers.

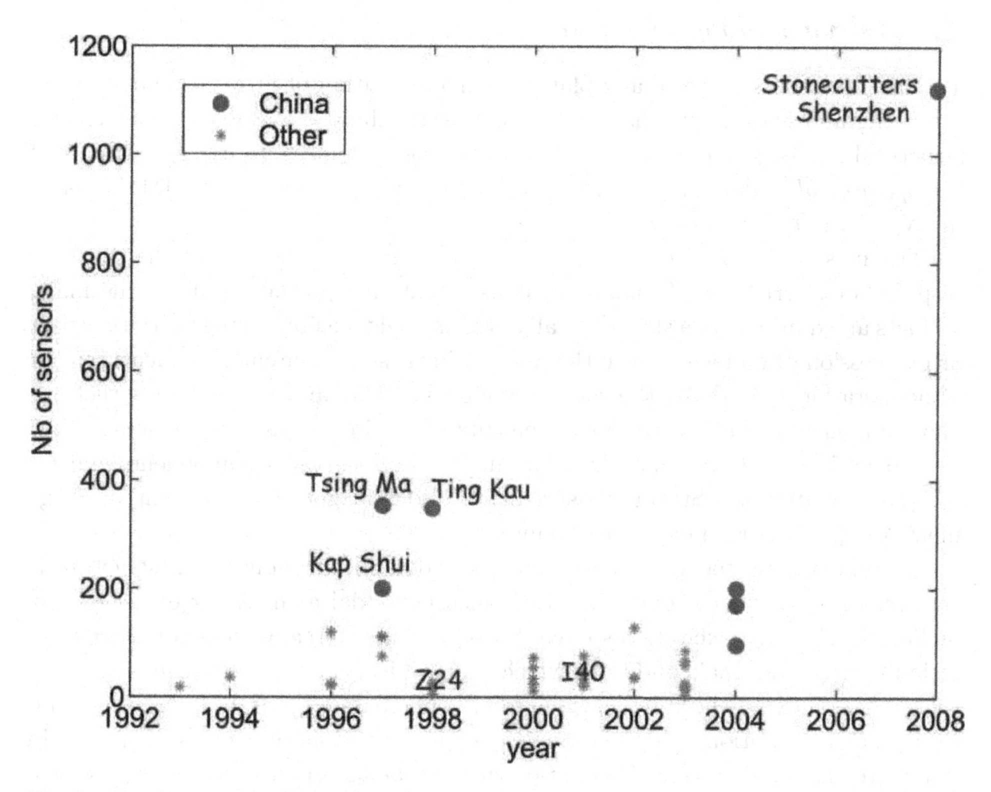

Fig. 1. Developments in the number of sensors on instrumented bridges over the last two decades.

2. SHM in Practice

The many benefits of SHM will come hand in hand with many challenges that must be overcome. The fundamental problem at the heart of SHM is that of how a measure of structural condition can be gained from an automated process. No sensor can measure damage directly [Worden *et al.* (2007)] and so this fundamental problem breaks down into a number of separate issues; what can be measured that correlates to damage, how to measure it and, importantly, how to use the raw measurements to make inferences and decisions about structural condition. Finally, before an SHM system can be relied upon, it must be proven to work, and issues such as how to cope with sensor failures, for example, must be addressed. In the following, each of these issues is addressed separately. As an introduction to SHM, this article certainly does not aim be exhaustive, instead for a comprehensive review of practices in SHM readers are referred to [Doebling *et al.* (1996); Sohn *et al.* (2002); Carden and Fanning (2004)].

2.1. *Instrumentation for SHM*

The first questions asked of any planned implementation of SHM, as alluded to in the paragraph above, are what is the most useful thing that can be measured for structural assessment purposes, and how can one best measure it? This is part of the *operational evaluation stage* in the four-stage implementation of SHM discussed in [Worden *et al.* (2007)].

The most common measurements sought by far in SHM are of the dynamic response of a structure. Dynamic responses contain information about the mass, stiffness and damping of a structure, all of which could feasibly change with the onset or progression of damage, hence the interest in these measurements. Measurement of acceleration is perhaps the most common in SHM and is used for structures and components of all sizes. Measurements of strain are also very common, but are currently most commonly used for small scale structures/components and for composite materials. Strain measurements also find good employment in usage monitoring, where load cycles are counted.

A dominating issue when considering the dynamic response of a structure is that many schemes developed for SHM, such as modal analysis for example, rely on knowledge of the excitation source. Structures in the real world experience excitation from operational conditions which, in practice, cannot be measured, such as the excitation experienced by a bridge from traffic passing over it. In these circumstances, an assumption as to the properties of an excitation source must be made. In other circumstances, artificial excitation that can be measured is introduced, with a hammer, a shaker or an electrical pulse, for example. As knowledge of an excitation source is desirable, much research effort is currently focused on sensor systems that can provide their own measurable excitation source [Lee and Sohn (2006)]. One particular area of interest where this is relevant is the use of higher frequency guided waves for damage assessment [Raghaven and Cesnik (2007)]. Guided waves have mainly been used for the detection of damage in plates and pipes, and are therefore arguably of most interest to the aerospace and process industries, although recently a growing interest in SHM for wind turbines has directed research into guided waves in that direction as well.

There are many reasons that the measurement types discussed above are the most commonly occurring in SHM, the main one being that the theory behind why these measurements are useful is well founded, and in many circumstances proven. A consequence, however, of the general acceptance of these measurement types as routine, is that they are commonly employed in an SHM system, perhaps, without consideration as to whether they would be the most useful for the scenario of interest. If these measurement types are not optimal for the damage type or location of interest, the consequence is that more effort may be required for data analysis, in order for the measurements to be useful (see Section 2.2.1). Of course, researchers must work within their means and make the most of readily available measurements, and of course, the most commonly occurring measurements may indeed be the most informative ones available. Some other measurements that have been found useful for SHM are measurement of acoustic emissions which occur with damage initiation and progression (for example, this has been investigated for monitoring individual cable snapping in the main cables of suspension bridges [Rizzo and di Scalea (2001)]), and measurement of electrical impedance, which has been found to correspond to the mass, stiffness and damping properties of a structure [Park and Inman (2007)].

One of the largest concerns in SHM is *how* to obtain the measurements that will be most useful; research into suitable instrumentation for SHM probably takes up more interest in the community than any other single topic. A complete SHM monitoring system will more than likely require a large number of different sensors of different types in order to monitor all components and be able to identify different damage scenarios. If suitable sensors are available, a number of important questions need to be addressed for a useful (or ideally optimal) monitoring system, these include; where sensors are best placed, how many are needed, how they can be powered, where an excitation source will come from and how data will be transmitted.

Most monitoring systems in place now, especially on civil infrastructure, use wired sensors, both for a power source and for data transferral. For numerous reasons, however, using wired sensors for the monitoring structures outside the laboratory proves to be difficult; the amount of wiring necessary to instrument whole structures quickly becomes infeasible for large scale structures, further to this, the addition of a large amount of wires is often very unappealing to operators (due to, for example, the extra weight or the increased lightning conductivity that a network of wires may introduce). Wires are also susceptible to the weather and may not work well in extreme conditions. For these reasons, wireless sensing has become a hot topic of research over the last few years, and is seen by some within the community as the future for SHM [Lynch (2007)].

Sensing wirelessly naturally introduces a whole new set of problems to overcome, the most pressing of which are how to power the sensor and data telemetry. To overcome some of the powering and telemetry issues, it is thought that some on-board processing of data at the sensor before transmission would be of great benefit. Self-powering sensors (energy harvesting) are also an emerging field of interest [Park *et al.* (2008)], as well as other novel techniques for power and data transferral, such as the use of remotely controlled vehicles.

An additional monitoring issue that has more recently occurred, concerns the management of large amounts of data collected by a monitoring system. Nowadays, many monitoring campaigns, especially for civil infrastructure, involve dense sensor arrays from which terabytes of data are collected (see Figure 1). Consequently the development of systems for storage of, and importantly, access to large amounts of data has become important. Aside from all the necessary signal processing inherent in monitoring campaigns, efficient management of data is essential for a successful SHM system.

2.2. *Assessment of structural condition from measurements*

The question of how to infer structural condition from different measurements is at the heart of SHM. Once measurements with some correlation with damage have been obtained, the process for arriving at a judgement on structural condition can be divided between two major tasks. The manipulation of measurements from a structure in order to create a usable variable that can give an indication of structural condition is often named *feature extraction.* The use of extracted features to make decisions (such as damaged or not damaged) is the second major challenge that must be faced, and one which has been identified by many as a problem in statistical pattern recognition [Farrar and Worden (2007)].

2.2.1. *Feature Extraction*

As previously stated, no sensor is available that can measure any type of damage directly, instead measurements can at best be correlated with the damage type one is interested in. A raw measurement is also unlikely to be directly useful for damage detection and assessment; on a practical level this is simply often because a raw measurement provides too much data/information than is feasible to work with (high dimensionality), and can also often be difficult to interpret. Often the pattern recognition techniques that are used to infer structural condition from data can only work well in a low dimension. For these reasons, feature extraction is used to create useful metrics from raw measurements which are often of a lower dimension than the raw data. Simple features that can be extracted from raw measurements include, for example, statistics from a signal such as the mean and variance. Where the aim of feature extraction is purely to reduce the dimension of measurements, approaches such as principal component analysis can be used, which acts to transform data in such a way that redundancy is simple to identify and remove. Many other feature extraction methods rely on working in the frequency domain of a signal provided through use of a Fourier transform. In the frequency domain, the spectrum of a signal is particularly useful as it is often of low dimension and can be easily interpreted.

Damage sensitive features in the form of modal properties extracted from acceleration measurements are probably the most frequently occurring features used in SHM. These include, but are not limited to, natural frequencies, mode shapes, and mode shape curvatures. The modal approach is attractive due to the interpretability of the features and additionally the fact that typically only a low number of sensors is required in order to extract these features. An important point to consider, however, is that modal features are global indicators for structural condition, meaning that any inference on condition applies to the whole structure. Because of this fact, and the fact that modal analysis can be carried out with a small number of sensors and therefore with relatively little hassle, these approaches have been found useful for the provision of a good one-off general assessment of a structure

([Wenzel (2009b)]). The well known disadvantage associated with modal properties is that they have been found to be insensitive to structural degradation on a local scale. Further to this, it is also very difficult to obtain accurate modal properties corresponding to higher frequencies for larger structures, and so often only information on the low frequency modes are available. Unfortunately, lower frequency modes have been found to show low sensitivity to damage.

Other approaches to feature extraction seek to fit measurements to mathematical models or functions (other than the Fourier transform) and use parameters from these models as features. A common example of this is fitting an ARMA type model to measurement data and using the model coefficients as features [Sohn *et al.* (2001)]. Another way in which fitting models to raw data can be used as a feature extraction methodology, is to use the residual error of a predictive model as a feature [Sohn *et al.* (2001)]. For more details on feature extraction and selection see [Worden *et al.* (2011)].

2.2.2. *Pattern Recognition for inference on structural condition from features*

Once a particular feature has been extracted from raw measurements, a decision process is needed to infer structural condition from this feature. As previously mentioned, this is essentially a problem in pattern recognition, where a feature will be classified according to whether it has arisen from a damaged or undamaged structure, and at higher levels of SHM, classified as to the location, type and severity of the damage if present. Pattern recognition in this form relies on one of two different approaches, the first is a supervised learning approach, the second relies on novelty detection [Worden and Dulieu-Barton (2004)].

Supervised learning for SHM is any procedure for the classification of a feature which is informed with data from all states of interest. In terms of the lowest level of Rytter's hierarchy this simply means that data must be available from the damaged and undamaged state of a structure. Techniques that use supervised learning can include all algorithms capable of classification, such as neural networks, support vector machines and Gaussian processes [Worden *et al.* (2011)]. A supervised learning approach is considered to be necessary where identification of different damage types and locations are required (see Section 2.4). Unfortunately, data collected from the damaged state of a structure is rarely available (let alone data from multiple damage scenarios), as, naturally, introducing damage to a structure to inform an SHM decision process is not an acceptable option. In cases where supervised learning is necessary, future advancements in the field may rely on physics based, or high fidelity models that can accurately simulate the response of a structure in a damaged state. It is possible that physical proxies for damage can also be found [Papatheou *et al.* (2010)].

Indeed, research into the use of high fidelity models (i.e. finite element) for SHM is popular. Model updating is used as a tool for inference on structural condition which, in simple terms, is the use of measurements from a real structure to update a physics based model, which can then be used, via an inverse problem, to predict the current state of the structure. Research into model updating approaches is reviewed comprehensively in [Friswell and Mottershead (1995)]. Such approaches are philosophically different to the data-based approach based on pattern recognition espoused in this article.

Novelty detection algorithms are required when data from the damaged condition of a structure are not available (which is most often the case). A decision process reliant on novelty detection will aim to define a baseline, with data from the undamaged condition of a structure, that represents the normal response of the structure in its undamaged condition. A non-normal response is then detected by any significant departure from this baseline. Common examples of techniques for novelty detection include outlier analysis and the use of statistical process control charts, both of which rely on the selection of a threshold which if crossed, signifies that a structure is not responding in a normal way [Fugate *et al.* (2001)]. A disadvantage of novelty detection approaches is the unavoidable fact that an indication that a structure has departed from its normal condition is uninformative as to what may have caused this departure. In this case, further investigation after a novel response has been detected will be necessary in order to eliminate the possibility that change has occurred for a benign reason e.g. because of a temperature change. However, in the authors' opinion, novelty detection must be considered for successful SHM, at least at the lowest level of SHM, simply due to the fact that a supervised learning approach is limited to scenarios that can be anticipated, or have occurred before and for which data are available. Novelty detection must be incorporated into any comprehensive SHM system to safe guard against unforeseen circumstances (Black Swan events [Taleb (2011)]), an indication that a structure is responding in a non-normal way can then be investigated further.

The pattern recognition problem as a whole is further complicated by the fact that many features undergo variability caused by operational and environmental conditions (as alluded to above), which must also be accounted for by the inference procedure. The influence of environmental and operational variation can make classification problems in supervised learning very complex, as the data may become separable in a large number of ways according to different operating conditions. Novelty detection is also compromised if external conditions produce a novel structural response from an undamaged structure. To overcome this, preprocessing of a feature before the decision making stage may be necessary [Sohn (2007)].

2.3. *Validation of SHM systems*

Before any of the benefits of SHM outlined in Section 1.2 can be realised, a proposed system for inference on structural condition must be rigorously proven or validated. A system that fails to detect serious or dangerous faults (referred to as a *false negative* detection of damage), or conversely detects faults where there are none (*false positive* detection of damage), would have serious life-safety or negative economic consequences respectively, which would completely undermine the reasons for embarking on SHM in the first place.

This question of validation is, however, one of the hardest faced by those working in the field of SHM, especially due to the earlier mentioned fact that data from the damaged condition of a structure are hard to come by, which would be one natural way to validate a decision making tool. The problem of validation in the civil infrastructure context is one of the reasons why benchmark structures as described in Section 4 assume such importance. It seems to the authors that the question of validation is not commonly addressed in current research in the field, perhaps because it is a premature one in relation to current progress towards the aims of SHM. Research that does address the issue of validation tends to focus on, for example, how scenarios arising from sensor failures can be dealt with. The lack of a real solution to the issue of validation has the consequence that, currently, an SHM system will not be able to completely replace timed visual inspections or routine maintenance of safety critical components, which is perhaps disappointing in view of the fundamental aims of SHM.

2.4. *Fundamental axioms of SHM*

From the practical progress towards the goals of SHM, naturally, much experience and knowledge has been gained. Reflecting the work in the field and the lessons learned over the last few decades a number of fundamental axioms have been formulated [Worden *et al.* (2007)], which are listed below:

Axiom I: All materials have inherent flaws or defects.

Axiom II: The assessment of damage requires a comparison between two system states.

Axiom III: Identifying the existence and location of damage can be done in an unsupervised learning mode, but identifying the type of damage present and the damage severity can generally only be done in a supervised learning mode.

Axiom IVa: Sensors cannot measure damage. Feature extraction through signal processing and statistical classification are necessary to convert sensor data into damage information.

Axiom IVb: Without intelligent feature extraction, the more sensitive a measurement is to damage, the more sensitive it is to changing operational and environmental conditions.

Axiom V: The length and time scales associated with damage initiation and evolution dictate the required properties of the SHM sensing system.

Axiom VI: There is a trade-off between the sensitivity to damage of an algorithm and its noise rejection capability.

Axiom VII: The size of damage that can be detected from changes in system dynamics is inversely proportional to the frequency range of excitation.

The axioms are intended to provide guidance on how to specify an SHM system in practice. Although they were initially formulated with a bias towards aerospace problems, in most cases the authors would say they apply well in a civil infrastructure context, some of them with greater or lesser emphasis in that domain. Some particular issues with civil structures will be discussed here.

Very few civil or structural engineers would argue with Axiom I. In the aerospace context, the 'flaws or defects' are usually associated with imperfections in the microstructure of metals like voids or dislocations. Mechanical engineering design has involved in such a way that the impact of defects on structural performance are understood and the design process will accommodate the imperfections, perhaps through the use of over-engineering via application of safety factors. One of the benefits of SHM mentioned earlier is that, through its use a transition from safe-life design to damage-tolerant design may be possible. In the civil engineering context, concrete is an extremely complex material and is almost always understood to have inherent flaws which are dealt with in the design process through the careful use of highly evolved codes.

Axiom III provides some interesting points for discussion in the context of civil infrastructure. Because only the lowest global modes are generally excited in practice for say a bridge, the modes will usually involve large scale coherent motions. If the structure has a dense sensor network, a breakdown in correlations between neighbouring sensors may give information about the location of damage. This means that location could be possible although data from damage states are not available. Severity of damage could also be inferred from the extent of the breakdown of correlation. Another possibility is that within the highly heterogenous structure that is a bridge, changes in local modes e.g. those of cables, may also allow some localisation of damage.

Axiom IVb is very pertinent for SHM of civil infrastructure as almost all structures will operate *in situ* at the mercy of the elements. Wind, rain, humidity, temperature will all induce benign variations in structural properties on various time scales. In addition, operational variations like traffic loading on a bridge are usually inescapable. While it is conceivable that an automated inspection of an aerospace structure could be carried out in a controlled environment (a hangar), this will not be an option for buildings, dams or bridges. The problem is even more complicated by the fact that civil structures are often so big that the environmental variations may vary themselves over the extent of the structure. This means for example, that many temperature measurements may be needed in order to account for the effects. As an example of this, one can cite the study of the Alamosa Canyon Bridge [Farrar *et al.* (2002)] where it was found that the latent variable driving the environmental effects was not a single ambient temperature, but the temperature gradient across the bridge in the east-west direction. Finally, one should also take note of the disparate time scales that environmental and operational variations operate on. A bridge will be subject to daily variations in structural characteristics as a result of temperature and traffic loading, will be subject to weekly variations due to traffic (usage patterns differ between working days and the weekend) and finally will be subject to seasonal variations in temperature. All of this means that the 'intelligent' feature extraction mentioned in Axiom IVb has to particularly intelligent in the context of civil infrastructure.

Although one could discuss each axiom in turn, the section will conclude here with some comments on Axiom V. Axiom V is the reason why operational evaluation is a critical stage in the development of any SHM system. The design of the sensor system must be tailored to the time and length scales that one expects damage to accumulate on. In the context of certain aerospace alloys, a critical crack size smaller than a millimetre means that confident detection of the crack almost immediately after initiation is critical. In the context of civil infrastructure it will usually be the case that damage will accumulate on longer time scales and will have a greater spatial extent before catastrophic failure would occur. Even so, these anticipated time and length scales must be considered very carefully in the specification for the monitoring system; because civil structures are so much larger, for confident detection sensor densities may well need to be relatively high.

3. Civil Infrastructure and SHM

It is often thought that aerospace applications monopolise, or are the most common concerns of, research carried out in SHM. Looking back through the SHM literature, however, shows that many studies are based on civil infrastructure, or have used it as a base for development and validation of algorithms [Sohn *et al.* (2002)]. This view, therefore, perhaps originates from the fact that many of the considered SHM successes have come from aerospace applications. Some technologies developed for aerospace have in fact made the jump from a research interest to applied technology in industry, the most notable example of which is the Health and Usage Monitoring System (HUMS) [Chronkite (1993)] that has been accepted into the rotorcraft industry as a matter of legislation (in some countries at least). There are a number of clear reasons why this success has, so far, not been mirrored in civil applications.

One potential reason may be contributed to a lack of general consensus within the civil community on how SHM should be approached, which can in turn be attributed to the fact that civil infrastructure is often privately owned, where owners answer to no single regulatory body. In comparison, in the aerospace industry a concerted effort is currently being made by industry and regulatory partners to coordinate and shape the development of SHM for aircraft. An Aerospace Industry Steering Committee (AISC) has been formed for this express purpose [Foote (2011)], where the aim is to develop guidelines and certification requirements and to encourage the use of SHM. Without a parallel effort for civil infrastructure dissemination of good SHM practice is difficult.

A separate issue that will also play a large role in the uptake of SHM for civil infrastructure, is the fact that most civil structures are one-offs, and as such are completely unique [Brownjohn (2007)]. Consequently, an SHM system developed for one structure will not be directly applicable to another, as is possible with fleets in aerospace. Finally, an additional complication comes from the challenge that the sheer size and complexity of some civil infrastructure provides. Detecting damage in a single component of a long-span bridge, for example, becomes a very difficult task, on top of the complications that arise from instrumenting a structure of that size.

Due to these challenges, a slightly different outlook on SHM seems to be emerging within the civil community. For some the belief seems to be, that for now certainly, automated condition assessment on a local level, e.g. detection of cracks on structural elements, is unobtainable. The focus, has therefore become, for some, more on monitoring practices, studying global response, and monitoring for performance. Many papers which are published in the name of SHM, solely focus on the monitoring effort of civil infrastructure, which is, of course, not inconsiderable. This is reflected by the fact that the state of the art in civil infrastructure SHM is commonly thought to be the large monitoring campaigns occurring in the East [Ko and Ni (2005)], where long-span bridges are instrumented with thousands of sensors (see Figure 1). The authors would argue that, despite the sophistication, until processes are developed to analyse the measurements obtained in order to make inferences on structural condition, so far only structural monitoring has occurred as mentioned in Section 1.3. The danger in this current trend lies, not in the development of instrumentation, which will always be useful, but in the loss of sight of the fundamental aims of SHM, a monitoring campaign cannot be useful without a process in place that utilises the information obtained.

This having been said, the idea of what constitutes SHM for civil applications is perhaps more blurred than in other areas due to the fact that monitoring is often undertaken for slightly different purposes than for other SHM applications. For large scale civil infrastructure, for example, monitoring is often carried out at the start of a structure's life and during construction to ensure that the structure responds in an expected manner to its environmental and operational conditions (especially to wind conditions). This is motivated by the problems (and disasters no less) caused by the phenomena of self-excited oscillations, the most famous example of which being the collapse of the Tacoma Narrows bridge [Larsen (2000)]. Although many of these examples motivated the beginnings of research into SHM, the current practice used to safe guard against these events doesn't align exactly with the view of SHM presented in this article, as the monitoring is generally not intended to extend into the general assessment of the structure's health throughout its life.

4. Benchmarks

One of the issues for SHM research in the context of civil infrastructure is the sheer size and expense of many of the structures. Whereas, with aerospace SHM it is possible to conceive and implement benchmark structures with the representative complexity of aerospace structures, it is very rare that structures on the scale of bridges or dams become available for comprehensive testing, let alone allow deliberate damage to be introduced. Over the last couple of decades however, some landmark benchmarking tests have been carried out. Arguably, the most comprehensive of these and those for which the data have spread most widely, are the I-40 and Z24 bridge tests, and the tests on the Steelquake structure at the JRC (Joint Research Centre) in Ispra, Italy. These benchmarks are described briefly in the following.

4.1. *The I-40 Bridge*

The I-40 bridge over the Rio Grande consisted of twin spans (there are separate bridges for each traffic direction) made up of a concrete deck supported by two welded-steel plate girders and three steel stringers [Farrar *et al.* (1994)]. Prior to its demolition this bridge was destructively tested for the purpose of developing and validating SHM procedures on an *in situ* structure.

Loads from the stringers were transferred to the plate girders by floor beams located at 6.1-m (20-ft) intervals. Cross-bracing was provided between the floor beams. Figure 2 shows an elevation view of the portion of the bridge that was tested. The cross-sectional geometries of each of the two bridges comprising the I-40 crossing are shown in Figure 3, and Figure 4 shows the actual substructure of the bridge.

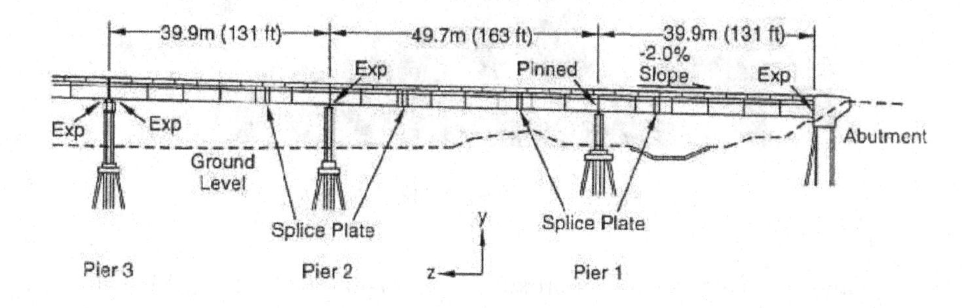

Fig. 2. Elevation view of the portion of the eastbound bridge that was tested.

Each bridge was made up of three identical sections. Except for the common pier located at the end of each section the sections are structurally independent. A section has three spans; the end spans are of equal length, approximately 39.9 m

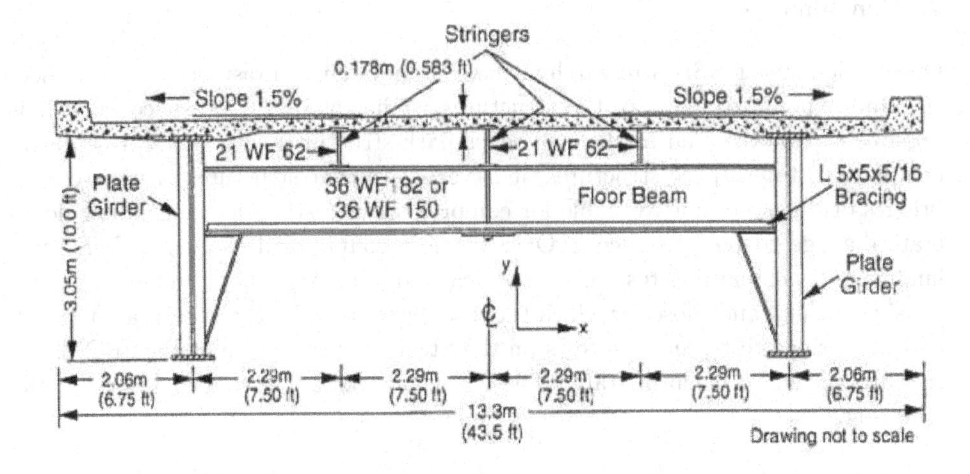

Fig. 3. Typical cross-section geometry of the I-40 Bridge.

Fig. 4. I-40 Bridge substructure.

(131 ft), and the centre span is approximately 49.7-m (163-ft) long. Five plate girders are connected with four bolted splices to form a continuous beam over the three spans. The portions of the plate girders over the piers have increased flange dimensions, compared with the midspan portions, to resist the higher bending stresses at these locations. Connections that allow for longitudinal thermal expansion were located where the plate girders attach to the abutment and where the plate girders attach to piers 2, 3 of the section that was tested (Figure 2). A connection that

prevented longitudinal translation was located at the base of each plate girder where they attached to pier 1.

The damage that was introduced was intended to simulate fatigue cracking that has been observed in plate girder bridges. This type of cracking results from out of plane bending of the plate girder web and usually begins at welded attachments to the web such as the seats supporting the floor beams (shown in Figure 3). Four levels of damage were introduced to the middle span of the north plate girder close to the seat supporting the floor beam at midspan. Damage was introduced by making various torch cuts in the web and flange of the girder. The first level of damage consisted of a 0.61-m-long (2-ft-long) cut through the web approximately 9.5-mm wide (3/8-in wide) centred at mid-height of the web. Next, this cut was continued to the bottom of the web. The flange was then cut half way in from either side directly below the cut in the web. Finally, the flange was cut completely through leaving the top 1.2 m (4 ft) of the web and the top flange to carry the load at this location. The various levels of damage designated E-1 - E-4 are shown in Figure 5(b).

The testing was carried out in a number of separate campaigns; various means of analysis were carried out by researchers at Los Alamos National Laboratories with a number of publications resulting [Farrar *et al.* (1994); Farrar and Jauregui (1998a,b); Doebling and Farrar (1998)]. The report [Farrar *et al.* (1994)] also discusses tests on another real bridge structure the Alamosa Canyon Bridge, this was one of the bridges where the effects of temperature on damage detection capability were first made clear. The paper [Farrar and Jauregui (1998a)] reports comparisons between five different SHM methods applied to the I-40 data.

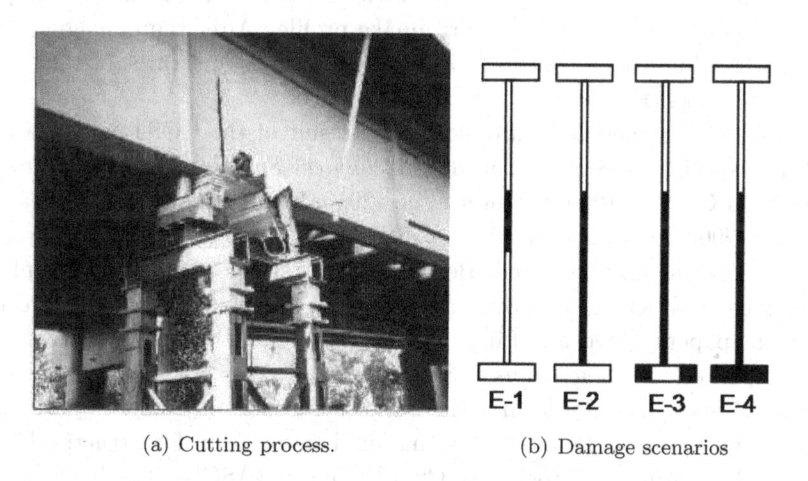

(a) Cutting process. (b) Damage scenarios

Fig. 5. Damage that was introduced into the I-40 Bridge in four increments.

4.2. *The Steelquake Structure*

The next two benchmarks were the subject of extensive study in a pan-European action on structural dynamics between the years of 1997 and 2001 — the COST F3 Action [Golinval and Link (2003)]. The COST action was focused on three main areas of structural dynamics: finite element model updating, structural health monitoring and nonlinear system identification. Three working groups were set up in order to conduct various tasks including the proposal of benchmarks; SHM was the subject of the Working Group 2 (WG2) effort [Worden (2003)]. The working group was granted valuable data from projects conducted elsewhere. In the first case from the Steelquake structure constructed and tested at the European Laboratory for Structural Assessment (ELSA) at the Joint Research Centre (JRC) at Ispra, Italy. In the second case from the Z24 bridge in Switzerland tested as part of the SIMCES project funded by Brite Euram [Roeck (2003)]. The Z24 bridge is discussed in the next section.

The Steelquake structure was designed as a large-scale benchmark, being a two-storey steel beam-column structure with concrete slab floors. The main dimensions were 9m by 6m by 3m. The decks were actually corrugated steel under the concrete slabs and this meant that the decks were composite with orthotropic material properties (this proved to one of the difficulties in attempting to model the structure). The structure was connected to a large reaction wall by four pistons which were used for excitation during a sequence of non-damaging and then damaging tests (Figure 6). Modal tests of the structure were carried out using hammer excitation with responses measured at 15 accelerometers; an attempt was made to place the accelerometers optimally using the Effective Independence Method(EIM). Substantial damage was introduced into the structure by driving it at large amplitude from the pistons with a recorded earthquake profile. After the damaging test, all of the beam-column joints showed plastic deformations and three of them showed substantial cracks (Figure 7).

The research immediately produced as a result of the COST action was published in a special issue of the journal *Mechanical Systems and Signal Processing* [Bodeux and Golinval (2003); Mevel *et al.* (2003a); Goerl and Link (2003); Fritzen and Bohle (2003); Zapico *et al.* (2003)], although a number of papers analysing the data have subsequently appeared. Because of the relative geometrical simplicity of the Steelquake structure, it proved amenable to FE modelling and updating and some of the papers [Goerl and Link (2003); Fritzen and Bohle (2003); Zapico *et al.* (2003)] exploited this as a means of SHM.

A steel frame structure formed the basis of another major benchmark exercise conducted under the aegis of the International Association for Structural Control (IASC) and the American Society of Civil Engineers (ASCE). The benchmark was based on a real structure which was proposed as the basis for a blind test of SHM methods during the 1998 International Modal Analysis Conference (IMAC) [Black and Ventura (1998)]. Although the initial blind test idea was not extensively pur-

Fig. 6. General view of Steelquake structure.

Fig. 7. Crack in the beam web, level 1, Column SE: Bottom view.

sued, the structure was adopted as the the basis of a major IASC-ASCE benchmarking exercise. Various phases of the benchmark followed with most of the exercises being based on simulated data generated from models of the original structure [Johnson *et al.* (2004); Lam (2003)], although experimental data were acquired and released [Dyke *et al.* (2003)]. The benchmark is not discussed here in detail as the structure in question was much smaller than the Steelquake structure; also, the damage scenarios for the IASC-ASCE benchmark were arguably less realistic than that imposed on the Steelquake structure. That said, the benchmark exercise was undeniably influential and the curious reader may consult the references given for further information.

4.3. *The Z24 Bridge*

The Z-24, a pre-stressed concrete highway bridge in Switzerland (Figure 8), was subject to a comprehensive monitoring campaign under the 'SIMCES project' [Roeck (2003)], prior to its demolishment in the late 1990s, it has since become a landmark benchmark study in SHM. The monitoring campaign, which spanned a whole year, tracked modal parameters and included extensive measurement of the environmental factors affecting the structure, such as air temperature, soil temperature, humidity etc. The Z24 monitoring exercise was an important study in the

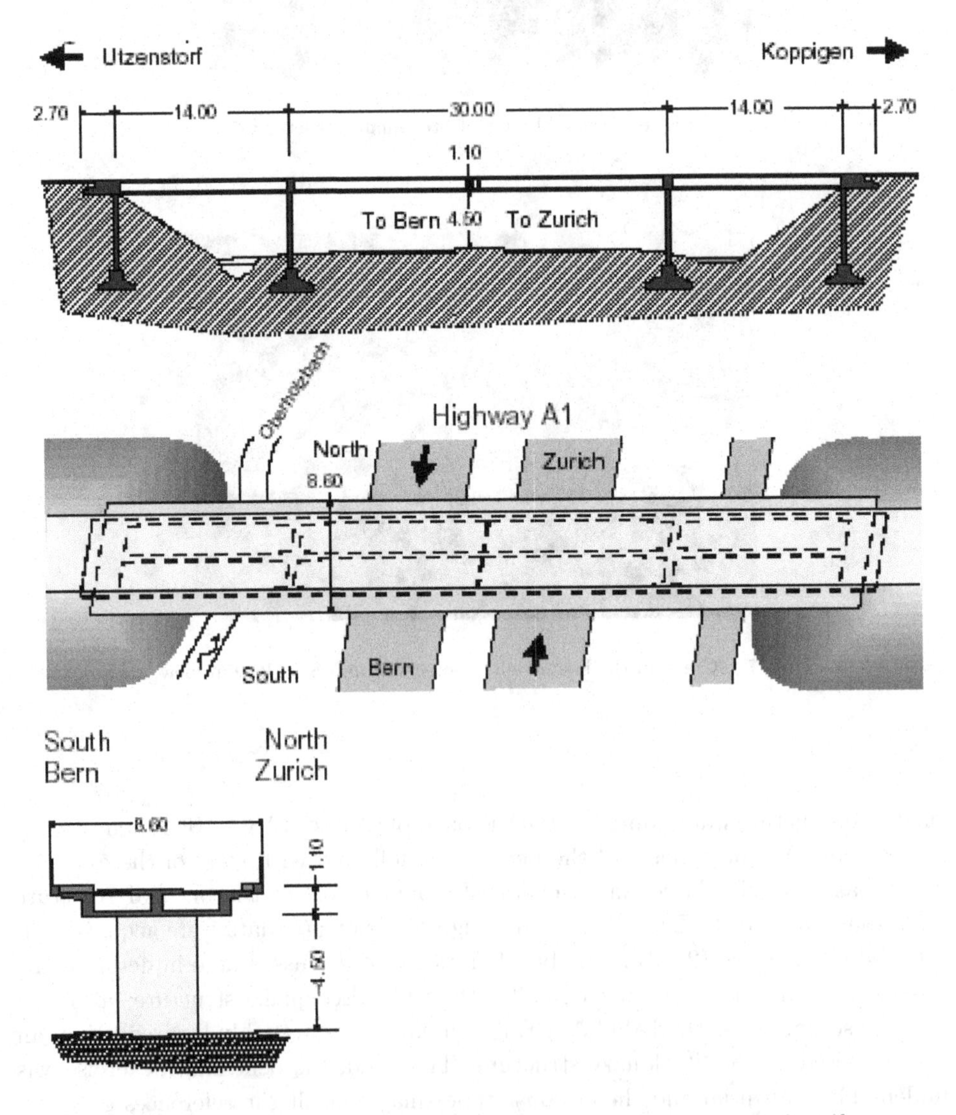

Fig. 8. Z24 bridge longitudinal section and top view [Kramer *et al.* (1999)].

history of SHM developments because towards the end of the monitoring campaign researchers were able to introduce a number of realistic damage scenarios to the structure. In order, these scenarios were [Yan *et al.* (2005)]:

- Pier settlement
- Tilt of foundation followed by settlement removal
- Concrete spalling
- Landslide
- Concrete hinge failure
- Anchor head failure
- Tendons rupture.

The papers immediately produced as a result of the COST action were [Maeck and Roeck (2003); Mevel *et al.* (2003b); Garibaldi *et al.* (2003); Kullaa (2003)], although many have followed. The bridge is discussed in more detail as a case study in the following section.

5. Case Study: Z24 Bridge

In this section a case study is presented showing how, for a real structure undergoing environmental variations, SHM can be accomplished. The structure of interest will be the Z24 bridge introduced in the last section; this choice is motivated by the fact that real damage was introduced in the later stages of the experimental campaign.

Of specific interest here are the natural frequencies of the bridge which were tracked over the period of a year and additionally over the period where the bridge was damaged according to the various scenarios. Modal properties of the bridge were extracted from acceleration data [Peeters and Roeck (2001)]. Figure 9 shows a time history of the four natural frequencies between 0-12 Hz of the bridge. The solid vertical line marks the start of the period where the different levels of damage (starting with pier lowering) were introduced. Gaps where the monitoring system failed have been removed. On inspection of Figure 9, the natural frequencies of the bridge are by no means stationary. There are some large fluctuations in the first half of the time history before the introduction of any damage. These fluctuations occurred during periods of very cold temperatures and have been associated with an increase in stiffness caused by freezing of the asphalt layer on the bridge deck. The natural frequency time histories are, therefore, a good illustrative example of damage sensitive parameters also sensitive to environmental variations, in this case temperature.

As the natural frequencies in their current form would not be suitable to monitor as a damage sensitive feature some action must be taken to remove the variable set's sensitivity to temperature. Various methods have been proposed in the literature as a means of removing environmental variations from feature data, a good survey can be found in [Sohn (2007)]. As discussed in [Sohn (2007)], linear projec-

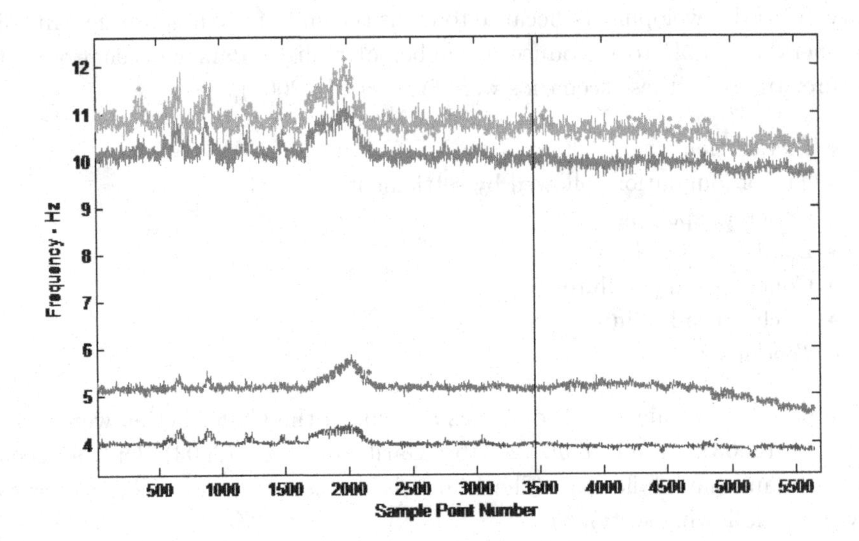

Fig. 9. Time histories of the extracted natural frequencies of the Z-24 Bridge, monitored over one year including a period when damage was introduced.

tion methods like Principal Component Analysis (PCA) and Factor Analysis (FA) have proved valuable for removing environmental and operational effects; however, these approaches are arguably suboptimal for SHM for the following reason. If one considers PCA for example, the projection on the data is made on the basis of preserving maximal signal power in the reduced dimensions; however, if the effects of damage actually manifest themselves in one of the minor components they will be removed. The motivation for PCA is not concordant with the requirements of SHM. An alternative approach to projection is the *cointegration* algorithm which projects on the basis of providing the most *stationary* linear combination of features [Cross *et al.* (2011)]. While this is still not perfectly consistent with the aims of SHM, it is arguably closer in spirit than PCA. The usual formulation of the cointegration is linear-algebraic [Cross *et al.* (2011)] and is based on a maximum likelihood algorithm due to Johansen [Johansen (1995)]. If each variable (natural frequency) is linearly related to temperature, cointegration is an ideal tool to remove temperature induced trends [Cross *et al.* (2011)]. Unfortunately in the Z24 case, the modal properties of the bridge are nonlinearly dependent on temperature (as an example, Figure 10 plots how the first natural frequency changes with temperature), which means that the Z24 data strictly requires the use of some nonlinear variant of cointegration [Dufrénot and Mignon (2002)].

The first sensible step when exploring the ideas of cointegration in this nonlinear context is to look at the results of using Johansen's linear procedure. Figure 11 shows the four natural frequencies of the Z24 bridge projected onto the 'best' (i.e. most stationary) cointegrating vector found by the Johansen procedure, when

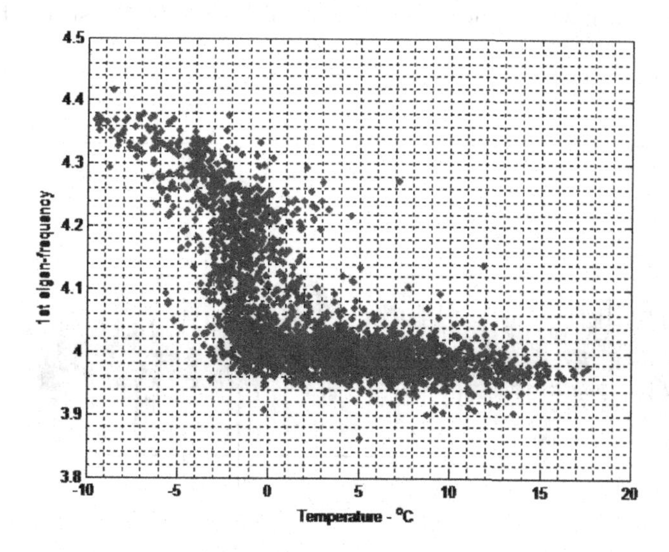

Fig. 10. First Natural frequency nonlinearly dependent on temperature.

trained on data points 1-500 shown in Figure 9. Such projections onto cointegrating vectors are commonly referred to as *residuals*. In the figure, the dotted horizontal lines indicate confidence intervals at $\pm 3\sigma$ of the residual from the training period, while the vertical solid line indicates the beginning of the period where damage was introduced. In the terminology of *Statistical Process Control* (SPC), Figure 11 is referred to as an *X-chart*; if Gaussian statistics are assumed for the residual the signal will only cross the $\pm 3\sigma$ *control limits* 0.3% of the time as as result of statistical fluctuations.

It is clear from Figure 11 that the first cointegrating vector has successfully detrended the data set and furthermore some indication of damage is visible towards the end of the data. It is perhaps unexpected that the linear Johansen procedure is able to remove nonlinear trends; looking at the relationships between the natural frequencies and how the Johansen procedure has combined them, however, sheds light on how this has been achieved. Although each frequency is related nonlinearly to temperature, which drives the frequency fluctuations, some of the frequencies (although not all) are linearly related to each other, which means that the Johansen procedure can successfully combine them to remove their common trends. The way in which the natural frequencies from the Z24 relate to each other is noted in Table 1. On studying the cointegrating vector in question, the residual in Figure 11 predominantly results from a combination of the first and third frequencies, which effectively removes the temperature dependent trends, the combination only contains very small contributions from the second and fourth frequencies. Although this is a successful removal of the temperature dependent trends, two of the variables have effectively not been included in the analysis. This is not an ideal situation as the loss of two variables reduces the chances of being able to successfully detect damage.

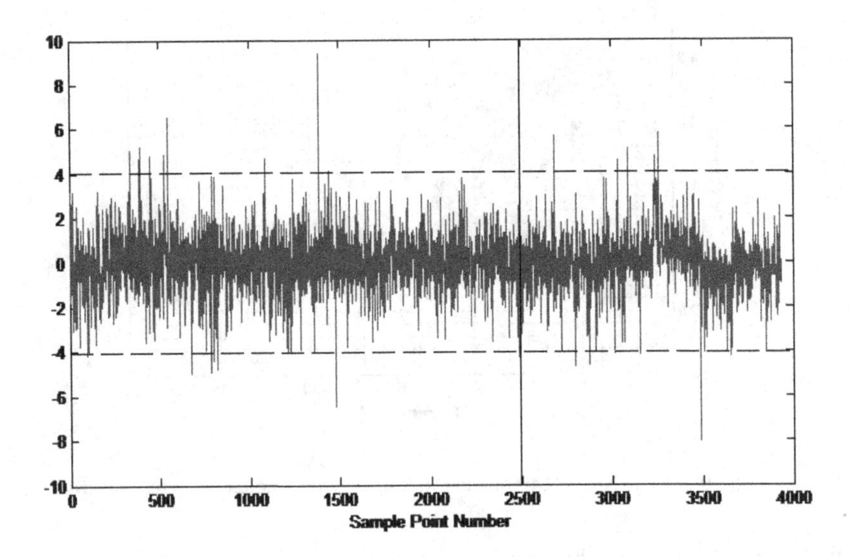

Fig. 11. Linear combination of the Z24 data set; data projected onto the first cointegrating vector found by the Johansen procedure.

Table 1. Nature of observed relationship between the deck natural frequencies of the Z24 Bridge.

Frequency	1	2	3	4
1	-	Nonlinear	Linear	Linear
2	Nonlinear	-	Nonlinear	Nonlinear
3	Linear	Nonlinear	-	Linear
4	Linear	Nonlinear	Linear	-

While cointegration has successfully removed the temperature effects, it has not produced a good indicator of damage. One immediate explanation for this is that the data are actually nonlinearly dependent on temperature and so a nonlinear variant of cointegration is needed. While a number of approaches to nonlinear cointegration based on optimisation principles have been explored in the recent past [Cross and Worden (2010, 2011a,b)], in this instance a more direct possibility presents itself. Because the dependence on temperature is essentially a 'switch' in linear characteristics as the temperature crosses the freezing point (as shown in Figure 10), one way around the problem of nonlinearity in this circumstance could be to use 'locally linear models'. Although overall the Z24 modal frequencies vary nonlinearly with temperature, each of them displays a linear relationship with temperatures above zero degrees, in other words there is a 'locally' linear relationship between temperature and each modal frequency. By only looking at data from the locale of the linear regime, cointegration principles once again become valid. The idea of using locally linear projections has already been explored successfully in the context of the Z24 data in [Yan *et al.* (2005)] using a locally linear variant of PCA.

To illustrate a simple locally linear adaption of cointegration, the Johansen procedure was re-implemented on the four Z24 natural frequency variables, this time with all data points removed if they occurred at temperatures below $1°C$. The first residual produced by the Johansen procedure for this modified data set was almost identical to that shown in Figure 11, in other words the Johansen procedure suggested the same combination to purge the temperature trends. As this particular combination is still not very sensitive to damage a better candidate may come from the other cointegrating vectors calculated by the Johansen procedure (for n variables, the Johansen procedure will produce $n - 1$ cointegrating vectors [Cross *et al.* (2011)]). Figure 12 shows the residual of the second cointegrating vector found by the Johansen procedure for the modified data set; it has successfully been purged of temperature dependence and also becomes nonstationary after the introduction of damage. The reason that the second cointegrating vector has proved more successful at detecting damage can be explained geometrically. The normal condition data will usually occupy some compact region of a subspace of the feature space and this region will encompass any temperature variations. In order to remove the temperature effects, one projects onto a lower-dimensional subspace which is orthogonal (insensitive) to temperature. Each cointegration vector corresponds to a different direction in this subspace and if there are different types of damage (as in the Z24 case) some cointegration projections will be more sensitive to certain types of damage than others. The first type of damage introduced in the case of the Z24 bridge was pier settlement, so it appears that the second cointegrating vector is more sensitive to this damage mode even though it is slightly less stationary under temperature variations. One can see that the first cointegrating vector is actually sensitive (although not very) to the later damage modes (Figure 11).

Even though the cointegrated residual in Figure 12 clearly becomes nonstationary when the damage occurs, the residual is so noisy that the alarm limits on the X-chart are barely breached. Fortunately, this issue is easily dealt with — one passes to an *X-bar chart* [Montgomery (2009)]. Rather than plotting individual values of the residual, one plots sequential averages of small groups of points. The X-bar chart for a subgroup size of 5 is shown in Figure 13; detection of damage is very clear in the figure. However, one should also note that the residual returns to control when the settlement is repaired and the bridge is taken through different damage scenarios. One simple way to deal with multiple damage types is to combine multiple cointegrated residuals into a multivariate feature and then conduct a multivariate SPC analysis (and this is what was carried out in the PCA case in [Yan *et al.* (2005)]; however, this is not shown here as it is considered the relevant points about damage detection have been made with the univariate features.

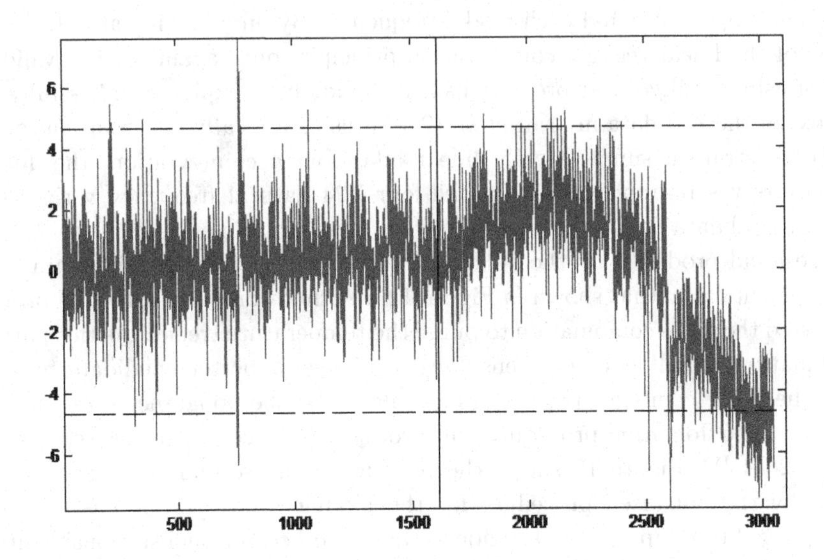

Fig. 12. Data points occurring above 1°C combined using the second cointegrating vector from the Johansen procedure.

In summary then, a case study has been presented for a real bridge where the effects of damage were masked by environmental variations; however, it proved possible to project out those variations and effectively perform SHM.

6. Continuing Challenges in SHM

Reflecting the difficulty of the challenge, it is an unavoidable fact that, despite the maturity of the research field, little of the SHM practice developed has made it into application in the real world. There are a number of diverse reasons why this might be (some touched on previously in this article), and a number of challenges are still to face before SHM can become a reality.

On a practical level, some challenges still remain for the instrumentation of a comprehensive monitoring system, such as how to maintain continual data collection on a large scale (this includes consideration of power sources, data transferral and storage, sensor failures, etc.). In practice, it is a considerable challenge to keep an in-place monitoring system working continuously. A separate issue arises from the transfer of successful SHM technology developed and tested in laboratory conditions to scenarios in the real world. As discussed earlier in this article, often many features monitored for their sensitivity to damage are also sensitive to changes caused in the structure by environmental and operational conditions. A successful SHM system must be able to distinguish between changes caused by innocent ambient variations, such as temperature fluctuations, and those caused by damage. This continues to

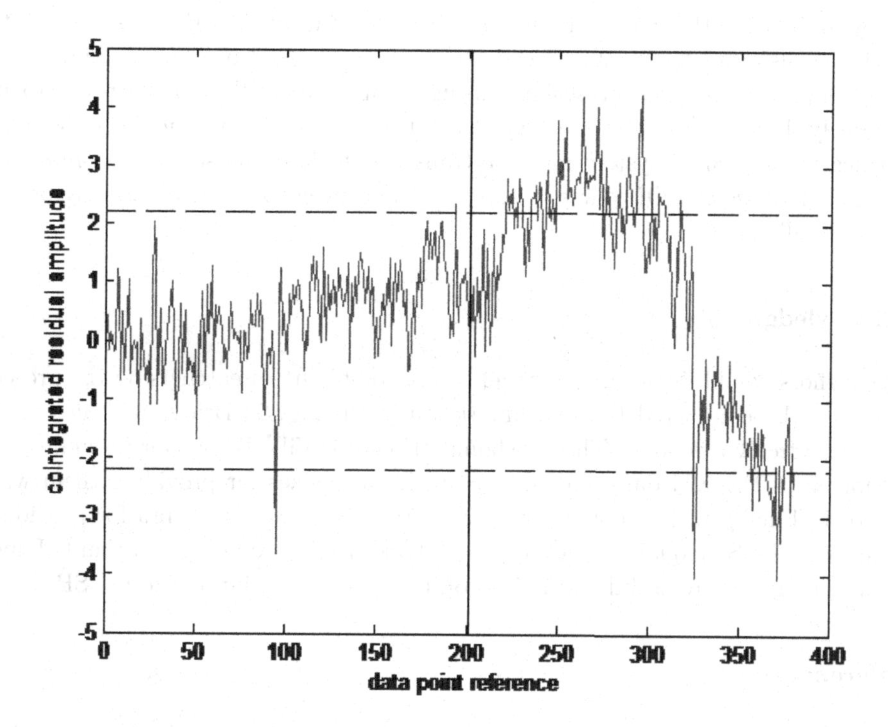

Fig. 13. Data points occurring above 1°C combined using the second cointegrating vector from the Johansen procedure: X-bar chart.

be a challenge for SHM practitioners and a particular one for civil infrastructure which is unavoidably embedded in an uncontrolled environment.

One general reason why SHM uptake has been slow could be put down to the fact that aerospace and civil operators have yet to be convinced by any technology developed so far. The most obvious explanation for this is that the developed technology is not yet up to the challenge and must be developed further. This explanation, is perhaps, however not the only contributing factor to why SHM uptake has not been fast. Other explanations concern general negative attitudes towards SHM; while some believe that SHM is an unobtainable vision, others may believe that change is unnecessary. Understandably, operators will be reluctant to accept an automated unproven system that may be expensive and unreliable (issues which must be addressed within the SHM community), but a general negativity towards SHM can be very unhelpful as this attitude will make it difficult in fact to even put test systems in place on commercial structures. Thankfully, however, this attitude is becoming less common; sceptics who where critical of the HUMS system mentioned earlier for rotorcraft, now find that they are reliant on it, and the aerospace industry are now making a concerted effort to shape the future of SHM.

Tying in with the earlier discussion about validation for SHM systems, certification is also an important issue that must be addressed by the community. So far, gaining certification for systems using complex algorithms has proved to be extremely difficult (certainly so for those labelled as black box models). To gain certification any SHM system must be transparent, based on sound reasoning and as simple as possible. Of course it must also be proven to work, which may be the biggest challenge yet.

Acknowledgments

The authors would like to express their gratitude to the researchers on the project SIMCES who conducted the benchmark study on the Z24 Bridge and also made their data freely available. Thanks should also go to Ville Lämsä for providing the authors with the Z24 data and to Arnaud Deraemaeker for providing them with Figure 1. Finally the authors thank Javier Molina of the JRC Ispra for providing pictures of the Steelquake structure and EJC and KW would like to thank James Brownjohn for many useful discussions on matters of civil infrastructure SHM.

References

Black, C. and Ventura, C. (1998). Blind test on damage detection of a steel frame structure, in *Proceedings of the 16th International Modal Analysis Conference (IMAC XVI)* (Santa Barbara, CA), pp. 623–629.

Bodeux, J.-B. and Golinval, J.-C. (2003). Modal identification and damage detection using the data-driven stochastic subspace and armav methods, *Mechanical Systems and Signal Processing* **17**, pp. 83–89.

Brownjohn, J. (2007). Structural health monitoring of civil infrastructure, *Philosophical Transactions of the Royal Society A: Mathematical, Physical and Engineering Sciences* **365**, 1851, p. 589.

Carden, E. and Fanning, P. (2004). Vibration based condition monitoring: a review, *Structural Health Monitoring* **3**, 4, p. 355.

Chronkite, J. (1993). Practical application of health and usage monitoring (hums) to helicopter rotor, engine and drive system, in *Proceedings of the 49th Forum of the American Helicopter Society*.

Cross, E. J. and Worden, K. (2010). An approach to nonlinear cointegration with a view towards SHM, in *Proceedings of 5th European Workshop on Structural Health Monitoring* (Sorrento, Italy), pp. 875–880.

Cross, E. J. and Worden, K. (2011a). Approaches to nonlinear cointegration with a view towards applications in SHM, in *Proceedings of 8th International Conference on Damage Assessment - DAMAS 2011* (Oxford, UK).

Cross, E. J. and Worden, K. (2011b). Nonlinear cointegration as a combinatorial optimisation problem, in *Proceedings of 8th International Workshop on Structural Health Monitoring* (Palo Alto, CA), pp. 87–94.

Cross, E. J., Worden, K. and Chen, Q. (2011). Cointegration; a novel approach for the removal of environmental trends in structural health monitoring data, *Proceedings of the Royal Society - Series A* **467**, pp. 2712–2732.

Doebling, S. and Farrar, C. (1998). Statistical damage identification techniques applied to the i-40 bridge over the rio grande, in *Proceedings of 16th IMAC* (Santa Barbara, CA).

Doebling, S., Farrar, C., Prime, M. and Shevitz, D. (1996). Damage identification and health monitoring of structural and mechanical systems from changes in their vibration characteristics: a literature review, Tech. rep., Los Alamos National Laboratories.

Dufrénot, G. and Mignon, V. (2002). *Recent Developments in Nonlinear Cointegration with Applications to Macroeconomics and Finance* (Kluwer Academic Publishers, Dordrecht).

Dyke, S., Bernal, D., Beck, J. and Ventura, C. (2003). Experimental phase ii of the structural health monitoring benchmark problem, in *Proceedings of the 16th ASCE Engineering Mechanics Conference* (Reston, VA).

Farrar, C., Baker, W., Bell, T., Cone, K., Darling, T. and Duffey, T. (1994). Dynamic characterization and damage detection in the i-40 bridge over the rio grande, Tech. Rep. LA-12767-MS, Los Alamos National Laboratories, Los Alamos, NM.

Farrar, C., Cornwell, P., Doebling, S. and Prime, M. (2002). Structural health monitoring studies of the alamosa canyon and i-40 bridges, Tech. Rep. LA-13635-MS, Los Alamos National Laboratories, Los Alamos, NM.

Farrar, C. and Jauregui, D. (1998a). Comparative study of damage identification algorithms applied to a bridge: I. experiment, *Smart Materials and Structures* **7**, pp. 704–719.

Farrar, C. and Jauregui, D. (1998b). Comparative study of damage identification algorithms applied to a bridge: Ii. numerical study, *Smart Materials and Structures* **7**, pp. 720–731.

Farrar, C. and Worden, K. (2007). An introduction to structural health monitoring, *Philosophical Transactions of the Royal Society A: Mathematical, Physical and Engineering Sciences* **365**, 1851, p. 303.

Foote, P. (2011). The aerospace industry steering committee on structual health monitoring management (aisc shm): Progress on shm guidelines for aerospace, in *Proceedings of the 8th International Workshop on Structural Health Monitoring.*

Friswell, M. and Mottershead, J. (1995). *Finite element model updating in structural dynamics*, Vol. 38 (Springer).

Fritzen, C.-P. and Bohle, K. (2003). Global damage identification of the 'steelquake' structure using modal data, *Mechanical Systems and Signal Processing* **17**, pp. 111–117.

Fugate, M., Sohn, H. and Farrar, C. (2001). Vibration-based damage detection using statistical process control, *Mechanical Systems and Signal Processing* **15**, pp. 707–721.

Garibaldi, L., Marchesiello, S. and Bonisoli, E. (2003). Identification and up-dating over the z24 benchmark, *Mechanical Systems and Signal Processing* **17**, pp. 153–161.

Goerl, E. and Link, M. (2003). Damage identification using changes of eigenfrequencies and mode shapes, *Mechanical Systems and Signal Processing* **17**, pp. 103–110.

Golinval, J.-C. and Link, M. (2003). Cost action f3 'structural dynamics' (1997-2001) - a european co-operation in the field of science and technology, *Mechanical Systems and Signal Processing* **17**, pp. 3–7.

Johansen, S. (1995). *Likelihood-based Inference in Cointegrated Vector Autoregressive Models* (Oxford University Press, Oxford).

Johnson, E., Lam, H.-F., Katafygiotis, S. and Beck, J. (2004). Phase i iasc-asce structural health monitoring benchmark problem using simulated data, *ACSE Journal of Engineering Mechanics* **130**, pp. 3–15.

Ko, J. and Ni, Y. (2005). Technology developments in structural health monitoring of large-scale bridges, *Engineering structures* **27**, 12, pp. 1715–1725.

Kramer, C., Smet, C. D. and Roeck, G. D. (1999). Z24-bridge damage detection tests, in *Proceedings of 17th IMAC* (Kissimmee, FL), pp. 1023–1029.

Kullaa, J. (2003). Damage detection of the z24 bridge using control charts, *Mechanical Systems and Signal Processing* **17**, pp. 163–170.

Lam, H.-F. (2003). Phase iie of the iasc-asce benchmark study on structural health monitoring, in *Proceedings of the 21st International Modal Analysis Conference (IMAC XXI)* (Kissimmee, FL).

Larsen, A. (2000). Aerodynamics of the tacoma narrows bridge - 60 years later, *Structural Engineering International* **10**, 4, pp. 243–248.

Lee, S. and Sohn, H. (2006). Active self-sensing scheme development for structural health monitoring, *Smart Materials and Structures* **15**, 6, pp. 1734–1746, cited By (since 1996) 11.

Lynch, J. (2007). An overview of wireless structural health monitoring for civil structures, *Philosophical Transactions of the Royal Society A: Mathematical, Physical and Engineering Sciences* **365**, 1851, pp. 345–372.

Maeck, J. and Roeck, G. D. (2003). Damage assessment using vibration analysis on the z24 bridge, *Mechanical Systems and Signal Processing* **17**, pp. 133–142.

Mevel, L., Basseville, M. and Boursat, M. (2003a). Stochastic subspace-based structural identification and damage detection - application to the steelquake benchmark, *Mechanical Systems and Signal Processing* **17**, pp. 91–101.

Mevel, L., Goursat, M. and Basseville, M. (2003b). Stochastic subspace-based structural identification and damage detection and localisatio - application to the z24 bridge benchmark, *Mechanical Systems and Signal Processing* **17**, pp. 143–151.

Montgomery, D. (2009). *Introduction to Statistical Quality Control* (John Wiley and Sons).

Ni, Y., Xia, H. and Ko, J. (2008). Structural performance evaluation of tsing ma bridge deck using long-term monitoring data, *Modern Physics Letters B* **22**, pp. 875–880.

Papatheou, E., Manson, G., Barthorpe, R. and Worden, K. (2010). The use of pseudo-faults for novelty detection in shm, *Journal of Sound and Vibration* **329**, pp. 2349–2366.

Park, G. and Inman, D. (2007). Structural health monitoring using piezoelectric impedance measurements, *Philosophical Transactions of the Royal Society, Series A* **365**, pp. 373–392.

Park, G., Rosing, T., Todd, M., Farrar, C. and Hodgkiss, W. (2008). Energy harvesting for structural health monitoring sensor networks, *Journal of Infrastructure Systems* **14**, p. 64.

Peeters, B. and Roeck, G. D. (2001). One-year monitoring of the z24-bridge: environmental effects versus damage events, *Earthquake Engineering and Structural Dynamics* **30**, pp. 149–171.

Raghaven, A. and Cesnik, C. (2007). Review of guided-wave structural health monitoring, *Shock and Vibration Digest* **39**, pp. 91–114.

Randall, R. (2011). *Vibration-Based Condition Monitoring: Industrial, Aerospace and Automotive Applications* (Wiley-Blackwell).

Rizzo, P. and di Scalea, F. (2001). Acoustic emission monitoring of carbon-fiber-reinforced-polymer bridge stay cables in large-scale testing, *Experimental mechanics* **41**, 3, pp. 282–290.

Roeck, G. D. (2003). The state-of-the-art of damage detection by vibration monitoring: the simces experience, *Journal of Structural Control* **10**, pp. 127–134.

Rytter, A. (1993). *Vibration-based inspection of civil engineering structures*, Ph.D. thesis, Department of Building Technology and Structural Engineering, University of Aalborg, Denmark.

Sohn, H. (2007). Effects of environmental and operational variability on structural health monitoring, *Philosophical Transactions of the Royal Society A: Mathematical, Physical and Engineering Sciences* **365**, 1851, pp. 539–561.

Sohn, H., Farrar, C., Hemez, F. and Czarnecki, J. (2002). A review of structural health review of structural health monitoring literature 1996-2001. Tech. rep., Los Alamos National Laboratories.

Sohn, H., Farrar, C., Hunter, N. and Worden, K. (2001). Structural health monitoring using statistical pattern recognition techniques, *ASME Journal of Dynamics, Measurement Systems and Control* **123**, pp. 706–711.

Taleb, N. (2011). *The Black Swan: The Impact of the Highly Improbable* (Penguin).

Wenzel, H. (2009a). The character of shm in civil engineering, in C. Boller, F.-K. Chang and Y. Fujino (eds.), *Encyclopedia of Structural Health Monitoring* (Wiley, Chichester, UK), pp. 2031–2037.

Wenzel, H. (2009b). *Health Monitoring of Bridges* (John Wiley and Sons, Ltd).

Worden, K. (2003). Cost action f3 on structural dynamics: benchmarks for working group 2 - structural health monitoring, *Mechanical Systems and Signal Processing* **17**, pp. 73–75.

Worden, K. and Dulieu-Barton, J. (2004). Damage identification in systems and structures, *International Journal of Structural Health Monitoring* **3**, pp. 85–98.

Worden, K., Farrar, C. R., Manson, G. and Park, G. (2007). The fundamental axioms of structural health monitoring, *Proceedings of the Royal Society - Series A* **463**, pp. 1639–1664.

Worden, K., Staszewski, W. and Hensman, J. (2011). Natural computing for mechanical systems research: A tutorial overview, *Mechanical Systems and Signal Processing* **25**, 1, pp. 4–111.

Yan, A.-M., Kerschen, G., Boe, P. D. and Golinval, J.-C. (2005). Structural damage diagnosis under varying environmental conditions part ii: local pca for non-linear cases, *Mechanical Systems and Signal Processing* **19**, pp. 865–880.

Zapico, J., Gonzaález, M. and Worden, K. (2003). Damage assessment using neural networks, *Mechanical Systems and Signal Processing* **17**, pp. 119–125.

Chapter 2

Enhanced Damage Locating Vector Method for Structural Health Monitoring

S. T. Quek[1], V. A. Tran[2], and N. N. K. Lee[3]

[1]*Professor, National University of Singapore,*
[2]*Director, SICOM Investment Construction JSC,*
[3]*Research Engineer, National University of Singapore,*
E-mail: [1]*ceeqst@nus.edu.sg,* [2]*anh.tv@sicom.com.vn,* [3]*ceenlmk@nus.edu.sg*

1. The DLV Method

1.1. *General concept*

Consider a linear elastic structure with *ns* sensors attached. Let F_u and F_d denote the (*ns* × *ns*) flexibility matrices constructed with respect to the *ns* sensor locations for the reference (undamaged or original) and damaged structures, respectively. If a non-zero (*ns* × 1) static load vector P exists and is applied to both structures such that their work done are equal, then

$$0.5P^T(F_dP) = 0.5P^T(F_uP) \quad \text{or} \quad (F_d\text{-}F_u)P = F_\Delta P = 0 \tag{1}$$

This implies two possibilities: (i) $\mathbf{F}_\Delta = \mathbf{0}$, or (ii) $\mathbf{F}_\Delta$ is rank deficient and **P** is a basis of the null space of $\mathbf{F}_\Delta$. The first case implies that the structure is unaltered and the latter implies that **P** effectively does not induce any additional stress (or energy) on the damaged elements, if any, in the structure. This feature facilitates the locating of structural damage and hence **P** is known as DLV. The DLV method was first developed by Bernal (2002).

By performing singular value decomposition (SVD) on $\mathbf{F}_\Delta$ gives

$$F_\Delta = U\Sigma V^T = [U_1 \quad U_0]\begin{bmatrix}\Sigma_1 & 0 \\ 0 & \Sigma_0(=0)\end{bmatrix}[V_1 \quad V_0]^T \tag{2}$$

where $\mathbf{\Sigma}$ is the singular value matrix, $\mathbf{\Sigma}_1$ is the sub-matrix of $\mathbf{\Sigma}$ containing all non-zero singular values (NZV), $\mathbf{\Sigma}_0$ is the sub-matrix of $\mathbf{\Sigma}$ containing all zero singular values; and $\mathbf{U}$ and $\mathbf{V}$ are the left and right singular matrices, respectively. Post-multiplying equation (2) by $\mathbf{V}$ on both sides and using the orthonormal properties of $\mathbf{V}$ yields

$$F_\Delta[V_1 \quad V_0] = [F_\Delta V_1 \quad F_\Delta V_0] = [U_1\Sigma_1 \quad U_0\Sigma_0] = [U_1\Sigma_1 \quad 0] \tag{3}$$

This gives $\mathbf{F}_\Delta \mathbf{V}_0 = \mathbf{0}$, implying that each column in $\mathbf{V}_0$ is a feasible solution to equation (1). This provides a mean to obtain sets of DLVs since the work done by this force on the structural changes is zero. The columns of $\mathbf{V}_0$ constitute a set of DLVs where the number of columns in $\mathbf{V}_0$ (denoted as *ndlv*) is less than *ns*.

1.2. *Normalized cumulative energy (NCE)*

If column i of $\mathbf{V}_0$ is applied to the reference structure, the energy induced on element j, denoted as Ξ_{ji}, is given by

$$\Xi_{ji} = \int_{L_j} \frac{M_{ji}^2}{2E_jI_j} ds + \int_{L_j} v\frac{Q_{ji}^2}{2G_jA_j} ds + \int_{L_j} \frac{N_{ji}^2}{2E_jA_j} ds \tag{4}$$

where M_{ji}, Q_{ji}, N_{ji} are the internal moment, shear and axial force, respectively within element j; L_j is its length; v is its Poisson's ratio; and E_jI_j, G_jA_j, E_jA_j are its flexural, shear and axial stiffness, respectively, in which E_j is its Young's modulus and G_j is its shear modulus.

The Normalized Cumulative Energy (NCE) of element j is defined as

$$\overline{\Xi}_j = \frac{\sum_{k=1}^{ndlv} \Xi_{jk}}{\max_{\text{all } j}(\Xi_j)} = \frac{\Xi_j}{\Xi_{max}} \tag{5}$$

Physically, the NCE of element j is the normalized sum of internal work done on the element j by the complete set of DLVs. In equation (5), the normalization is done using the maximum sum of internal work on an element in the structure. An alternative normalization base is the sum of the total internal work done by all elements. In such case, the NCE of the element is the fraction contributed by this element to the total work done on the structure.

Theoretically, potential damage elements (PDEs) are classified as elements with $\overline{\Xi}_j = 0$. In reality, the NCE of damage elements may not be exactly zero due to the presence of noise, uncertainties and round-off errors. Thus, a non-zero threshold is needed for practical application. Based on study by Quek *et al.* (2009), NCE threshold of 0.01 may be used.

The above approach uses the flexibility matrix for the SVD, and the resulting DLV corresponds to the physical quantity of force. If stiffness matrix is used instead, a set of DLVs can be similarly obtained by performing SVD; the resulting DLVs correspond to the physical quantity of displacement. For both cases, applying these DLVs to the reference structure as force (flexibility matrix) or displacement (stiffness matrix), damaged elements can be identified through the NCE of the elements.

2. Identifying Actual Damage Elements

2.1. *Intersection scheme*

In the DLV method, if limited numbers of sensors are being employed, the set of PDEs identified would contain some undamaged elements (Bernal, 2002). To filter out the actual damage elements from the set of PDEs, all potential sets derived using data from various combinations of sensors are compared. The actual damaged elements are the common elements across these different sets of PDEs. This is basically an intersection scheme and its flow diagram is illustrated in Fig. 1.

Starting with *ns* sensors, a set of PDEs is computed and denoted as the current intersected damage set (IDS). Denote the number of elements contained in the IDS as *ne*. The next step is to use only data from *ns* – 1 sensors, from which another set of PDE can be identified. The current IDS is updated by taking the common elements from the computed PDEs and the previous IDS. This procedure can be repeated using a different combination of *ns* – 1 sensors to update the IDS. If the new updated IDS is identical to the previous IDS, the elements in the IDS are identified as the actual damage elements and the identification process is considered complete. The process is also terminated if the new IDS is a null set, implying that there is no damaged (or altered) element in the structure. If two consecutive IDSs are not identical and all possible combinations of *ns* – 1 sensors are exhausted, then combinations of *ns* – 2 sensors are next considered until the criterion of two consecutive identical IDSs is met. To ensure robustness of the scheme, the termination criterion can be changed by increasing the requirement of number of consecutive identical IDSs at the expense of computational cost.

The scheme works provided that *ns* is greater than two since at least two measurements are needed to form a matrix before any SVD can be performed to compute the DLVs. With $ns = 2$, only one set of PDEs can be computed and no subsequent combination of sensors is available to filter out the actual damaged elements.

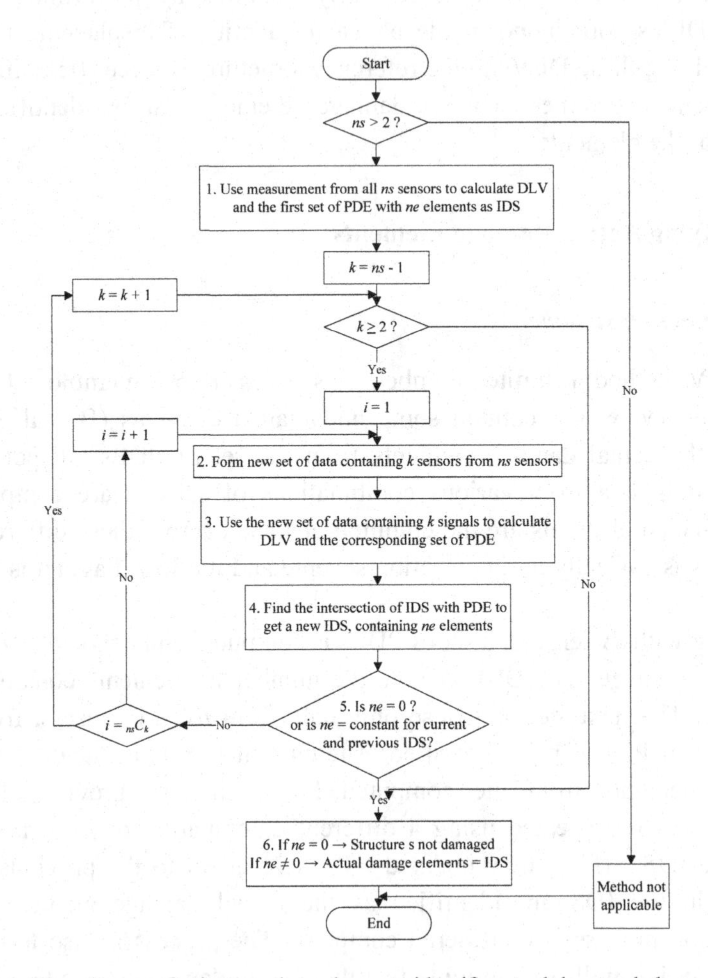

Fig. 1. Flow diagram for intersection scheme to identify actual damaged elements.

3. Formulation of Flexibility Matrix at Sensor Location

Before the DLV method can be implemented, the flexibility or stiffness matrices of the reference and damaged structures need to be constructed based on measurement data. Depending on the type of actuation (static or dynamic), different techniques are required and a brief summary is provided.

3.1. *Forming flexibility matrix using static responses*

3.1.1. *Static responses with load of known magnitude*

To formulate the flexibility matrix using static response, displacement measurements at targeted locations are needed. This can be done, for example, by employing non-contact laser instruments or displacement transducers with fixed reference points. When a load P is applied at DOF j (where there is sensor mounted) of an n-DOF structure, the compatibility conditions at all DOF of the structure can be expressed as

$$\mathrm{Ff} = \mathrm{d} \tag{6}$$

where $\mathbf{F}$ is the $(n \times n)$ structural flexibility matrix; $\mathbf{f} = \mathbf{B}_2 P$ is the applied force vector which $\mathbf{B}_2$ is the $(n \times 1)$ input matrix to map P to the structural DOF, and $\mathbf{d}$ is the $(n \times 1)$ nodal displacement vector. Extracting the compatibility conditions at ns sensor locations in equation (6)

$$\begin{bmatrix} d_{1\,j} \\ d_{2\,j} \\ \vdots \\ d_{i\,j} \\ \vdots \\ d_{ns\,j} \end{bmatrix} = \begin{bmatrix} \delta_{1\,1} & \delta_{1\,2} & \cdots & \delta_{1\,j} & \cdots & \delta_{1\,ns} \\ \delta_{2\,1} & \delta_{2\,2} & \cdots & \delta_{2\,j} & \cdots & \delta_{2\,ns} \\ \vdots & \vdots & & \vdots & & \vdots \\ \delta_{i\,1} & \delta_{i\,2} & \cdots & \delta_{i\,j} & \cdots & \delta_{i\,ns} \\ \vdots & \vdots & & \vdots & & \vdots \\ \delta_{ns\,1} & \delta_{ns\,2} & \cdots & \delta_{ns\,j} & \cdots & \delta_{ns\,ns} \end{bmatrix} \begin{bmatrix} 0 \\ 0 \\ \vdots \\ P_j \\ \vdots \\ 0 \end{bmatrix} \tag{7}$$

where $d_{i\,j}$ is the displacement at DOF i resulting from load P_j being applied at DOF j and $\delta_{i\,j}$ is the displacement at DOF i due to a unit load being applied at DOF j.

$$\begin{bmatrix} d_{1\,j} \\ d_{2\,j} \\ \vdots \\ d_{i\,j} \\ \vdots \\ d_{ns\,j} \end{bmatrix} = \begin{bmatrix} \delta_{1\,j}\,P_j \\ \delta_{2\,j}\,P_j \\ \vdots \\ \delta_{i\,j}\,P_j \\ \vdots \\ \delta_{ns\,j}\,P_j \end{bmatrix} \text{ or } \begin{bmatrix} d_{1\,j}\,/\,P_j \\ d_{2\,j}\,/\,P_j \\ \vdots \\ d_{i\,j}\,/\,P_j \\ \vdots \\ d_{ns\,j}\,/\,P_j \end{bmatrix} = \begin{bmatrix} \delta_{1\,j} \\ \delta_{2\,j} \\ \vdots \\ \delta_{i\,j} \\ \vdots \\ \delta_{ns\,j} \end{bmatrix} \tag{8}$$

By shifting the load P through each location of the ns sensors and measuring the displacements at the sensor locations, the corresponding flexibility matrix ($\mathbf{F}_s$) can be assembled as

$$\mathbf{F}_s = \begin{bmatrix} \delta_{1\,1} & \delta_{1\,2} & \cdots & \delta_{1\,ns} \\ \delta_{2\,1} & \delta_{2\,2} & \cdots & \delta_{2\,ns} \\ \vdots & \vdots & & \vdots \\ \delta_{ns\,1} & \delta_{ns\,2} & \cdots & \delta_{ns\,ns} \end{bmatrix} = \begin{bmatrix} \frac{d_{1\,1}}{P_1} & \frac{d_{1\,2}}{P_2} & \cdots & \frac{d_{1\,1}}{P_{ns}} \\ \frac{d_{2\,1}}{P_1} & \frac{d_{2\,2}}{P_2} & \cdots & \frac{d_{2\,ns}}{P_{ns}} \\ \vdots & \vdots & & \vdots \\ \frac{d_{ns\,1}}{P_1} & \frac{d_{ns\,2}}{P_2} & \cdots & \frac{d_{ns\,ns}}{P_{ns}} \end{bmatrix} \tag{9}$$

For ease of implementation and computation, the static load P is usually kept constant. It should also be noted that the formulation of flexibility matrix is still valid when moments are applied instead of static forces at the sensor locations if nodal rotation measurements are available.

3.1.2. *Static responses with load of unknown magnitude*

When the applied static load is of unknown magnitude, the flexibility matrices can still be formulated using displacement responses with some modifications. Shifting each of the two static loads, P_u and P_d, through the ns sensor-attached DOFs of the reference and damaged structures, respectively, the displacement responses at the ns sensor locations are measured. The flexibility matrices with respect to the sensor locations of the reference ($\mathbf{F}_u$) and damaged ($\mathbf{F}_d$) structures can be written as

$$\mathrm{F_u} = \frac{1}{\mathrm{P_u}} \begin{bmatrix} \mathrm{d^u_{1\,1}} & \mathrm{d^u_{1\,2}} & \cdots & \mathrm{d^u_{1\,n}} \\ \mathrm{d^u_{2\,1}} & \mathrm{d^u_{2\,2}} & \cdots & \mathrm{d^u_{2\,ns}} \\ \vdots & \vdots & & \vdots \\ \mathrm{d^u_{ns\,1}} & \mathrm{d^u_{ns\,2}} & \cdots & \mathrm{d^u_{ns\,ns}} \end{bmatrix} = \frac{1}{\mathrm{P_u}} \mathrm{U_u} \tag{10}$$

$$\mathrm{F_d} = \frac{1}{\mathrm{P_d}} \begin{bmatrix} \mathrm{d^d_{1\,1}} & \mathrm{d^d_{1\,2}} & \cdots & \mathrm{d^d_{1\,n}} \\ \mathrm{d^d_{2\,1}} & \mathrm{d^d_{2\,2}} & \cdots & \mathrm{d^d_{2\,ns}} \\ \vdots & \vdots & & \vdots \\ \mathrm{d^d_{ns\,1}} & \mathrm{d^d_{ns\,2}} & \cdots & \mathrm{d^d_{ns\,ns}} \end{bmatrix} = \frac{1}{\mathrm{P_d}} \mathrm{U_d} \tag{11}$$

where $d^u_{i\,j}$ and $d^d_{i\,j}$ are the displacements at DOF i due to load P_u and load P_d at DOF j corresponding to reference and damaged states, respectively. $\mathbf{U}_u$ and $\mathbf{U}_d$ are the displacement matrices with respect to the sensor locations of the reference and the damaged structures, respectively.

Assuming there is a constant α such that

$$\mathrm{P_u} = \alpha \mathrm{P_d} \tag{12}$$

Multiplying equation (1) by P_u and combining with equation (13) gives

$$(\alpha \mathrm{P_d F_d}\text{-}\mathrm{P_u F_u})\mathrm{P} = 0 \rightarrow (\alpha \mathrm{U_d}\text{-}\mathrm{U_u})\mathrm{P} = \mathrm{U_\Delta P} = 0 \tag{13}$$

where

$$\mathrm{U_\Delta} = \alpha \mathrm{U_d}\text{-}\mathrm{U_u} = \mathrm{P_u F_\Delta} \tag{14}$$

From equation (13), performing SVD on $\mathbf{U}_\Delta$ and $\mathbf{F}_\Delta$ produces the same right singular matrix. Hence, the set of DLVs can also be computed by replacing $\mathbf{F}_\Delta$ with $\mathbf{U}_\Delta$ in equations (2) and (3) and the remaining requirement is to estimate α such that equation (12) is satisfied.

The distinguishing feature of α which satisfies equation (12) is that: (i) $\mathbf{U}_\Delta = \mathbf{F}_\Delta = \mathbf{0}$ if the structure is not damaged; (ii) $\mathbf{F}_\Delta$ is rank-deficient, if the structure is damaged. Hence, $\mathbf{U}_\Delta$ is also rank-deficient. In both cases, α produces the smallest summation of singular values of $\mathbf{U}_\Delta = \alpha \mathbf{U}_d - \mathbf{U}_u$. Thus, α can be estimated by minimizing the sum of singular values of $\mathbf{U}_\Delta$; that is, minimize the function z_α given by

$$\mathrm{z}_\alpha = \textstyle\sum_{\mathrm{i}=1}^{\mathrm{ns}} \mathrm{s}_\Delta(\mathrm{i}) \tag{15}$$

where $s_\Delta(i)$ is the ith singular value of $\mathbf{U}_\Delta$. The minimization can be executed in MATLAB using the function fminbnd. Upon estimating α, DLVs can be computed by replacing $\mathbf{F}_\Delta$ by $\mathbf{U}_\Delta$ in equations (2) and (3).

3.2. *Forming flexibility matrix using dynamic responses*

While the formulation of flexibility matrices based on static responses is simple conceptually, it is not widely used due to practical limitations as direct displacement measurements are hard to come by. To measure a nodal displacement using displacement transducer, a fixed reference point within close proximity to that node is required. Employing non-contact laser instruments may solve the problem but they do not come cheap. In addition, large static load is generally required to produce measureable displacements. Moreover, moving the heavy load from point to point to generate sufficient readings is time consuming. The results are also sensitive to the presence of noise and uncertainties. Hence, the formulation of flexibility matrices using dynamic (acceleration) responses is preferred. In formulating flexibility matrices based on dynamic responses, depending on whether the input excitations are measured, different formulation techniques are needed.

3.2.1. *Dynamic responses with known excitation*

When input excitations are measured and there is at least one collocated sensor-actuator pair, the mass normalized mode shapes can be obtained from the measured data (Alvin and Park, 1994) and, hence, the flexibility matrix can be computed. Bernal and Gunes (2004) proposed using of eigensystem realization algorithm (ERA) (Juang and Pappa, 1985; Juang, 1994) to construct the flexibility matrices at sensor locations. The three major steps of this procedure are as follows.

Step 1: Compute structural flexibility matrix in term of complex modes
Consider the equation of motion for the free vibration of an n-DOF linear structure

$$\mathrm{M}\ddot{\mathrm{d}} + \mathrm{D_d}\dot{\mathrm{d}} + \mathrm{Kd} = 0 \tag{16}$$

where $\mathbf{M}$, $\mathbf{D}_d$ and $\mathbf{K}$ are the ($n \times n$) mass, damping and stiffness matrices, respectively; $\ddot{\mathbf{d}}$, $\dot{\mathbf{d}}$ and $\mathbf{d}$ are the ($n \times 1$) acceleration, velocity and displacement vectors, respectively. A state-space representation of the structure can be expressed as

$$\begin{bmatrix} \mathbf{D}_d & \mathbf{M} \\ \mathbf{M} & \mathbf{0} \end{bmatrix} \dot{\mathbf{x}} = \begin{bmatrix} -\mathbf{K} & \mathbf{0} \\ \mathbf{0} & \mathbf{M} \end{bmatrix} \mathbf{x} \tag{17}$$

where $\mathbf{x} = [\mathbf{d} \quad \dot{\mathbf{d}}]^T$ is the state vector. Solving equation (17), the eigenvalue and eigenvector matrices are

$$\Lambda_1 = \begin{bmatrix} \Lambda & 0 \\ 0 & \Lambda^* \end{bmatrix} \quad \text{and} \quad \Phi = \begin{bmatrix} \psi & \psi^* \\ \psi\Lambda & \psi^*\Lambda^* \end{bmatrix} \tag{18}$$

where '*' denotes complex conjugate. Let

$$\mathrm{D}_g^{-1} = \Phi^{\mathrm{T}} \begin{bmatrix} -\mathrm{K} & 0 \\ 0 & \mathrm{M} \end{bmatrix} \Phi \tag{19}$$

Based on the orthogonality of the eigenvectors,

$$\begin{bmatrix} -\mathrm{K}^{-1} & 0 \\ 0 & \mathrm{M}^{-1} \end{bmatrix} = \begin{bmatrix} -\mathrm{F} & 0 \\ 0 & \mathrm{M}^{-1} \end{bmatrix} = \Phi \mathrm{D_g} \Phi^{\mathrm{T}} \tag{20}$$

From equation (18),

$$F = \Upsilon D_g \Upsilon^T \tag{21}$$

where $\Upsilon = [\psi \quad \psi^*]$ is the normalized complex mode shapes of the structure and D_g is the modal normalization constant.

Step 2: Partition flexibility matrix from state-space realization

After obtaining the state-space representation of the structure using ERA, the discrete time state-space representation identified from measurement data can be converted to continuous time as

$$\dot{\boldsymbol{x}} = \boldsymbol{A}_c \boldsymbol{x} + \boldsymbol{B}_c \boldsymbol{u}$$

$$\mathbf{y} = \mathbf{Cx} + \mathbf{Du} \tag{22}$$

where $\mathbf{A}_c$ is the state matrix; $\mathbf{B}_c$ is the input influence matrix; $\mathbf{C}$ is the output influence matrix; $\mathbf{D}$ is the direct transmission matrix; $\mathbf{x}$ is the state vector; $\boldsymbol{u}$ is the input excitation vector; and $\mathbf{y}$ is the output vector. Performing Fourier Transformation on equation (22) yields

$$y(\omega) = \left\{C\left[I \cdot i\omega - A_c\right]^{-1} B_c + D\right\} u(\omega) \tag{23}$$

The Fourier Transformation of the output vector can be expressed in terms of the displacement vector y_D as

$$y_D(\omega) = \frac{1}{(i\omega)^p}\left\{C\left[I \cdot i\omega - A_c\right]^{-1} B_c + D\right\} u(\omega) \tag{24}$$

where $p = 0$, 1 or 2 corresponding to the output measurement of displacement, velocity or acceleration, respectively.

The flexibility matrix relates the inputs to the outputs at $\omega = 0$. By defining $\mathbf{F}_f$ as a suitable partition of the flexibility matrix $\boldsymbol{F}$ such that

$$y_D(0) = F_f u(0) \tag{25}$$

then

$$\mathbf{F}_f = \lim_{\omega \to 0}\left[\frac{1}{(i\omega)^p}\{\mathbf{C}[\mathbf{I} \cdot i\omega - \mathbf{A}_c]^{-1}\mathbf{B}_c + \mathbf{D}\}\right] = -\mathbf{C}\mathbf{A}_c{}^{-(p+1)}\mathbf{B}_c \tag{26}$$

Step 3: Compute diagonal matrix containing modal constant ($\mathbf{D}_g$)

Bernal and Gunes (2004) shows that the complex eigenvector ($\mathbf{\Upsilon}_s$) for the reduced system, defined with respect to DOF at the sensor locations, can obtained via the following relationship

$$\Upsilon_s = C\Upsilon_c \tag{27}$$

The flexibility matrix of the reduced system, defined at the sensor locations, can be expressed as

$$F_s = -\Upsilon_s D_g \Upsilon_s^T \tag{28}$$

Expressing $\mathbf{A}_c$ in terms of its eigenvalue ($\mathbf{\Lambda}_c$) and eigenvector ($\mathbf{\Upsilon}_c$) matrices, equation (26) becomes

$$F_f = -C\Upsilon_c\Lambda_c^{-(p+1)}\Upsilon_c^{-1}B_c = -\Upsilon_s\Lambda_c^{-(p+1)}\Upsilon_c^{-1}B_c \tag{29}$$

Extracting from the columns of $\mathbf{B}_c$ and $\mathbf{\Upsilon}_s^T$ corresponding to the DOFs where the actuators and sensors are collocated and define the extracted columns as $\mathbf{B}_{cc}$ and $\mathbf{\Upsilon}_{sc}^T$, respectively. By comparing the collocated columns of the flexibility matrices in equations (28) and (29), we have

$$-\Upsilon_s D_g \Upsilon_{sc}^T = -\Upsilon_s\Lambda_c^{-(p+1)}\Upsilon_c^{-1}B_{cc}$$

or

$$D_g \Upsilon_{sc}^T = \Lambda_c^{-(p+1)}\Upsilon_c^{-1}B_{cc} \tag{30}$$

From equation (30), the modal normalized constant matrix $\mathbf{D}_g$can be computed as

$$\mathbf{D}_g = \mathbf{\Lambda}_c^{-(p+1)}\mathbf{\Upsilon}_c^{-1}\mathbf{B}_{cc}\begin{bmatrix} \Upsilon_{sc}(1) & & & 0 \\ & \Upsilon_{sc}(2) & & \\ & & \ddots & \\ 0 & & & \Upsilon_{sc}(2n) \end{bmatrix}^{-1} \tag{31}$$

where $\Upsilon_{sc}(k)$ is the kth component of $\mathbf{\Upsilon}_{sc}^T$. Once $\mathbf{D}_g$ is estimated, the flexibility matrix with respect to the sensor locations can be calculated via equation (28).

3.2.2. *Dynamic responses with unknown excitation*

To formulate the flexibility matrix from measured accelerations with unknown excitation, Bernal (2004) proposed adding a known disturbance mass to the structure to evaluate the change in eigensolution by assuming that the changes in structural frequency and mode shape are small. Using the information from the change of eigensolution due to a known disturbed mass, the mass normalized constants are evaluated and, hence, the flexibility matrix can be computed.

Alternatively, Gao *et al.* (2007) assumed that the change in mass normalized constants is negligible when the structure changes in the event when a structural element is damaged. Hence, the flexibility matrix can be computed based on the mass normalized constants at the reference state. While the proposal by Bernal (2004) may not be convenient in practice, the proposal by Gao *et al.* (2007) may not be accurate when the damage to the element is severe.

In a third method, the stiffness matrix of the structure is computed based solely on the acceleration measurements at the sensor locations, with an assumption that the mass matrix is known. Consider an n-DOF structure with ns sensors attached. Assuming r input excitations of unknown magnitude are acting on the structure, nc (≥ 0) of which are collocated with the sensors. The locations of the sensors and excitations are also assumed known. The equation of motion of the system can be expressed as

$$\mathbf{M}\ddot{\mathbf{d}} + \mathbf{D}_d\dot{\mathbf{d}} + \mathbf{K}\mathbf{d} = \mathbf{P} = \mathbf{B}_2\mathbf{u} \tag{32}$$

where the definitions of $\mathbf{M}$, $\mathbf{D}_d$, $\mathbf{K}$, $\ddot{\mathbf{d}}$, $\dot{\mathbf{d}}$ and $\mathbf{d}$ are identical to those in equation (16); $\mathbf{P}$ is the $(n \times 1)$ applied force vector; $\mathbf{B}_2$ is the input influence matrix to map the $(r \times 1)$ input force vector $\mathbf{u}$ to the structural DOFs.

If a sampling time interval of Δt is used, then the structural displacement and velocity responses at time step j can be estimated through Newmark- β method as

$$\dot{\mathrm{d}}_{\mathrm{j}} = \dot{\mathrm{d}}_{\mathrm{j-1}} + (1\text{-}\gamma)\Delta t\ddot{\mathrm{d}}_{\mathrm{j-1}} + \gamma\,\Delta t\ddot{\mathrm{d}}_{\mathrm{j}}$$

$$\mathrm{d}_{\mathrm{j}} = \mathrm{d}_{\mathrm{j-1}} + \Delta t\dot{\mathrm{d}}_{\mathrm{j-1}} + (0.5\text{-}\beta)\,\Delta t^2\ddot{\mathrm{d}}_{\mathrm{j-1}} + \beta\,\Delta t^2\ddot{\mathrm{d}}_{\mathrm{j}} \tag{33}$$

where γ and β are the integration constants, taken as $\gamma = 0.50$ and $\beta = 0.25$ corresponding to the case of constant average acceleration within Δt. Knowing the initial conditions $\mathbf{d}_1$ and $\dot{\mathbf{d}}_1$, equation (32) can be rewritten as

$$\dot{\mathrm{d}}_{\mathrm{j}} = \dot{\mathrm{d}}_1 + \mathrm{B}_1\sum_{\mathrm{i}=1}^{\mathrm{j-1}}(\ddot{\mathrm{d}}_{\mathrm{i}} + \ddot{\mathrm{d}}_{\mathrm{i+1}})$$

$$\mathrm{d}_j = \mathrm{d}_1 + (j\text{-}1)\Delta t\dot{\mathrm{d}}_1 + C_1 \sum_{i=1}^{j-1}(\ddot{\mathrm{d}}_i + \ddot{\mathrm{d}}_{i+1})$$
$$+B_1\Delta t \sum_{i=1}^{j-1}[(j-i-1)(\ddot{\mathbf{d}}_i + \ddot{\mathbf{d}}_{i+1})] \quad (34)$$

where $B_1 = 0.5\Delta t$ and $C_1 = 0.25\,\Delta t^2$.

Theoretically, with k_1 time steps of measured accelerations, $k_1 n$ equations can be formulated using equation (32) as

$$\boldsymbol{M}\ddot{\boldsymbol{d}}_1 + \boldsymbol{D}_d\dot{\boldsymbol{d}}_1 + \boldsymbol{K}\boldsymbol{d}_1 = \boldsymbol{P}_1$$
$$\boldsymbol{M}\ddot{\boldsymbol{d}}_2 + \boldsymbol{D}_d\dot{\boldsymbol{d}}_2 + \boldsymbol{K}\boldsymbol{d}_1 = \boldsymbol{P}_2$$
$$\cdots\cdots\cdots\cdots\cdots\cdots\cdots\cdots\cdots\cdots$$
$$\mathbf{M}\ddot{\mathbf{d}}_{k_1} + \mathbf{D}_d\dot{\mathbf{d}}_{k_1} + \mathbf{K}\mathbf{d}_{k_1} = \mathbf{P}_{k_1} \quad (35)$$

where subscripts 1, 2, ⋯, k_1 denote the time step. Substituting equation (34) into equation (35), the unknowns in equation (35) include: (i) $2n$ initial conditions; (ii) k_1 $(n - ns)$ accelerations which are not measured; (iii) $n(n + 1)/2$ entries in each of the symmetric $\mathbf{D}_d$ and $\mathbf{K}$ matrices; and $k_1 r$ unknown input forces. To solve these unknowns, the minimum number of time steps (k_1) of acceleration data needed is

$$k_1 n \geq 2n + k_1(n\text{-}ns) + n(n+1) + k_1 r \;\Rightarrow\; k_1 \geq \frac{n(n+3)}{ns\text{-}r} \quad (36)$$

assuming that (i) $ns > r$ and; (ii) stiffness and damping do not change within the k_1 time steps.

There are a few limitations in employing this method. If $ns \leq r$ or the location of the excitation is not known, this method is not applicable. For cases where the structure has many DOFs and r is not much smaller than ns, then, k_1 may be extremely large causing this method to be impractical. If stiffness or damping changes rapidly within the k_1 time steps, the computed $\mathbf{D}_d$ and $\mathbf{K}$ matrices using this method may not accurately represent those of the damaged structure. However, if the stiffness and damping changes slowly (less than 0.1% within k_1 time steps), then the values computed are average values.

Observing equation (36), the number of time steps k_1 required to solve equation (35) does not depend on the number of collocated sensor-actuator pairs (nc). Hence, this method overcomes the requirement of at least one collocated sensor-actuator pair in the original DLV method.

The system of nonlinear equations in equation (35) can be solved by Secant method (Wolfe, 1959) or Newton-Raphson method (Bathe, 1996). Since Jacobian matrix can be computed directly, the Newton-Raphson method can be employed using the structural parameters at the reference state as an initial guess for the unknown $\mathbf{D}_d$ and $\mathbf{K}$ matrices to obtain the solution iteratively.

In cases where measured time history is long, then the data from consecutive time segments of k_1 time steps can be used to monitor the changes in the D_d and K matrices with time. In fact, the evolution of damage with time can be captured and whether the damage has stabilized may be deduced.

In the event where the structure undergoes significant change in the mass matrix, then the mass matrix needs to be estimated as well. Hence, the new minimum number of time steps k_1 of acceleration data needed is

$$k_1 n \geq 2n + \frac{k_1(n-ns)}{2} \times 3 + n(n+1) + k_1 r \ \Rightarrow \ k_1 \geq \frac{n(3n+7)}{2(ns-r)} \tag{37}$$

4. Lost Data Reconstruction for Wireless Sensors

The quality of sensors affects the reliability of the DLV method in detecting damage. Traditionally, sensors of a structural health monitoring system are connected to the base station using lengthy wire. The installation and maintenance costs of such system are usually high and the wires may interfere with the normal operations of the structure (Lynch, 2004).

With the development of wireless sensing technology, response measurements may become more convenient and pervasive. Although various enhancements in wireless sensing for structural damage detection have been reported in recent years, such as improvement in the sampling rate (Hou et al., 2008) and transmission topology (Mechitov *et al.*, 2006), there are still challenges with regards to implementation of wireless sensor technology for structural health monitoring. One challenge is the issue of lost of data in wireless transmission. To address this problem, an algorithm for lost data reconstruction performed for signals of each sensor at the base station will be presented in relation to the DLV method for damage detection.

4.1. *Lost data reconstruction algorithm*

In general, structural response signals can be decomposed into various discrete frequencies via Fourier transformation. If the proportion of lost data is not significant, the lost data can be reconstructed by using the available signals (says m points from a total of ξ points; that is, $\xi - m$ points are lost, their magnitudes are set to zero) to identify the significant frequencies. By classifying significant frequencies as those frequencies with power spectra value (PSV) normalized by the maximum PSV exceeding the threshold of 1%, a set of significant frequencies (say $nfreq$) can be filtered out, where $nfreq < 0.5\xi$. (This threshold of 1% is chosen on the basis that if the signals have no lost data, the reconstructed signals using all significant frequencies and the corresponding Fourier coefficients will yield less than 1% error relative to the exact signals.) Using these frequencies together with Fourier coefficients A_k determined by the least square fit of m measured values (requiring that $m > nfreq$ or $m > 0.5\xi$) an approximate complete signal is reconstructed using the following equation

$$x_n = \frac{1}{\xi}\sum_{k=1}^{nfreq}\left[A_k \exp\left(jn\frac{2\pi(k-1)}{\xi}\right)\right] \tag{38}$$

where j denotes the imaginary unit. The reconstructed signals, in principle, do not have lost package. Performing Fourier transform on the reconstructed signals, a new set of significant frequencies which may differ from the previously identified significant frequencies is calculated. The Fourier coefficients based on the least square fit of m measured data points are obtained using the new set of frequencies. This can then be used to generate a new set of signals and compared with the previous reconstructed signals. The relative difference ($Rerr$) between two consecutive reconstructed lost portions of the signals is defined as

$$\text{Rerr} = \sqrt{\frac{\sum_{k=1}^{\xi}\left(g_k^i - g_k^{i-1}\right)^2}{\sum_{k=1}^{\xi}\left(g_k^i\right)^2}} \times 100\% \tag{39}$$

where g_k^i and g_k^{i-1} are the estimated lost values at the ith and $(i\text{-}1)$th iterations, respectively. The iteration terminates when Rerr is less than 1%. The algorithm is illustrated in Fig. 2.

5. Numerical and Experimental Examples

The use of NCE as damage indicator and the intersection scheme to identify the actual damaged element are illustrated through a numerical example of a 2-D warehouse frame; while, the lost data reconstruction algorithm is demonstrated through an experimental example of a 3-D modular truss structure.

5.1. *Numerical example: 2-D warehouse frame structure*

A 2-D warehouse frame shown in Fig. 3 is first considered. The properties of the structural elements are listed in Table 1. Two scenarios are shown: (i) element 14 is damaged; and (ii) element 7 and 14 are damaged. The damage in column element 7 is simulated by imposing a reduction of 20% to the flexural stiffness (*EI*) whereas for truss member 14, 20% reduction of the axial stiffness (*EA*) is assumed. The structure is excited horizontally by a zero-mean white random load with root mean square (RMS) of 300N at node 9. Horizontal and vertical acceleration responses at nodes (4, 6, 8, 9, 10, 13) and (5, 7, 11, 12), respectively, are computed and sampled at a rate of 1 kHz.

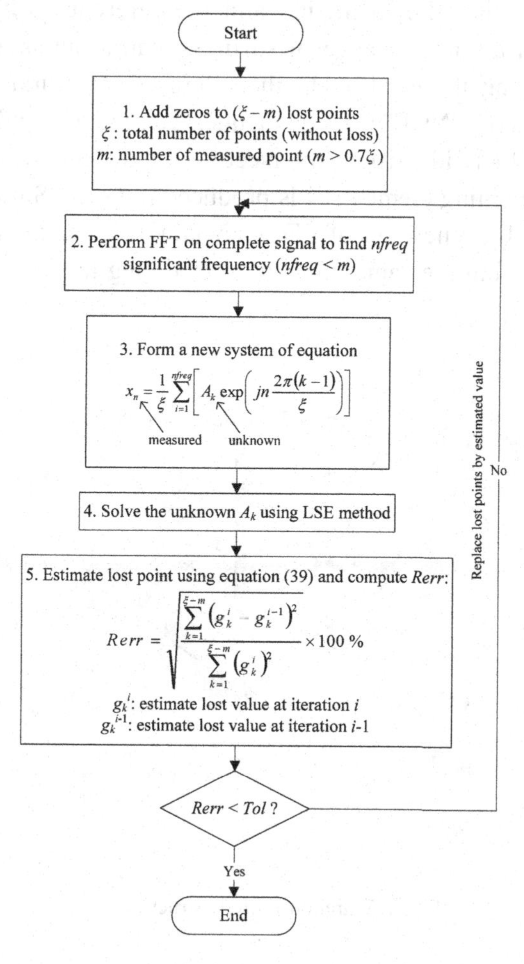

Fig. 2. Flow diagram for lost data reconstruction algorithm.

Performing the procedure outlined in section 1.3.2.2, the stiffness matrices for all three states (one healthy case and two damaged cases) are formulated. For the first damage case (element 14 damaged), the change in stiffness matrix is computed by comparing the identified and the reference (healthy) stiffness matrices. Performing SVD on the change in stiffness matrix, a set of nine DLVs is obtained. By applying these DLVs to the reference structural model as displacement vectors, the NCE of all elements are computed and the set of PDEs which includes elements (14, 17) is identified. Therefore, the current IDS contains elements (14, 17) and *ne* = 2. By omitting the sensor reading at node 4, the stiffness matrix at the remaining nine sensor locations is computed using the same procedure in section 1.3.2.2. Again, the change in stiffness matrix is computed by comparing the identified and the reference stiffness matrices. Performing SVD on the new change in stiffness matrix, another eight DLVs is identified. By applying these DLVs to the reference structural model as nodal displacement vectors, the NCE of all elements are computed and the set of PDE containing element 14 is identified. Intersecting the set of PDE and the current IDS, a new IDS containing element 14 is produced (*ne* = 1). Similar, by omitting sensor data at node 10, a new set of PDEs containing elements (12, 14, 16) can be obtained by repeating the same procedure. Updating the current IDS with the

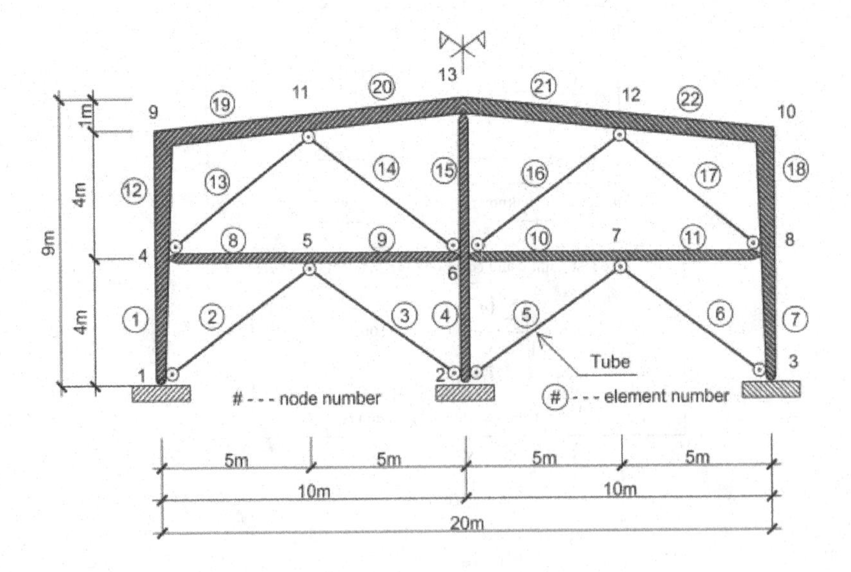

Fig. 3. Warehouse frame structure.

Table 1. Specification for members of warehouse structure.

Element numbers		Height (mm)	Width (mm)	Flange thickness (mm)	Web thickness (mm)	Young's modulus ($10^{11}Nm^{-2}$)	Moment of inertia ($10^{-8}m^4$)	Cross-sectional area ($10^{-4}m^2$)
1, 7	end 1	300	300	16	10	2.10	20,982	122.8
	end 2	450	300	16	10	2.10	51,312	137.8
12, 18	end 1	450	300	16	10	2.10	51,312	137.8
	end 2	600	300	16	10	2.10	97,145	152.8
4, 15		300	300	16	10	2.10	20,982	122.8
19, 20, 21, 22		500	300	16	10	2.10	64,784	142.8
8, 9, 10, 11		300	300	16	10	2.10	20,982	122.8
2, 3, 5, 6, 13, 14, 16, 17		Tubular sections				2.10	4	2.0

identified PDEs gives a new IDS containing only element 14 ($ne = 1$). Since two consecutive iterations produce identical IDS, the iteration is terminated and the actual damage element 14 is identified. Similarly for the second case where elements (7, 14) are damaged, the same procedure is applied and results in Table 3 confirm the feasibility of the method.

In the lower portion of Table 2, the remaining eight combinations resulting from omitting a sensor reading at a time are considered and the identified PDEs shown confirm that the IDS contains only element 14.

Table 2. Damage detection of warehouse (element 14 damaged).

	Set of sensor includes sensor at nodes	No. of DLV	PDEs	Eliminated elements	IDS	ne
$ns = 10$	[4, 5, 6, 7, 8, 9, 10, 11, 12, 13]	9	[14, 17]		[14, 17]	2
$k = 9$						
$i = 1$	[5, 6, 7, 8, 9, 10, 11, 12, 13]	8	[14]	17	[14]	1
$i = 2$	[4, 5, 6, 7, 8, 9, 11, 12, 13]	8	[12, 14, 17]		[14]	1
$i = 3$	[4, 6, 7, 8, 9, 10, 11, 12, 13]	8	[4, 14]		[14]	1
$i = 4$	[4, 5, 7, 8, 9, 10, 11, 12, 13]	8	[1, 2, 7, 14, 20]		[14]	1
$i = 5$	[4, 5, 6, 8, 9, 10, 11, 12, 13]	8	[12, 14]		[14]	1
$i = 6$	[4, 5, 6, 7, 9, 10, 11, 12, 13]	8	[14, 16]		[14]	1
$i = 7$	[4, 5, 6, 7, 8, 10, 11, 12, 13]	8	[14]		[14]	1
$i = 8$	[4, 5, 6, 7, 8, 9, 10, 12, 13]	8	[12, 14, 16]		[14]	1
$i = 9$	[4, 5, 6, 7, 8, 9, 10, 11, 13]	8	[12, 14]		[14]	1
$i = 10$	[4, 5, 6, 7, 8, 9, 10, 11, 12]	8	[14, 17]		[14]	1

To illustrate the effect of noise on the performance of the DLV method, zero-mean white noises with RMS of (i) 5%; and (ii) 10% of the RMS of the response signal are added to contaminate the acceleration response signal. For the case of 5% noise, from the contaminated responses of ten sensors, the same procedure

Table 3. Damage detection of warehouse (elements (7, 14) damaged)

	Set of sensor includes sensor at nodes	No. of DLV	PDEs	Eliminated elements	IDS	ne
$ns = 10$	[4, 5, 6, 7, 8, 9, 10, 11, 12, 13]	9	[1, 4, 7, 13, 14]		[1, 4, 7, 13, 14]	2
$k = 9$						
$i = 1$	[4, 5, 6, 7, 8, 9, 11, 12, 13]	8	[7, 12, 14]	1, 4, 13	[7, 14]	1
$i = 2$	[4, 5, 6, 7, 8, 10, 11, 12, 13]	8	[1, 7, 13, 14, 17]		[7, 14]	1

yields a set of PDEs containing six elements (4, 6, 12, 13, 14, 17). These six elements are assigned to the current IDS (*ne* = 6). By omitting readings of the sensor at node 4 and repeating the same procedure using only data from the remaining nine sensor locations, a PDE set containing elements (1, 8, 10, 14) is identified.

Taking the intersection between the set of PDEs and the current IDS, a new IDS containing only element 14 is obtained (*ne* = 1). Similarly, by omitting the readings of the sensor at node 10 instead of the sensor at node 4, another set of

Table 4. Damage detection of warehouse with 5% noise added (element 14 damaged).

	Set of sensor includes sensor at nodes	No. of DLV	PDEs	Eliminated elements	IDS	ne
$ns = 10$	[4, 5, 6, 7, 8, 9, 10, 11, 12, 13]	7	[4, 6, ,12, 13, 14, 17]		[14, 17]	2
$k = 9$						
$i = 1$	[5, 6, 7, 8, 9, 10, 11, 12, 13]	5	[1, 8, 10, 14]	4, 6, 12, 13, 14,17	[14]	1
$i = 2$	[4, 5, 6, 7, 8, 9, 11, 12, 13]	5	[5, 14, 16]		[14]	1
$i = 3$	[4, 6, 7, 8, 9, 10, 11, 12, 13]	5	[14]		[14]	1
$i = 4$	[4, 5, 7, 8, 9, 10, 11, 12, 13]	5	[7, 14]		[14]	1
$i = 5$	[4, 5, 6, 8, 9, 10, 11, 12, 13]	5	[5, 6, 14]		[14]	1
$i = 6$	[4, 5, 6, 7, 9, 10, 11, 12, 13]	5	[10, 11, 12, 14, 15]		[14]	1
$i = 7$	[4, 5, 6, 7, 8, 10, 11, 12, 13]	5	[1, 8, 9, 14, 16, 21]		[14]	1
$i = 8$	[4, 5, 6, 7, 8, 9, 10, 12, 13]	5	[14, 15, 19, 20]		[14]	1
$i = 9$	[4, 5, 6, 7, 8, 9, 10, 11, 13]	5	[14, 18]		[14]	1
$i = 10$	[4, 5, 6, 7, 8, 9, 10, 11, 12]	5	[7, 14]		[14]	1

PDEs comprising elements (5, 14, 16) is identified. The intersection of the set of PDEs and the current IDS which contains only element 14 produces element 14 as the new IDS (*ne*=1). Since the IDS for the two consecutive steps are identical, the iteration is stopped and the identified damaged element 14 matches the actual case.

The procedure is summarized in the upper portion of Table 4. Performing the same procedure for the other eight combinatorial sets of sensors by dropping a sensor record each time, the results in the lower portion of Table 4 confirm that only element 14 is damaged. The same computation is performed for the case of 10% noise and the results in Table 5 support the feasibility of the proposed methodology.

It is observed that the number of DLVs computed from the contaminated data is less than that computed from the pure data. If the noise level goes beyond a certain threshold, no DLV may be computed.

5.2. *Experimental example: 3-D modular truss structure*

A 3-D modular truss structure (Fig. 4(a)) comprising twenty-three aluminum tubes and a pre-tensioned cable is used in this example. The geometrical and material properties of the truss members are listed in Table 6 whereas the

Table 5. Damage detection of warehouse with 10% noise added (element 14 damaged).

	Set of sensor includes sensor at nodes	No. of DLV	PDEs	Eliminated elements	IDS	ne
$ns = 10$	[4, 5, 6, 7, 8, 9, 10, 11, 12, 13]	5	[3, 8, 9, 13, 14, 16, 18, 20, 21, 22]		[3, 8, 9, 13, 14, 16, 18, 20, 21, 22]	10
$k = 9$						
$i = 1$	[5, 6, 7, 8, 9, 10, 11, 12, 13]	4	[1, 4, 7, 14, 17]	3, 8, 9, 13, 16, 18, 20, 21, 22	[14]	1
$i = 2$	[4, 5, 6, 7, 8, 9, 11, 12, 13]	4	[5, 14, 16]		[14]	1
$i = 3$	[4, 6, 7, 8, 9, 10, 11, 12, 13]	4	[14]		[14]	1
$i = 4$	[4, 5, 7, 8, 9, 10, 11, 12, 13]	4	[7, 14]		[14]	1
$i = 5$	[4, 5, 6, 8, 9, 10, 11, 12, 13]	4	[5, 6, 14]		[14]	1
$i = 6$	[4, 5, 6, 7, 9, 10, 11, 12, 13]	4	[10, 11, 12, 14, 15]		[14]	1
$i = 7$	[4, 5, 6, 7, 8, 10, 11, 12, 13]	4	[1, 8, 9, 14, 16, 21]		[14]	1
$i = 8$	[4, 5, 6, 7, 8, 9, 10, 12, 13]	4	[14, 15, 19, 20]		[14]	1
$i = 9$	[4, 5, 6, 7, 8, 9, 10, 11, 13]	4	[14, 18]		[14]	1
$i = 10$	[4, 5, 6, 7, 8, 9, 10, 11, 12]	4	[7, 14]		[14]	1

dimensions of the truss structure are shown in Fig. 4(b). Damage in the structure was simulated by cutting the pre-tensioned cable member midway through the test. A load simulated by zero-mean white random load with RMS of 50N was applied vertically to the truss at node 1 using a shaker and the acceleration responses of the truss were captured by six wireless sensors (sensor locations shown in Fig. 4(a)). Summary of results is shown in Table 7.

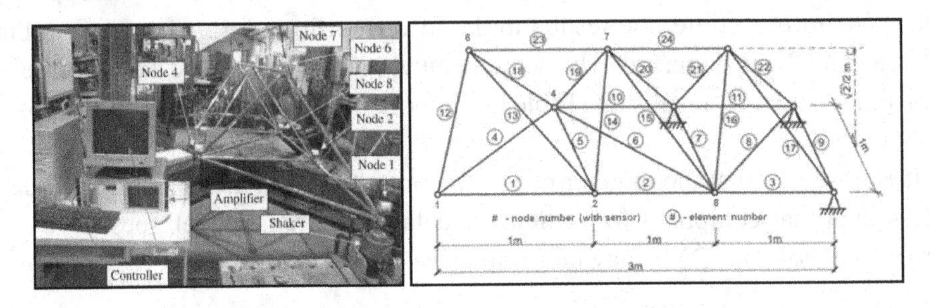

Fig. 4. (a) Experimental setup; (b) Element and node numbers for experimental truss structure.

Table 6. Geometric and material properties of truss members.

	Aluminum tubular members	Pre-tensioned cable members
Outer diameter (mm)	20.0	4.0
Thickness (mm)	1.0	-
Young's modulus (N/m^3)	6.8×10^{10}	1.6×10^{11}
Mass density (kg/m^3)	2690.0	7500.0
Pre-tensioned force (N)	-	2000.0

Table 7. Computational details for signal reconstruction at 6 sensor nodes.

	Node 1	Node 2	Node 4	Node 6	Node 7	Node 8
No. of lost points	352	380	392	376	304	372
Total no. of points	1600	1600	1600	1600	1600	1600
Lost percentage (%)	22.00	23.75	24.50	23.50	19.00	23.25
No. of iterations	12	14	14	11	14	10
No. of frequencies used (last iteration)	479	467	495	367	479	343
Relative error at the last iteration (%)	0.62	0.33	0.55	0.72	0.87	0.83

The lost percentages of the acceleration responses captured by the six wireless sensors indicate that the loss of data packets is in fact random and not synchronized which agree with findings by Nagayama (2007). To estimate lost data values, the reconstruction procedure is applied individually to the signals of each wireless sensor. The results for sensor node 4 are shown in Fig. 5 where Fig. 5(a) gives the signals captured by wireless sensor at node 4 including an insert to show an example of the lost data. Fig. 5(b) gives a better indication of the locations of the lost data points based on the 'corrected' values. Fig. 5(c) compares in frequency domain the raw signals against the reconstructed signals which show no drastic change in trend or general characteristics.

The reconstruction signals from each sensor are divided into fifteen segments, each of which contains 109 time steps, can be used to estimate the structural stiffness coefficients and selected results are plotted in Fig. 6. Of the six stiffness coefficients plotted in Fig. 6, only K_{88} shows significant reduction since it was contributed directly by the stiffness of the pre-tensioned cable member which was cut. When the pre-tensioned cable was cut midway through the experiment, the resulting transient oscillations towards dynamical equilibrium of the new system is manifested by the varying stiffness coefficients estimated for segments 7–9 (from 7 to 11 s). The identified stiffness coefficients stabilized from segment 10 (11th second) onwards. Based on the identified stiffness coefficients at segment 1 (first) and segment 15 (last), the change in the stiffness matrix is calculated and by performing SVD, four DLVs are identified. Applying these DLVs to the reference structural model as nodal displacement vectors, the NCE of all elements are computed and the set of PDEs comprising elements (1, 8) is identified. The current *IDS* therefore comprises elements (1, 8) and $ne = 2$.

By omitting data of the sensor at nodes 9, the data of the remaining five sensors are used to estimate the difference in structural stiffness matrices between the first and last time segments. Again, performing SVD on the change in stiffness matrix, a set of three DLVs is obtained and are applied to the reference model as nodal displacement vectors. The set of PDEs identified comprise elements (4, 5, 8, 12, 14). Taking intersection between the set of PDEs and the current IDS which contains elements (1, 8) gives the new IDS with element 8 as the only member ($ne = 1$). Similarly, by omitting the data of the sensor at node 1, another set of PDEs comprising elements (1, 4, 8, 12) is identified. Intersecting the set of PDEs and the current IDS which contains only element 8 gives element 8 as the new IDS (ne=1). Since the IDS for two consecutive steps are identical, the iteration is terminated, confirming that element 8 is damaged. The whole procedure is summarized in Table 8.

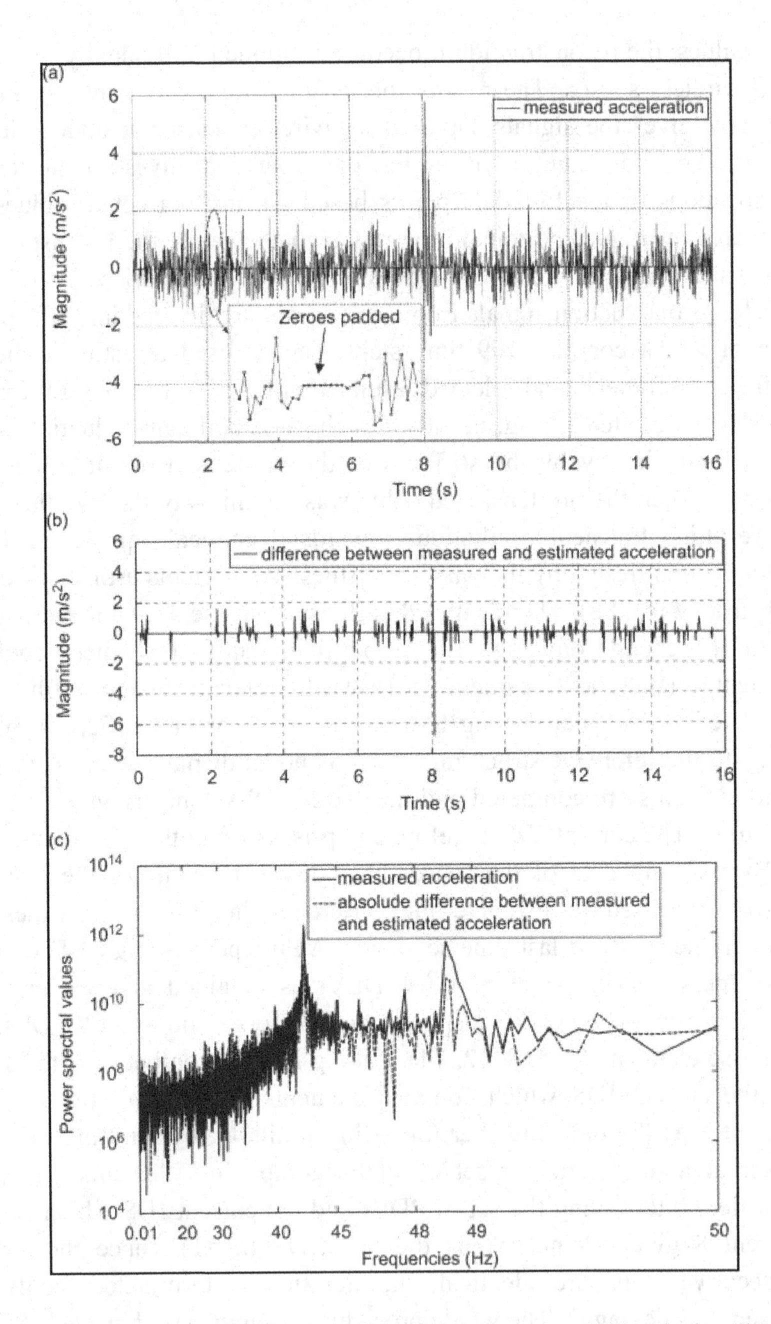

Fig. 5. Comparison between measured and estimated accelerations at sensor node 4 in time ((a) and (b)), and frequency (c)

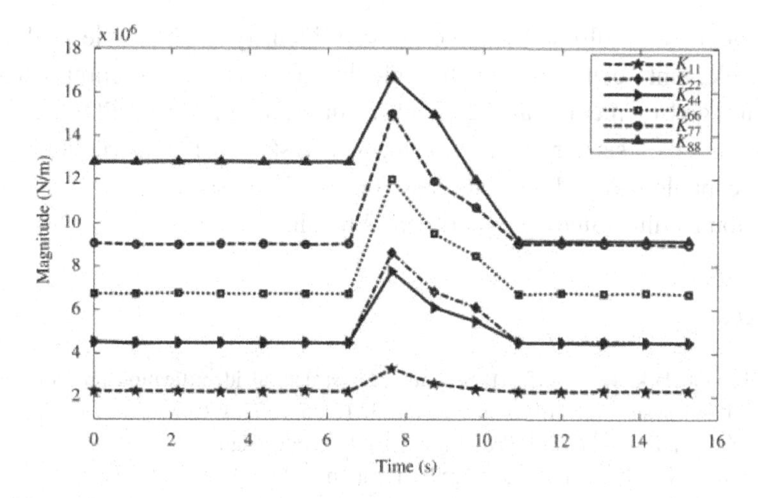

Fig. 6. Variation of structural stiffness coefficients with time

Table 8. Damage detection of experimental truss structure.

	Set of sensors includes sensors at nodes	No. of DLV	PDE	Eliminated elements	IDS	ne
$ns = 6$	[1, 2, 3, 5, 8, 9]	4	[1,8]		[1, 8]	2
$k = 5$						
$i = 1$	[1, 2, 3, 5, 8]	3	[4, 5, 8, 12, 14]	1	[8]	1
$i = 2$	[2, 3, 5, 8, 9]	3	[1, 4, 8, 12]		[8]	1

If the raw signals where lost values are padded with zeroes are used instead (that is, without correction), the stiffness matrix at all fifteen time segments can still be deduced. However, if SVD is performed on the difference in stiffness matrix between the first and the last time segments, no DLV can be computed. This confirms the necessity of the reconstruction procedure to improve the quality of the measured signals before damage identification procedure can be applied to obtain reliable results.

6. Concluding Remarks

Although conceptually simple and theoretically tractable, the DLV method has proven to be feasible for damage identification using either static or dynamic response measurements. The DLV method is shown to be robust even under the presence of reasonable level of measurement noise (up to 10%). Using the NCE as damage indicator, the application of DLV method is applicable for structures with elements in multi-stress state. It can identify both single and multiple damaged element scenarios and through the proposed intersection scheme, the actual damaged elements can be filtered out from the set of PDEs even when limited number of sensors is employed. The lost data reconstruction algorithm shown is capable of handling data loss (up to 25%) associated with the wireless transmission as illustrated by experimental results.

References

Alvin, K. F. and Park, K. C. (1994). Second-order structural identification procedure via state-space-based system identification, *AIAA J.*, 32 (2), pp. 397-406.

Bathe, K. J. (1996). *Finite Element Procedures*, (PTR Prentice Hall, NJ).

Bernal, D. (2002). Load vectors for damage localization, *J. Eng. Mech.*, 128 (1), pp. 7-14.

Bernal, D. (2004). Modal scaling from known perturbation, *J. Eng. Mech.*, 130 (9), pp. 1083-1088.

Bernal, D. and Gunes, B. (2004). Flexibility based approach for damage characterization: Benchmark application, *J. Eng. Mech.*, 130 (1), pp. 61-70.

Gao, Y., Spencer, B. F. and Bernal, D. (2007). Experimental verification of the flexibility-based damage locating vector method, *J. Eng. Mech.*, 133 (10), pp. 1043-1049.

Hou, X. Y., Quek, S. T. and Qin, F. (2008). A wireless sensor network with continuous high frequency sampling and time synchronization for structural dynamic testing, *Proc. Eur. Workshop Struct. Health Monit.*, pp. 630-637.

Juang, J. N. and Pappa, R. S. (1985). An eigensystem realization algorithm for modal parameter identification and model reduction, *J. Guid. Control Dyn.*, 8 (5), pp. 620-627.

Juang, J. N. (1994). *Applied System Identification*, (PTR Prentice Hall, NJ).

Lynch, J. P. (2004). Overview of wireless sensors for real time health monitoring of civil structures, Proc. 4th Intl. Workshop on Struct. Control Monitor, pp. 189-194.

Mechitov, K. A., Kim, W., Agha, G. A. and Nagayama, T. (2006). High-frequency distributed sensing for structural monitoring, *Proc. 1st Intl. Workshop on Networked Sensing System*, pp. 101-105.

Nagayama, T. (2007). Structural health monitoring using smart sensors, PhD thesis, University of Illinois Urbana-Champaign.

Quek, S. T., Tran, V. A., Duan, W. H. and Hou, X. Y. (2009). Structural damage detection using wireless sensors accounting for data loss, *Proc. SPIE Intl. Soc. Opt. Eng., SPIE*, pp. 7292-9.

Wolfe, P. (1959). The Secant method for simultaneous nonlinear equations, *Commun. ACM*, 2 (12), pp. 12-13.

Chapter 3

Dynamics-based Damage Identification

Pizhong Qiao[1] and Wei Fan[2]

[1,2]*Department of Civil and Environmental Engineering, Washington State University, Sloan 117, Spokane Street, Pullman, WA 99164-2910, USA*
E-mails: qiao@wsu.edu

1. Introduction

Structural systems including bridges, airplanes, pipelines, etc. are susceptible to material aging, deterioration and structural damage. Structural damage identification has gained increasing attention from the scientific and engineering communities because unpredicted structural failure may cause catastrophic economic and human life loss. A reliable and effective non-destructive damage identification technique is crucial to maintain safety and integrity of structures. When structures are designed with safe-life design philosophy, in which the structures are designed to work in their design life without any damage, it is important to detect the existence and location of the damage to warn the failure of designed safety. When structures are designed with damage-tolerant design philosophy, in which the structure are designed to work in their design life with damage within a certain extent, it is also important to monitor the damage extent for potential repair to ensure that structures can still function in a satisfactory state.

Most non-destructive damage identification methods can be categorized as either local or global damage identification techniques (Doebling *et al.*, 1996). Local damage identification techniques, such as ultrasonic methods and X-ray methods, require that the vicinity of damage is known a priori and readily accessible for testing, which cannot be guaranteed for most cases in civil or aerospace engineering. Hence, the dynamics-based damage identification method as a global damage identification technique is developed to overcome these difficulties. The dynamics-based damage identification is based on the idea that the damage-induced changes in the physical properties (mass, damping, and stiffness) will cause detectable changes in dynamic mechanical properties (e.g., vibration modal parameters, such as natural frequencies, modal damping and mode shapes). Therefore, it is intuitive that damage can be identified by analyzing the changes in dynamic features of the structure.

During the last three decades, extensive research has been conducted in dynamics-based damage identification, and significant progress has been achieved in this highlighted area. A broad range of techniques, algorithms and methods are developed to solve damage identification problems in structures, from basic structural components (e.g., beams and plates) to complex structural systems (e.g., bridges and buildings).

The state-of-art damage identification process can be divided into four stages. In general, these stages are with increasing difficulty and require the knowledge of former stages:

(i) Detect the existence of damage;
(ii) Locate the damage;
(iii) Identify the types of damage;
(iv) Quantify the damage severity.

The terminology "damage detection method" is often used when the damage identification method only targets problems in Stages (i) and (ii).

Doebling *et al.* (1996) presented an extensive review of vibration-based damage detection methods up to 1996. Sohn et al. (2003) then presented an updated version of this review on the literature up to 2001. In both the articles, the features extracted for identification were used to classify the damage identification methods. Following closely this classification, Carden and Fanning (2004) presented a literature survey with particular emphasis on the papers and articles published from 1996 to 2003. Based on an extensive literature, Worden *et al.* (2007) identified a set of fundamental axioms for damage identification:

(i) All materials have inherent flaws or defects.
(ii) The assessment of damage requires a comparison between two system states.
(iii) Identifying the existence and location of damage can be done in an unsupervised learning mode, but identifying the type of damage present and the damage severity can generally only be done in a supervised learning mode.

(iv) Sensors cannot measure damage. Feature extraction through signal processing and statistical classification is necessary to convert sensor data into damage information. Without intelligent feature extraction, the more sensitive a measurement is to damage, the more sensitive it is to changing operational and environmental conditions.

(v) The length- and time-scales associated with damage initiation and evolution dictates the required properties of the SHM sensing system.

(vi) There is a trade-off between the sensitivity to damage of an algorithm and its noise rejection capability.

(vii) The size of damage that can be detected from changes in system dynamics is inversely proportional to the frequency range of excitation.

Based on different criteria, damage identification methods can be classified into different categories.

Based on the domain in which the dynamic features for damage identification are extracted, the methods can be categorized into "time series-based methods", "frequency-based method" and "modal parameter-based method". Although in vibration test the excitation and response are always measured and recorded in the form of time history, it is usually difficult to directly examine the time domain data for damage identification. A more popular method is to extract the modal domain features through the modal analysis technique, in which the time domain data is transformed into the frequency domain, and then to the modal domain data. During the past three decades, great effort has been made in the researches within all three domains (i.e., time, frequency, and modal domains). The modal domain methods attract more attention than the other two and play a dominant role in the state-of-the-art of structural damage identification. The modal domain methods evolve along with the rapid development of experimental modal analysis technique, and they gain their popularity because the modal properties (i.e., natural frequencies, modal damping, modal shapes, etc.) have their physical meanings and are thus easier to be interpreted or interrogated than those abstract mathematical features extracted from the time or frequency domain.

Based on the dynamic features extracted for damage identification, the methods can be categorized as "natural frequency-based method", "mode shape-based method" and "strain energy-based method", etc.

Based on the assumption of the structure in healthy state, damage identification methods can also be classified as 'model-based method' or 'response-based method'. The model-based method assumes that the structure in healthy state can be simulated with a detailed numerical model; while the response-based method depends only on experimental response data from damaged and healthy structures for damage identification. It should be noted that some response-based methods claim to detect damage depending only on experimental response data from damaged state structures without the requirement of baseline data from healthy state. Recall the fundamental axioms (ii), the model-based methods adopt the detailed numerical model to define the healthy state of structures, while these damaged state response-based methods do not have a clear explicit definition of the healthy state of structures. But the damaged state response-based methods usually have a common embedded assumption that the structure should not have any local changes (irregularity) in its geometry and mechanical properties, which is used as an implicit definition of the healthy state structure.

This chapter is to present dynamics-based damage identification for beam-type and plate-type structures. A brief introduction to damage identification algorithms is provided. Based on finite element simulation and experimental test, a comparative study of several damage identification algorithms for plate-type structures is presented to illustrate the validity and effectiveness of the algorithms in various scenarios with different damage and sensor types.

2. Damage Identification Algorithms

2.1. *Literature review*

Natural frequency-based methods use the natural frequency change of structures as the basic feature for damage identification. The use of natural frequency changes for crack detection has been extensively studied in the past three decades. The choice of the natural frequency change is attractive because the natural frequencies can be conveniently measured from just a few accessible points on the structure and are usually less contaminated by experimental noise. But it has several limitations such as complexity in structural modeling and damage and non-uniqueness of the solutions. In 1997, Salawu (1997) presented an extensive review of publications before 1997 dealing with the detection of structural damage through frequency changes. It was concluded that the natural frequency changes alone may not be sufficient for a unique identification of the location of structural damage because cracks associated with similar crack lengths but at two different locations may cause the same amount of frequency change.

Compared to using natural frequencies, the advantages of using mode shapes and their derivatives as a basic feature for damage detection are obvious. First, mode shapes contain local information, which makes them more sensitive to local damages and enables them to be used directly in multiple damage detection. Second, the mode shapes are less sensitive to environmental effects, such as temperature, than natural frequencies (Farrar and James, 1997). However, their disadvantages are also apparent. First, measurement of the mode shapes requires a series of sensors. Second, the measured mode shapes are more prone to noise contamination than natural frequencies.

Many damage identification methods have been developed based on direct or indirect use of measured mode shapes. Most of these methods take mode shape data as a spatial-domain signal and adopt signal processing technique to locate damage by detecting the local discontinuity of mode shapes or their derivatives caused by damage. These methods can be roughly categorized into two types here: displacement mode shape-based methods and curvature mode shape-based methods.

It has been shown by many researchers that the displacement mode shape itself is not very sensitive to small damage, even with high density mode shape measurement (Khan *et al.*, 1999; Salawu and Williams, 1994; Huth *et al.*, 2005). As an effort to enhance the sensitivity of mode shape data to the damage, the curvature mode shape is investigated as a promising feature for damage identification. The curvature mode shape can be either obtained by direct measurement of strain mode shape or derived from displacement mode shape. When the displacement mode shapes are acquired from modal testing, the central difference method or the Chebyshev polynomial approximation can be applied to the measured displacement mode shape to derive the curvature mode shape. The curvature mode shapes are also closely involved in the modal strain energy-based method (Kim and Stubbs, 1995; Stubbs and Kim, 1996; Shi and Law, 1998; Shi *et al.*, 2000; Kim *et al.*, 2003).

Although the above-mentioned methods directly use a single type of basic modal characteristics (modal frequencies, mode shapes or modal damping) to identify damage, further research shows that the basic modal characteristics may not be the most effective and sensitive modal features for damage identification. Topole and Stubbs (1995) first presented a damage detection method to locate and size structural damage from both measured natural frequencies and mode shapes of the damaged structure without *a priori* knowledge of the modal characteristics of a baseline structure. However, its application is limited by several requirements which are difficult to meet in real experiment, such as the knowledge of baseline mass and stiffness, measurement of mode shapes from all DOFs and many modes. Since then, in order to find an effective and sensitive dynamic feature for damage identification, extensive research effort has also been put into damage identification methods utilizing a combination of the basic modal characteristics or modal parameters derived from these characteristics, such as modal strain energy (Kim and Stubbs, 1995; Stubbs et al., 1995; Stubbs and Kim, 1996; Shi and Law, 1998; Shi *et al*., 2000; Kim *et al*., 2003), modal flexibility (Pandey and Biswas, 1994; Zhang and Aktan, 1998) and uniform load surface (Wu and Law, 2004; Wu and Law, 2005; Wang and Qiao, 2007).

In the past three decades, extensive research efforts have been invested into the field of structural damage identification. Most research efforts on damage identification methods still focus on damage localization (the second damage identification stage), since it sets the foundation of further identifying damage type and severity (the third and fourth stages). Some researchers also explored a unified approach to identify damage location as well as quantification. Even though numerous theories and algorithms have been developed, up to the present time, most methods are developed in the context of 1-Dimensional beam-type structures (Fan and Qiao 2011). Only limited researches on damage identification of 2-D plate-type structures can be found in literature. For existing 2-D damage identification algorithms, most of them are generalized from their 1-D version algorithms.

A few established damage identification algorithms for damage identification of plate-type structures are presented here and will be investigated in the following section. The characteristics of these four algorithms are listed in Table 1.

2.2. *Two-dimensional gapped smoothing method (GSM)*

Ratcliffe (1997) developed a response-based gapped smoothing method (GSM) to locate the damage in beam-type structures. Yoon *et al*. (2005) generalized this one-dimensional GSM to two dimensional for plate-like structural applications.

In the one-dimensional GSM method, the displacement mode shapes are extracted from the damaged structure and then converted to curvatures. A gapped cubic polynomial is then fit to the curvature mode shape. The gapped polynomial at the ith grid C_i is represented by.

$$C_i = a_0 + a_1 x_i + a_2 x_i^2 + a_3 x_i^3 \tag{1}$$

where x_i is the distance between the ith grid point and the beam end. The coefficients $a_0, a_1, a_2,$ and a_3 are determined explicitly using the neighboring curvatures from the damaged structure.

Table 1. Characteristics of four damage identification algorithms

Algorithm	Type	Modal parameters requirement	Damage index
GSM	Response-based, damaged state response only	Experimental mode shapes in damaged plate	Defined at node
DIM	Model-based	Experimental mode shapes in damaged plate, and theoretical mode shape in healthy plate model	Defined at element
ULS	Model-based	Experimental mode shapes in damaged plate, and theoretical mode shape in healthy plate model	Defined at node
GFD	Response-based, damaged state response only	Experimental mode shape in damaged plate	Defined at node

The structural irregularity index is then calculated by the squared difference polynomial function and the curvatures as follows.

$$\delta_i = \left(\emptyset''_{i,d} - C_i\right)^2 \tag{2}$$

where $\emptyset''_{i,d}$ is the curvature at the ith grid point from the damaged structure, which can be calculated using a four-point backward/forward looking finite difference approximation that maintains a Bachmann-Landau order of magnitude of $O(h^2)$ as:

$$\emptyset''_i = (\emptyset_{i+1} + \emptyset_{i-1} - 2\emptyset_i)/h^2 \tag{3}$$

where $\emptyset''_i$ is the mode shape obtained at the ith grid point and h is the uniform separation of the test grid.

In the two-dimensional gapped smoothing method (2-D GSM), instead of using a gapped line smoothening algorithm, a gapped surface smoothening algorithm is employed. First, the curvature mode shape $\nabla^2\psi_{i,j}$ is calculated from the displacement mode shape ψ_{ij} by the central difference approximation at grid point (i, j) as follows:

$$\nabla^2\psi_{i,j} = (\psi_{i+1,j} + \psi_{i-1,j} - 2\psi_{i,j})/h_x^2 + (\psi_{i+1,j} + \psi_{i-1,j} - 2\psi_{i,j})/h_y^2 \tag{4}$$

In this equation, i and j are the location indicators in x and y directions, respectively, and h_x and h_y are the horizontal and vertical grid increments, respectively, for the grid points on the edges for which the forward or backward difference approximations are applied.

Then, the smoothed surface is obtained based on the curvature values at its neighboring grid points by surface fitting. Finally, the structural irregularity index is calculated as:

$$\delta_{i,j} = \left|\nabla^2 \psi_{i,j} - C_{i,j}\right| \tag{5}$$

where $C_{i,j}$ is the smoothed curvature at point (i, j).

2.3. *Strain Energy-based Damage Index Method (DIM)*

Stubbs and Kim (1996) first presented the damage index method based on the change in modal strain energy for 1-D beam-type structures. Cornwell et al. (1999) then extended this one dimensional strain energy method to two dimensional plate-like structures. This method requires that the mode shapes before and after damage be known. They subdivided the plate into N_x, N_y subdivisions in the x and y directions, respectively, as shown in Fig. 1, so that the damage index can be derived in each of them. Then, the fractional strain energy is assumed to remain relatively constant in undamaged sub-regions. In this case, the Young modulus and the Poisson's ratio are assumed to be essentially constant over the whole plate for both the undamaged and damaged modes, and the strain energy for a given mode shape $\psi_i(x, y)$ in a plate can be derived as:

$$U = \frac{D_{jk}}{2}\int_0^b \int_0^a \left(\frac{\partial^2 \psi_i}{\partial x^2}\right)^2 + \left(\frac{\partial^2 \psi_i}{\partial y^2}\right)^2 + 2\nu\left(\frac{\partial^2 \psi_i}{\partial x^2}\right)\left(\frac{\partial^2 \psi_i}{\partial y^2}\right)$$

$$+2(1-\nu)\left(\frac{\partial^2 \psi_i}{\partial x\,\partial y}\right)^2 dxdy \tag{6}$$

where D is the bending stiffness of the isotropic plate ($D = Eh^3/12(1-\nu^2)$).

Then the energy associated with sub-region j, k for the ith mode is:

$$U_{ijk} = \frac{D_{jk}}{2}\int_{b_k}^{b_{k+1}} \int_{a_j}^{a_{j+1}} \left(\frac{\partial^2 \psi_i}{\partial x^2}\right)^2 + \left(\frac{\partial^2 \psi_i}{\partial y^2}\right)^2 + 2\nu\left(\frac{\partial^2 \psi_i}{\partial x^2}\right)\left(\frac{\partial^2 \psi_i}{\partial y^2}\right)$$

$$+2(1-\nu)\left(\frac{\partial^2 \psi_i}{\partial x\,\partial y}\right)^2 dxdy \tag{7}$$

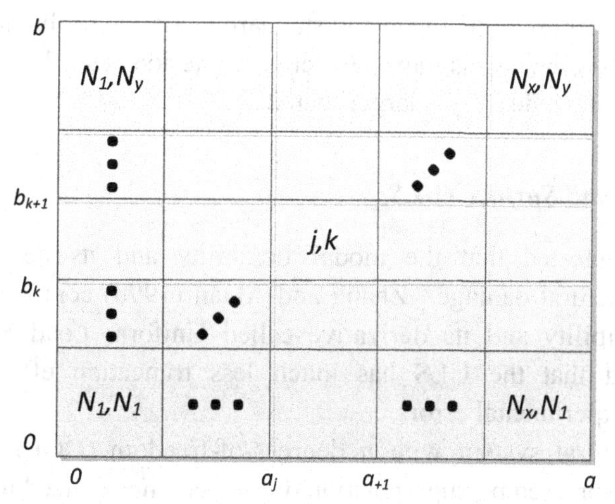

Fig. 1. A schematic illustrating a plate's $N_x \times N_y$ sub-regions (Cornwell *et al.*, 1999).

After summing over all the sub-regions, the total energy is:

$$U_i = \sum_{k=1}^{N_y} \sum_{j=1}^{N_x} U_{ijk} \tag{8}$$

The fractional energy at location j, k is defined to be:

$$F_{ijk} = U_{ijk}/U_i \quad \text{and} \quad \sum_{k=1}^{N_y} \sum_{j=1}^{N_x} F_{ijk} = 1 \tag{9}$$

Therefore, the fractional strain energy f_{ijk} in sub-region (j, k) for the ith mode is given as:

$$f_{ijk} = \frac{\int_{b_k}^{b_{k+1}} \int_{a_j}^{a_{j+1}} \left(\frac{\partial^2 \psi_i}{\partial x^2}\right)^2 + \left(\frac{\partial^2 \psi_i}{\partial y^2}\right)^2 + 2\nu \left(\frac{\partial^2 \psi_i}{\partial x^2}\right) \left(\frac{\partial^2 \psi_i}{\partial y^2}\right) + 2(1-\nu) \left(\frac{\partial^2 \psi_i}{\partial x \partial y}\right)^2 dxdy}{\int_0^b \int_0^a \left(\frac{\partial^2 w}{\partial x^2}\right)^2 + \left(\frac{\partial^2 \psi_i}{\partial y^2}\right)^2 + 2\nu \left(\frac{\partial^2 \psi_i}{\partial x^2}\right) \left(\frac{\partial^2 \psi_i}{\partial y^2}\right) + 2(1-\nu) \left(\frac{\partial^2 \psi_i}{\partial x \partial y}\right)^2 dxdy} \tag{10}$$

Following the same procedure for the damaged mode shapes indicated with an asterisk (*), the damage index β at a sub-region can be obtained from:

$$\beta_{jk} = \sum_{i=1}^{m} f_{ijk}^* \,/\, \sum_{i=1}^{m} f_{ijk} \tag{11}$$

A normalized damage index can be obtained for the purpose of comparison using:

$$Z_{jk} = \frac{\beta_{jk} - \bar{\beta}}{\sigma_\beta} \tag{12}$$

where $\bar{\beta}$ and σ_β represent the mean and standard deviation of the damage indices, respectively. Usually, a damage detection criterion can be set when the normalized damage index Z_{ij} is larger than 2.

2.4. *Uniform Load Surface (ULS)*

Researchers suggested that the modal flexibility and its derivative can be sensitive to structural damage. Zhang and Aktan (1998) comparatively studied the modal flexibility and its derivative called Uniform Load Surface (ULS). They suggested that the ULS has much less truncation effect and is less susceptible to experimental errors.

For a structural system with n degrees-of-freedom (DOF), the flexibility matrix can be expressed by superposition of the mass normalized modes ϕ_r

$$F = \sum_{r=1}^{n} \frac{\phi_r \phi_r^T}{\omega_r^2} \tag{13}$$

where ω_r is the rth natural frequency. It can be seen that as the frequency increases the flexibility decreases substantially, so that the flexibility matrix converges quickly with several lower modes, which make it easier to approximate the matrix with a few lower modes as:

$$F_T = [f_{k,l}] = \sum_{r=1}^{m} \frac{\phi_r \phi_r^T}{\omega_r^2} f_{k,l} \tag{14}$$

where the modal flexibility, $f_{k,l}$ at the kth point under the unit load at point l is the summation of the products of two related modal coefficients for each available mode

$$f_{k,l} = \sum_{r=1}^{m} \frac{\phi_r(k) \phi_r^T(l)}{\omega_r^2} \tag{15}$$

Then, the ULS is defined as the deflection vector of the structure under uniform load

$$U_T = \{u(k)\} = F_T . L \tag{16}$$

where $L = \{1 \quad 1 \quad \dots \quad \dots \quad 1\}_{1\text{x}n}^T$ is the unit vector representing the uniform load acting on the structure

$$U_T = \{u(k)\} = \begin{bmatrix} f_{1,1} & f_{1,2} & \cdots & f_{1,n} \\ f_{21} & f_{2,2} & & f_{2,n} \\ \vdots & & \ddots & \vdots \\ f_{n,1} & f_{n,2} & \cdots & f_{nn} \end{bmatrix} \begin{Bmatrix} 1 \\ 1 \\ \vdots \\ \vdots \\ 1 \end{Bmatrix} \tag{17}$$

$u(k)$ is the modal deflection at point k under uniform unit load all over the structure.

$$u(k) = \sum_{l=1}^{n} f_{k,l} = \sum_{r=1}^{m} \frac{\phi_r(k) \sum_{l=1}^{n} \phi_r^T(l)}{\omega_r^2} \tag{18}$$

Finally, the damage index can be formulated as follows:

$$d(x_i, y_j) = \left[\propto_{xx} |u_{xx}^D - u_{xx}| + \propto_{yy} |u_{yy}^D - u_{yy}| + \propto_{xy} |u_{xy}^D - u_{xy}|\right]^2 \tag{19}$$

where u_{xx}, u_{yy}, u_{xy} $u_{xx}^D, u_{yy}^D, u_{xy}^D$ are the curvature values of the intact structure and the damaged structure, respectively. $\propto_{xx}, \propto_{yy}, \propto_{xy}$ are the weights that can be set from 0 to 1 to account for the importance of the curvature in the corresponding direction.

2.5. *Generalized Fractal Dimension (GFD)*

The concept of a fractal and its relevant mathematical model were developed by Mandelbrot and Van Ness (1968). If we take an object residing in Euclidean dimension D and reduce its linear size by $1/r$ in each spatial direction, its measure (length, area, or volume) would increase to $N = r^D$ times the original. D is called the fractal dimension (FD) of a fractal curve, and it can be expressed as (Wang and Qiao, 2007; Hadjileontiadis and Douka, 2007):

$$FD(x) = \frac{\log n}{\log n + \log\left[\frac{d(x_i, M)}{L(x_i, M)}\right]} \tag{20}$$

$$L(x_i, M) = \sum_{j=1}^{M} \sqrt{\left(y(x_{i+j}) - y(x_{i+j-1})\right)^2 + \left(x_{i+j} - x_{i+j-1}\right)^2}$$

$$d(x_i, M) = \max_{1 \le i \le M} \sqrt{\left(y(x_{i+j}) - y(x_{i+j-1})\right)^2 + \left(x_{i+j} - x_{i+j-1}\right)^2}$$

where $x = \frac{1}{2}(x_i + x_{i+M})$, $n = \frac{1}{\alpha}$, α is the average distance between successive points, x_i and y_i are the coordinate values of curve. The term M represents the sliding window dimension length. The dimension of a smooth curve is 1, and the more fractions the curve has, the larger the dimension of the curve is. The sharp peak of the FD curve indicates the location of the damage. Therefore, the fractal dimension has the potential to serve as a damage index to reveal the irregularity introduced by local damage in the structure. However, when the higher mode shape is considered, the above FD approach may give some misleading peak information in the location of the maximum and minimum points in a curve. To overcome this shortcoming, a generalized fractal dimension (GFD) was defined by Wang and Qiao (2007) as a modification of the FD method:

$$GFD(x) = \frac{\log n}{\log n + \log\left[\frac{d_S(x_i,M)}{L_S(x_i,M)}\right]}$$

$$L_s(x_i, M) = \sum_{j=1}^{M} \sqrt{\left(y(x_{i+j}) - y(x_{i+j-1})\right)^2 + S^2\left(x_{i+j} - x_{i+j-1}\right)^2} \tag{21}$$

$$d_s(x_i, M) = \max_{1 \le i \le M} \sqrt{\left(y(x_{i+j}) - y(x_{i+j-1})\right)^2 + S^2\left(x_{i+j} - x_{i+j-1}\right)^2}$$

where s is a scale parameter, as inspired by the wavelet transformation. Compared to the mode shape itself, the irregularity caused by the damage on the deformation mode shape is local and smaller, and it can be filtered at the sharp peak value of FD introduced by the mode shape itself while keeping the one caused by the local damage through choosing a proper scale value s.

It should be noted that the FD/GFD method is still a 1-D method, and when applying it to a plate-like structure, the plate must be divided into slices. Each slice is treated as a 1-D beam. Then, the FD/GFD can be applied to each slice. Hadjileontiadis *et al.* (2007) successfully applied the FD method to a thin plate with a crack lying parallel to one side of the plate. They considered a signal consisting of spatial data (i.e., displacement of a 2-D structure) from a 2-D vibration mode. The FD operator was applied to succeeding horizontal, vertical and diagonal 1-D slices of the 2-D vibration mode. The method was shown to be able to identify the location and length of the crack.

3. Comparative Study

3.1. *Geometry of the composite plate*

A comparative study on the four aforementioned damage identification algorithms is conducted. The evaluation of the performance of the algorithms is based on both the numerical and experimental mode shape data from a sample composite plate. The pultruded E-glass Fiber Reinforced Plastic (FRP) plate is 486.2 mm long × 254 mm wide × 3.175 mm thick (19 in. ×10 in. × 0.125 in.) and manufactured by Creative Pultrusions. Inc., Alum Bank, PA. The damage is

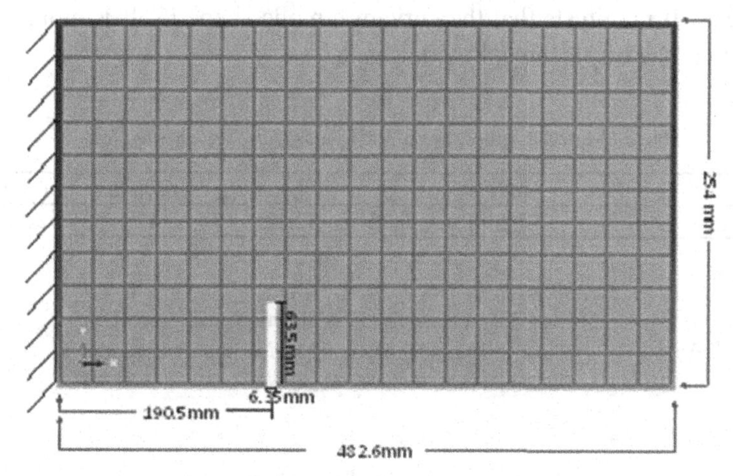

Fig. 2. Geometry of the cantilevered composite plate and the artificially induced saw-cut damage.

artificially induced as a 63.5 mm × 6.35 mm (2.5 in. × 0. 25 in.) through-thickness saw-cut, as shown in Fig. 2. The plate is clamped at one of the longitudinal edge and free at others (in a cantilever configuration).

3.2. *Numerical analysis*

The numerical modal analysis of the composite plate is performed using the commercial Finite Element Analysis (FEA) package ABAQUS version 6.10. ABAQUS offers a linear perturbation procedure to perform eigenvalue extraction to calculate the natural frequencies and the corresponding mode shapes of a system. The equivalent material properties of the GFRP composite plate adopted in the FEA are listed in Table 2. The plate is modeled with the cantilevered boundary condition, fixed at one edge and free at the others. The 4-node general-purpose reduced-integration shell element S4R is utilized in the analysis. Fine meshing is used to discrete the plate with the element size of 1.27 mm × 1.27 mm (0.05 in. × 0.05 in.), generating a total of 76581 elements for the healthy plate and fewer for the damaged plate. Fig. 3 shows the displacement and curvature mode shapes of the first and second vibration mode of the damaged plate. The displacement and curvature modes shapes are extracted from 18 × 10 nodes with a node spacing of 25.4 mm (1 in.). The grid lines/nodes are shown in Fig. 2. The fundamental vibration mode shapes of the damaged plate obtained from numerical simulation are shown in Fig. 3. As we can see in the figure, the damage can be directly noticed from curvature mode shape but not from displacement. It manifests that the curvature mode shape itself is more sensitive to damage than the displacement mode shape.

Table 2. Material .properties of the E-glass *fiber composite plate.*

ρ (kg/m^3)	E_x (GPa)	E_y (GPa)	v_{xy}	G_{xy} *(GPa)*	G_{xz} (GPa)	G_{yz} (GPa)
1,800	12.41	6.90	0.32	2.42	2.42	2.23

where

ρ	-	Density;
E_x, E_y	-	Modulus of elasticity in *x*- direction and *y*- direction;
G_{xy}, G_{xz}, G_{yz}	-	Shear modulus in the *xy, xz, and yz* planes respectively;
v_{xy}	-	Poisson's ratio in the *xy* plane.

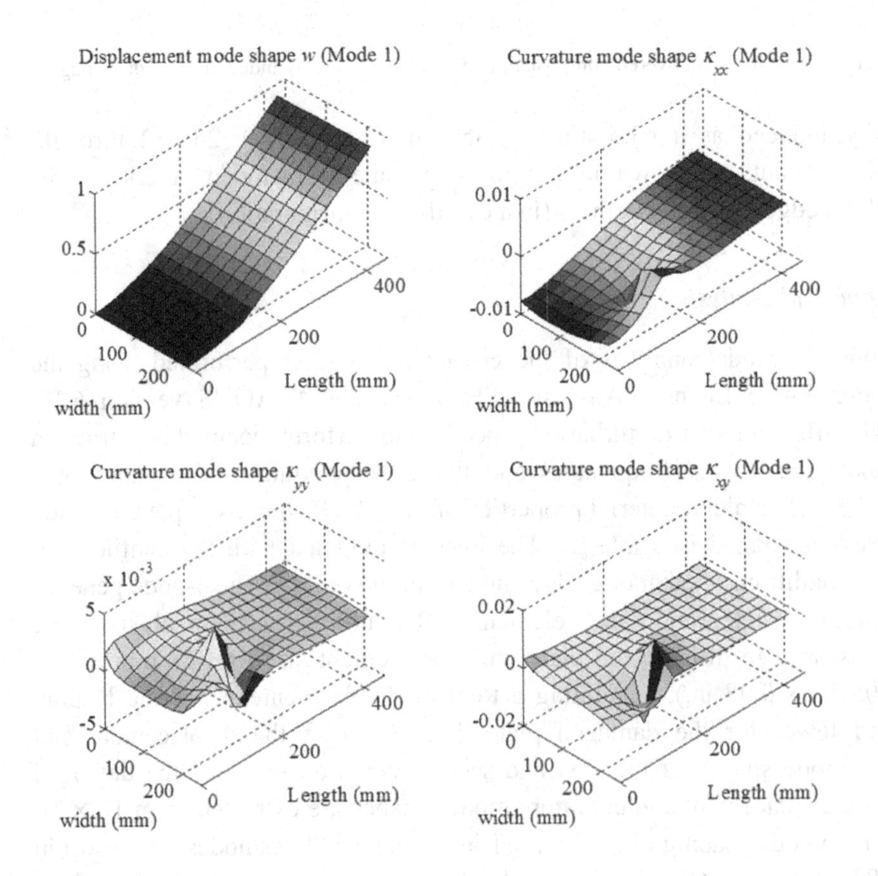

Fig. 3. Displacement and curvature mode shapes of the 1st mode of the damaged plate from numerical simulation.

3.3. *Damage identification based on numerical data*

First, the four typical damage identification algorithms aforementioned in Section 2 are applied to the numerically simulated displacement mode shapes for damage identification. In order to generate the curvature mode shape required by DIM from the displacement mode shape, the central difference method is adopted. It should be noted that the damage index is defined at nodes in GSM and ULS; while the damage index is defined at elements in DIM and GFD. When the GFD method is used, the slices are along longitudinal direction (x-direction) and a constant scale parameter $s = 1$ is chosen so that GFD is reduced to FD in the analysis. For comparison, the damage indices generated by these methods are all normalized using Eq. (12), and a criterion of $|DI| > 2$ is adopted for the indication of damage.

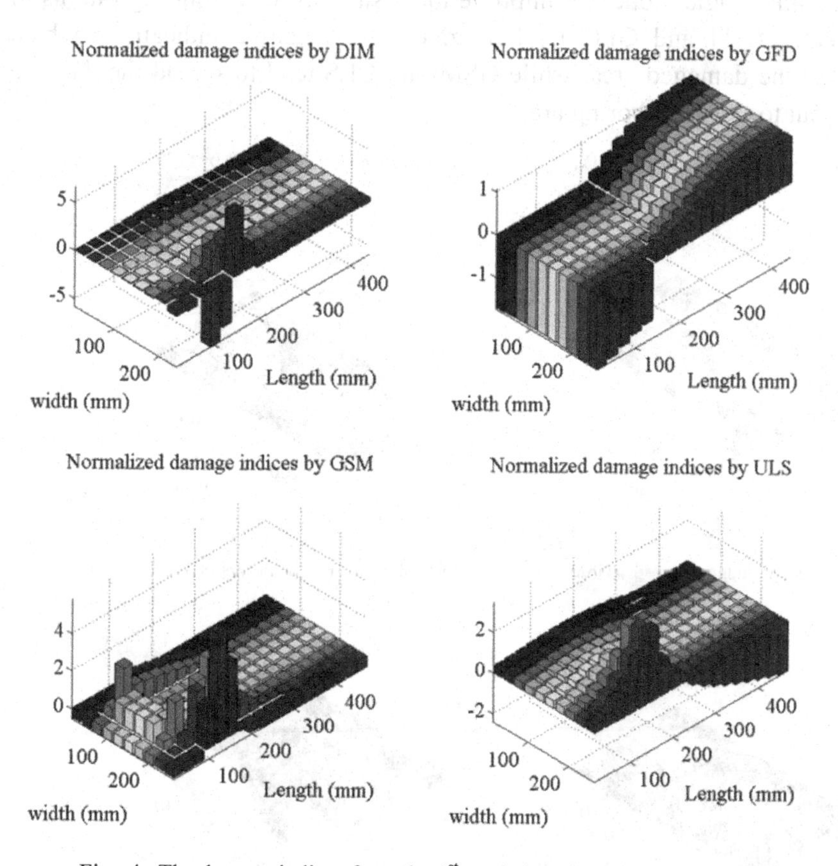

Fig. 4. The damage indices from the 1st mode based on displacement mode shape

When only the fundamental mode shape data are adopted, as suggested by some damage identification algorithms, the damage indices from four algorithms are shown in Fig. 4. It is shown that GFD fails to indicate any damage when only the fundamental mode is adopted. Further investigation shows that using other vibration modes or adjusting the scale parameter s may significantly improve the result. For GSM and ULS, there are obviously some false positive indications of damage far away from the actual damaged area although the actual damage can be identified.

To make full use of the accessible modes from experimental modal testing, a combination of several modes can also be adopted to derive the damage indices, e.g., Eq. (11). The damage indices based on all the modes below 250 Hz are shown in Fig. 0.5. It can be seen that all four algorithms can correctly approximate the location of the damage without any false positive indication far away from the damage. We can notice that using a combination of modes instead of using single mode can improve the results from all four algorithms to some extent. DIM and GFD tend to give false negative indications which underpredict the damaged area, while GSM and ULS tend to spread the damage indication out to the neighboring area.

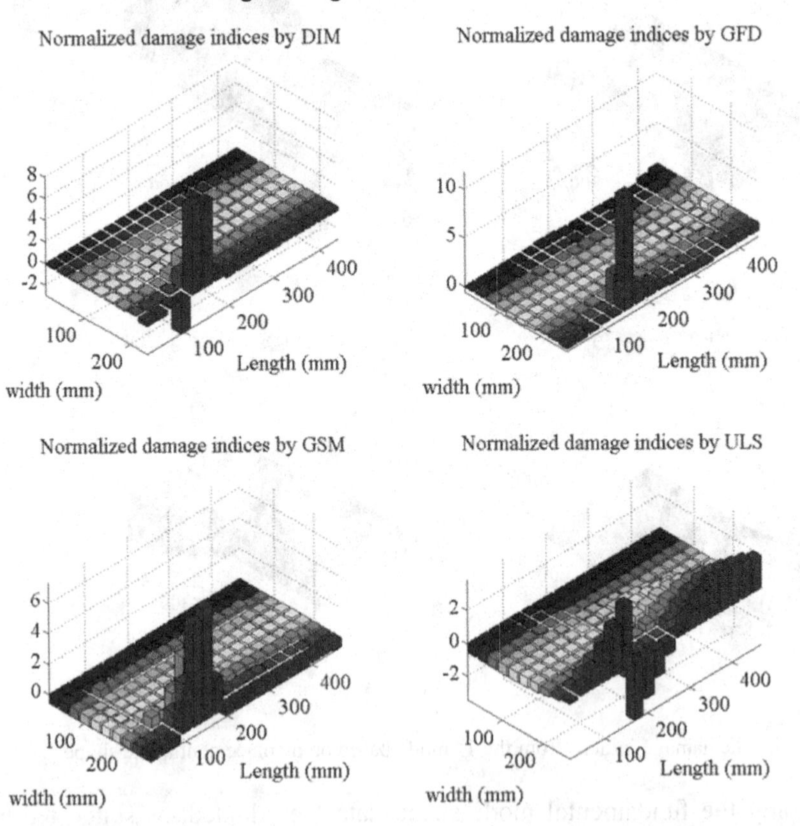

Fig. 5. The damage indices from the mode combinations based on displacement mode shape.

Then, the four damage identification algorithms are applied to the numerically simulated curvature mode shapes. It should be noted that the GFD and ULS methods are directly applied to the curvature mode shapes, although they are originally developed based on the displacement mode shape. The damage indices using curvature κ_{xx}, κ_{yy} and κ_{xy} are summed up and normalized to generate the combined damage indices for GFD and ULS.

The damage indices based on the fundamental mode and modes below 250 Hz are shown in Figs. 6 and 7, respectively. It can be seen that all four algorithms can correctly approximate the location of the damage when the curvature mode shapes are adopted for damage identification. It can also be noticed that the damage indices in the damaged area shown in Figs. 6 and 7 are much higher than those shown in Figs. 4 and 5. It further confirms that the curvature mode shapes are more sensitive to damage than the displacement mode shape. The comparison between Figs. 6 and 7 also reinforces the conclusion that using a combination of modes instead of using single mode can improve the damage identification results from all four algorithms.

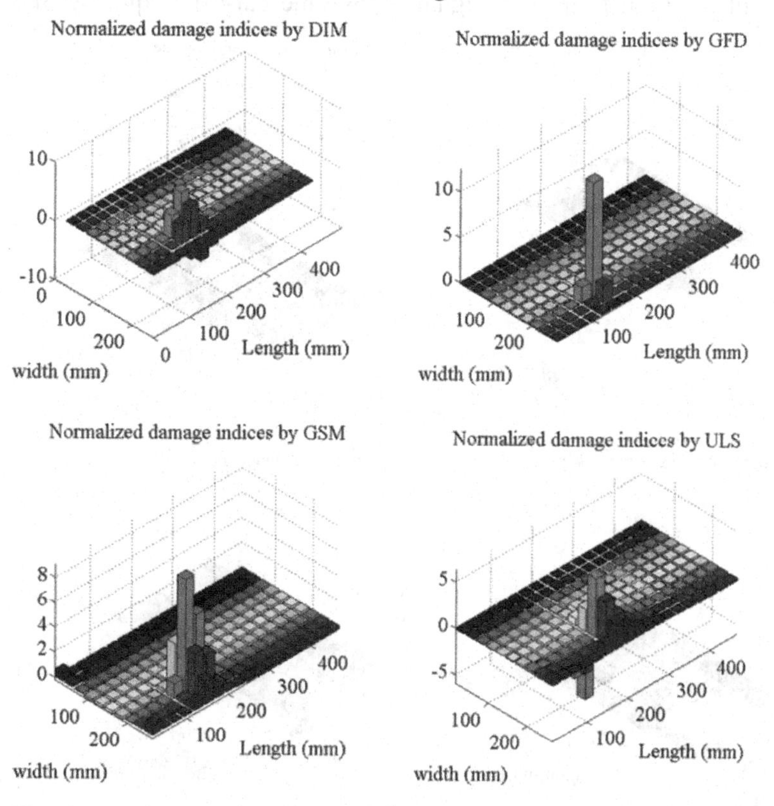

Fig. 6. The damage indices from the 1st mode based on curvature mode shape.

3.4. *Experimental program*

An experimental program was developed to test these four damage identification algorithms. A composite plate was set up in laboratory as described in Section 3.1. The experiment test is shown in Fig. 8.

The displacement mode shapes from the healthy and the damaged plate were extracted from a roving excitation test. The plate was uniformly divided into 19×11 elements by the grid lines as shown in Fig. 2. The plate was subjected to a dynamic pulse load applied at each grid point using modally tuned hammer (PCB 086C03). A total of 18×10 grid nodes were tested within the area of interest corresponding to a spatial sampling distance of 25.4 mm (1.0 in.). A 2.4 lb. modally-tuned ICP® impact hammer was used to excite the plate at each node. The response measurements were made using one accelerometer (PCB 352C68) fixed on the plate surface to record the structure response time histories, as shown in Fig. 8(a). The analog signals then passed a low-pass anti-aliasing filter to prevent aliasing. A Krohn-Hite 3382 8-pole dual channel filter was employed to filter out the high frequency signals above the cut-off frequency of 500 Hz.

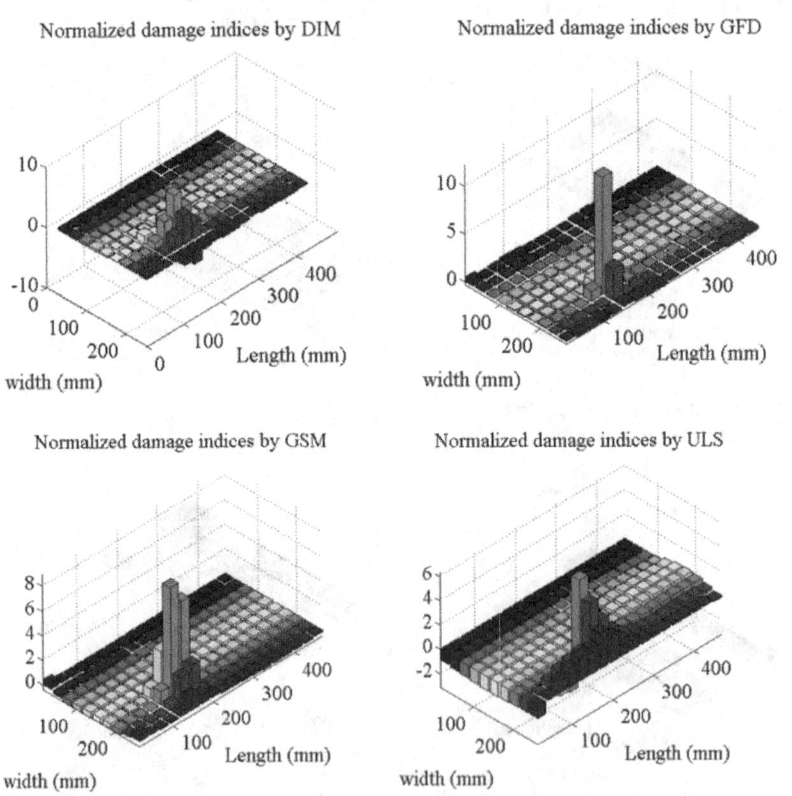

Fig. 7. The damage indices from the mode combinations based on curvature mode shape.

The filtered signals are then digitized and collected by the data acquisition system dSPACE CP1103 at a sampling frequency of 1000Hz. The measurements at each point are repeated 8 times, and the synchronized time histories from the excitation and response measurements were averaged to enhance the signal-to-noise ratio (SNR). Then, the frequency response functions (FRFs) of these tested points can be calculated using Fast Fourier Transform (FFT) on these excitation and response time histories. Then, these FRF curves were imported to the modal analysis program ME'Scope for curve fitting and modal extraction.

The curvature mode shapes from the healthy and the damaged plate can be extracted using a similar approach. However, a roving sensor test was adopted instead of roving excitation test to avoid the high cost and complexity of bonding PZT actuator on each node. A bidirectional sweep sine signal varying from 0 Hz to 1,000 Hz within 10 seconds was generated by the Digital-to-Analogue Converter (DAC) channel of the dSPACE system and then amplified by a power amplifier to excite a PZT patch actuator fixed at one node.

(a)

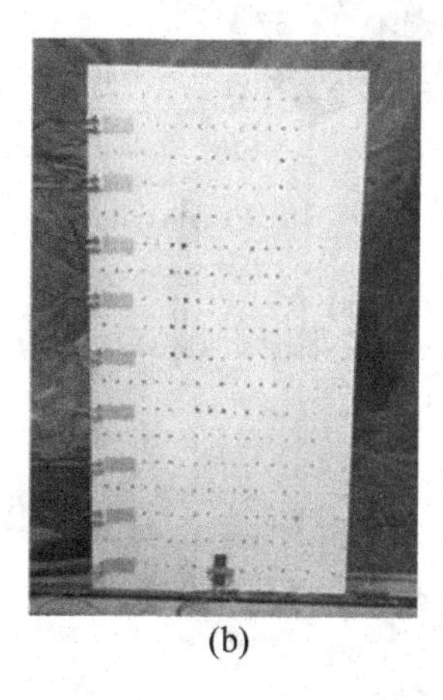

(b)

Fig. 8. FRP plate mounted on a steel I-beam fixture with (a) Accelerometer and (b) PZT and PVDFs attached to the surface.

The PVDF sensors were attached to all the nodes to measure the curvature (strain) response over the plate. Due to limitation of the available DAC channels, only one set of nine sensors is measured at a time, as shown in Fig. 8(b). The synchronized actuation and response time histories were simultaneously collected from PZT actuator and PVDF sensors. The measurements were repeated eight times. Then, the synchronized time histories from the excitation and response measurements are averaged and transformed into FRFs. The FRFs of each node can be obtained and imported into ME'Scope for modal analysis. Since the roving sensors were PVDFs which measure the strain responses on the plate surface, the experimentally-generated mode shapes are the strain mode shapes of the plate. The curvature mode shapes are considered as the same to the strain mode shape because the curvatures of the plate section are proportional to the strain of plate surface in the corresponding directions according to classic plate theory.

It should be noted that the accuracy of curvature mode shape measurement using the low-profile PVDF sensors is lower than the displacement mode shape measurement using accelerometer due to issues, such as electromagnetic noise immunity, sensor array calibration, etc.

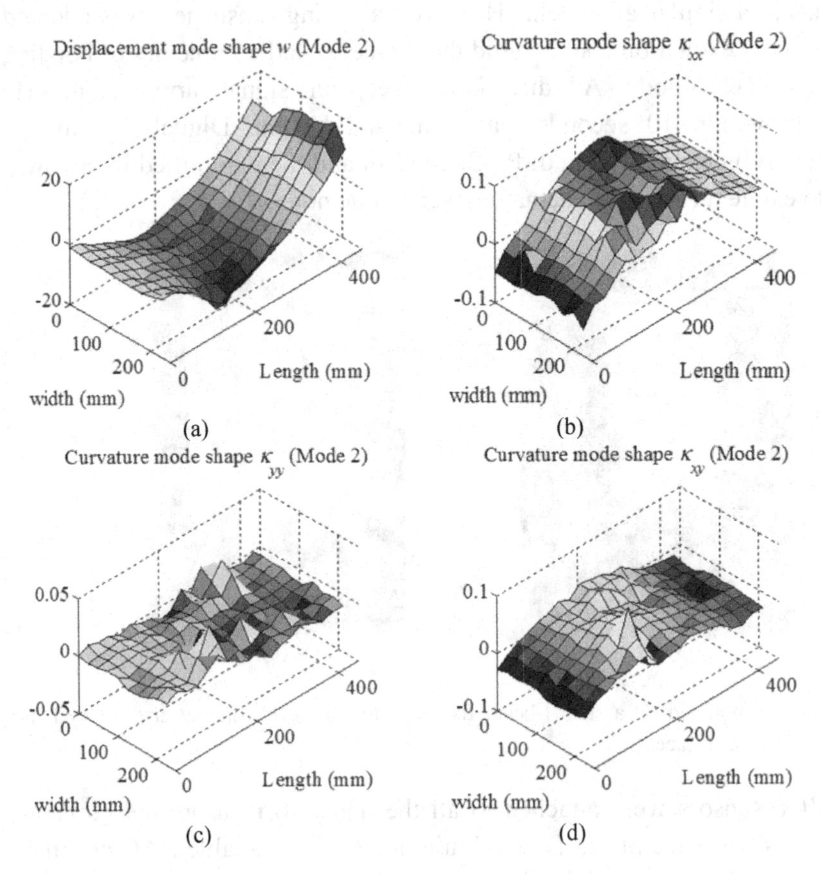

Fig. 9. Displacement and curvature mode shape of the 2nd mode of the damaged plate from experimental modal testing.

The 2nd mode shapes of the damaged plate obtained from the experimental modal testing are shown in Fig. 9. The 1st mode shapes are not shown here, and they are not adopted in damage identification because the experimentally obtained 1st mode shape is severely contaminated by the room noise (the 1st

modal frequency (= 5.2 Hz) is in the frequency range of room noise (< 20 Hz)). It is noticed that in the 2nd mode the curvature mode shape in *y*- direction κ_{yy} is corrupted. Since the 2nd vibration mode is primarily dominated by the longitudinal bending mode (*x*-direction bending mode), it is expected that the absolute value of the modal curvature in *y*- direction is small and leads to a low SNR.

3.5. *Damage identification based on experimental data*

Similar to the numerical analysis, the four damage identification algorithms are first applied to the experimentally-obtained displacement mode shapes for damage identification. According to the conclusions from the numerical data, only a combination of several modes below 250 Hz, instead of a single mode, is adopted in the damage identification. The 1st mode shapes are excluded due to severe noise contamination in the low frequency range. The damage indices are shown in Fig. 10.

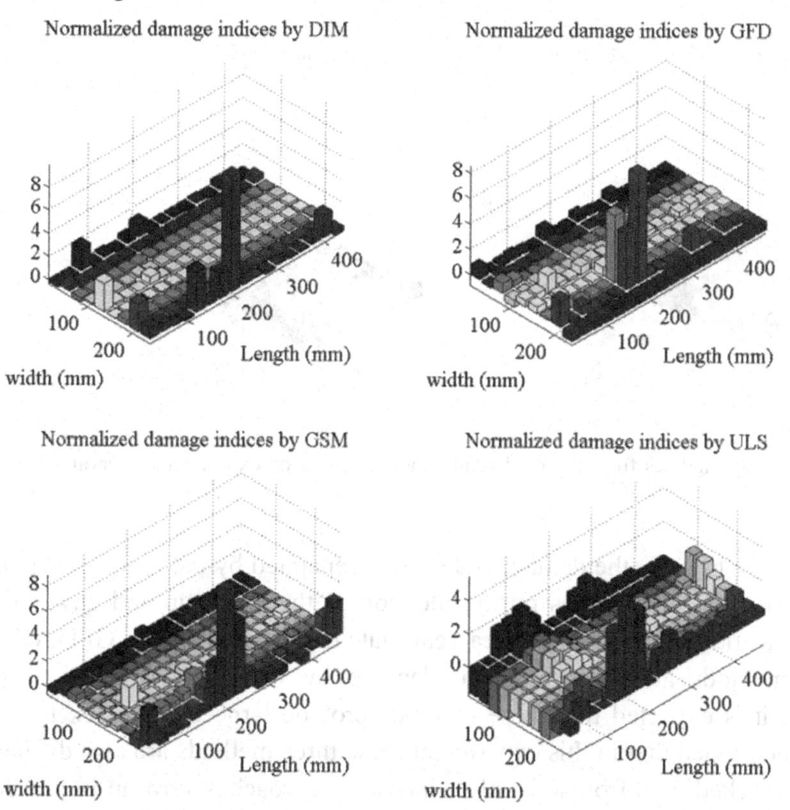

Fig. 10. The damage indices from the mode combinations based on experimental displacement mode shape.

Then, the four damage identification algorithms are applied to the curvature mode shapes. It is interesting to note that the damage identification results using curvature are comparable to or better than results using displacement, although the accuracy of the curvature mode shape measurement is lower than the displacement measurement.

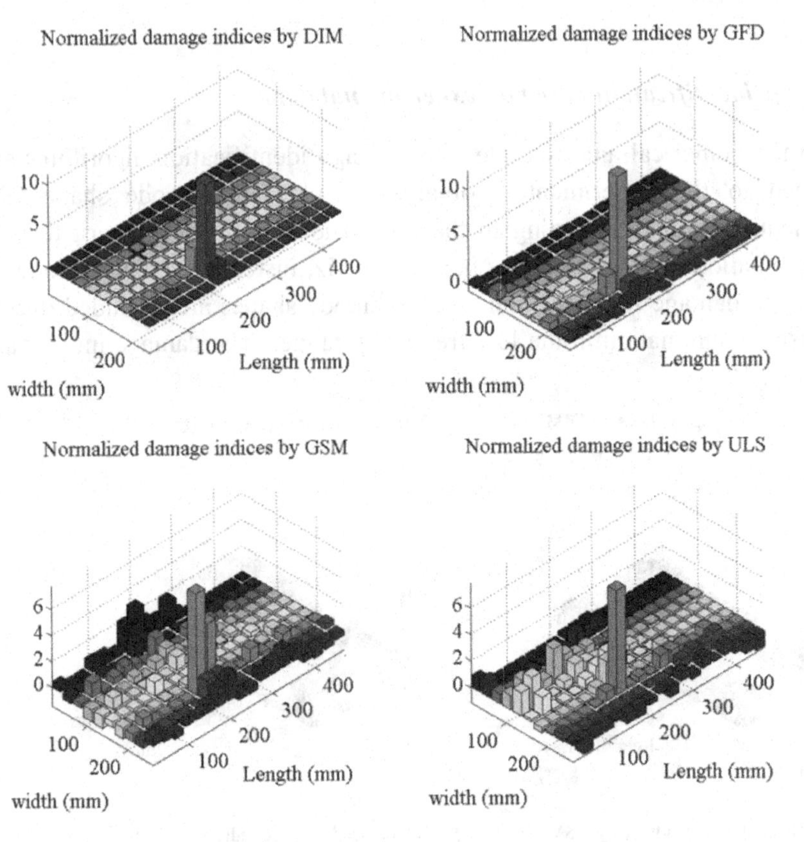

Fig. 11. The damage indices from the mode combinations based on experimental curvature mode shape.

As shown in Fig. 11, the damage index (DI) generated by the GFD, GSM and ULS methods only emphasizes on the location of the crack tip and gave false negative indication of the cracked area near plate edge. Because the GFD, GSM and ULS methods attempts to locate damage by locating the mode shape irregularity, it is expected that these methods provide large DI at crack tips. It should be mentioned that in this experiment these three methods are also difficult to find the cracked area from κ_{xx} only, since κ_{xx} approaches zero on both sides of the crack. While the DI generated by the strain energy-based DIM can correctly approximate the cracked area. From the comparison of damage indices from four algorithms, it is concluded that the DIM has the best overall performance for damage detection when applied to both numerical simulated data and experimental test data.

It is shown in Fig. 11 that both GSM and ULS give a few false positive damage indications. Note that these false positive indications can be eliminated if a higher threshold value (e.g. $|DI| = 3$) is adopted to indicate damage. However, increasing the threshold value will result in higher noise immunity and lower damage sensitivity. Therefore, the trade-off between the sensitivity to damage and the noise rejection capability should be carefully considered for the application before adjusting the criterion of DI for indicating damage.

4. Summary and Conclusions

This chapter has provided a brief introduction to dynamics-based (vibration-based) damage identification for plate-type structures. According to the dynamic features extracted for damage identification, the algorithms can be categorized as natural frequency-based method, mode shape-based method, modal strain energy-based method, modal flexibility-based method, etc.

Upto date, numerous dynamics-based methods have been proposed in the literature. However, most research efforts on damage identification methods still focus on damage localization in 1-D beam-type structures (Fan and Qiao, 2011). Some researchers also explored a unified approach to identify damage location and quantification. Limited research on damage identification of 2-D plate-type structures can be found in the literature. Some 1-D algorithms can be generalized to 2-D for application in damage identification of plate-type structures.

The rest part of this chapter is devoted to the introduction of four damage identification algorithms (GSM, DIM, ULS, GFD) and the comparative study among them for 2-D plate-type structures. The algorithms are investigated and compared based on numerically-simulated data and experimentally-obtained data. The effect of the use of displacement mode shape and curvature mode shape in damage identification is discussed. It is observed that not only the curvature mode shape itself is more sensitive to damage than the displacement mode shape, but also the curvature mode shape works better when used with all four investigated algorithms than the displacement mode shape. It can also be noticed that using a combination of several vibration modes instead of using single mode can improve the damage identification results from all four algorithms.

Acknowledgements

This research was supported by the Alaska University Transportation Center (AUTC), State of Alaska Department of Transportation & Public Facilities, and US Department of Transportation (Contract/Grant Number: DTRT06-G-0011). The involvement and assistance in the experimental portion of this research by the graduate student Hussam Saleem at the WSU-Smart Structures Lab are appreciated.

References

Carden, E.P., and Fanning, P. (2004). Vibration based condition monitoring: A review, *Structural Health Monitoring.*, 3, 355-377.

Cornwell, P., Doebling, S.W., and Farrar, C.R. (1999). Application of the strain energy damage detection method to platelike structures, *J Sound Vib.*, 224, 359-374.

Doebling, S.W., Farrar, C.R., Prime, M.B., and Shevitz, D.W. (1996). Damage Identification and Health Monitoring of Structural and Mechanical Systems from Changes in their Vibration Characteristics: A Literature Review in *Los Alamos National Laboratory report.*

Fan, W. and Qiao, P.Z. (2011). Vibration-based damage identification methods: A review and comparative study. *Structural Health Monitoring, An International Journal,* 10(1), 83-111. DOI: 10.1177/1475921710365419.

Farrar, C.R., and James, G.H. (1997). System identification from ambient vibration measurements on a bridge, *J Sound Vib.*, 205, 1-18.

Hadjileontiadis, L. J., and Douka, E. (2007). Crack detection in plates using fractal dimension, *Eng. Struct.*, 29, 1612-1625.

Huth, O., Feltrin, G., Maeck, J., Kilic, N., and Motavalli, M. (2005). Damage identification using modal data: Experiences on a prestressed concrete bridge, *J Struct Eng-ASCE.*, 131, 1898-1910.

Khan, A.Z., Stanbridge, A.B., and Ewins, D.J. (1999). Detecting damage in vibrating structures with a scanning LDV, *Optics and Lasers in Engineering.*, 32, 583-592.

Kim, J.T., Ryu, Y.S., Cho, H.M., and Stubbs, N. (2003). Damage identification in beam-type structures: frequency-based method vs mode-shape-based method, *Eng Struct.*, 25, 57-67.

Kim, J.T., and Stubbs, N. (1995). Model-Uncertainty Impact and Damage-Detection Accuracy in Plate Girder, *Journal of Structural Engineering-ASCE.*, 121, 1409-1417.

Mandelbrot, B. B., and Van Ness, J. W. (1968). Fractional Brownian Motions, Fractional Noises and Applications, *SIAM Review.*, 10, 422-437.

Pandey, A.K., and Biswas, M. (1994). Damage Detection in Structures Using Changes in Flexibility, *J. Sound Vib.*, 169, 3-17.

Ratcliffe, C.P. (1997). Damage detection using a modified laplacian operator on mode shape data, *J Sound Vib.*, 204, 505-517.

Salawu, O. S. (1997). Detection of structural damage through changes in frequency: A review, *Eng Struct.*, 19, 718-723.

Salawu, O.S., and Williams, C. (1994). Damage location using vibration mode shapes, *International Modal Analysis Conference*, 933-9.

Shi, Z.Y., and Law, S.S. (1998). Structural damage localization from modal strain energy change, *J Sound Vib.*, 218, 825-844.

Shi, Z.Y., Law, S.S., and Zhang, L.M. (2000). Damage localization by directly using incomplete mode shapes, *J Eng Mech-ASCE.*, 126, 656-660.

Sohn, H., Farrar, C.R., Hemez, F.M., Shunk, D.D., Stinemates, D.W., and Nadler, B.R. (2003). A Review of Structural Health Monitoring Literature: 1996-2001 in *Los Alamos National Laboratory report.*

Stubbs, N., and Kim, J.T. (1996). Damage localization in structures without baseline modal parameters, *AIAA J.*, 34, 1644-1649.

Stubbs, N., Kim, J. T., and Farrar, C.R. (1995). Field verification of a nondestructive damage localization and severity estimation algorithm, *Proceedings of 13th International Modal Analysis Conference.*, 1, 210-8.

Topole, K.G., and Stubbs, N. (1995). Nondestructive Damage Evaluation in Complex Structures from a Minimum of Modal Parameters, *Modal Analysis-the International Journal of Analytical and Experimental Modal Analysis.*, 10, 95-103.

Wang, J., and Qiao, P.Z. (2007). Improved damage detection for beam-type structures using a uniform load surface, *Structural Health Monitoring-an International Journal.*, 6, 99-110.

Wang, B. T. (1998). Structural modal testing with various actuators and sensors, *Mechanical Systems and Signal Processing.*, 12, 627-639.

Worden, K., Farrar, C.R., Manson, G., and Park, G. (2007). The fundamental axioms of structural health monitoring. Paper presented at the *P R Soc A.*

Wu, D., and Law, S.S. (2004). Damage localization in plate structures from uniform load surface curvature, *J. Sound Vib*, 276, 227-244.

Wu, D., and Law, S.S. (2005). Sensitivity of uniform load surface curvature for damage identification in plate structures, *Journal of Vibration and Acoustics-Transactions of the ASME.*, 127, 84-92.

Yoon, M.K., Heider, D., Gillespie, J.W., Ratcliffe, C.P., and Crane, R.M. (2005). Local damage detection using the two-dimensional gapped smoothing method, *J Sound Vib.*, 279, 119-139.

Zhang, Z., and Aktan, A.E. (1998). Application of modal flexibility and its derivatives in structural identification, *Research in Nondestructive Evaluation.*, 10, 43-61.

Chapter 4

Simulation Based Methods for Model Updating in Structural Condition Assessment

H. A. Nasrellah[1], B. Radhika[2], V. S. Sundar[2], and C. S. Manohar[2*]

[1]*Department of Civil Engineering*
Alzaiem Alazhari University, Khartoum, Sudan

[2]*Department of Civil Engineering, Indian Institute of Science*
Bangalore 560 012 India
Email[2]: manohar@civil.iisc.ernet.in*

1. Introduction

The problem of structural system identification constitutes an important and difficult class of inverse problems in the study of existing engineering structures. Various complicating features, such as, ill-posed governing equations, stochastic characterization of imperfections in measurements and structural modeling, structural nonlinearities, spatially incomplete measurements, lack of measurements when excited, model selection, criteria for identifiability, and the possible need for online identification of system parameters (in actively controlled structures, for example) make these problems very challenging. Beyond the identification step, the updating of models for applied forces, response sensitivities, and reliability are also of central importance in the condition assessment of structures.

The basic premise of the present study is that a mathematical model based on principles of mechanics is deducible for the structure under consideration. A set of model parameters are taken to be unknown and the model is deemed to be imperfect. Furthermore, a set of noisy measurements on a subset of structural responses and applied actions is taken to be available and the measured quantities are related to system states via mathematical models which, again, could be imperfect. Both the model and measurement imperfections are assumed to be amenable for probabilistic modeling. The basic problem here is that of drawing inferences on unknown model parameters, hidden system states, and unmeasured applied forces based on the available measurements. The solutions to these problems become increasingly difficult as systems behave nonlinearly, the size of the problem increases (in terms of degrees of freedom of the structural model, the number of unknown states, the parameters and applied actions to be estimated), or noises become multiplicative and (or) non-Gaussian.

The problem of determining system parameters can be tackled based on two alternative view points: the first treats the system parameters as being deterministic and determines them by maximizing a likelihood function; the second treats the parameters as a vector of random variables and employs Bayesian framework to determine the posterior probability density function (PDF) via a multi-fold integration. The application of Bayes' theorem in the latter case leads to formal solutions for the posterior PDFs only up to a constant multiplication factor. Complete closed-form solutions are possible only for a few special cases. In general, one has to resort to either analytical approximations or numerical solutions based on Monte Carlo simulations. The Monte Carlo simulation-based methods are versatile in terms of handling complicated nonlinearities and additive/multiplicative non-Gaussian noise models, and when combined with finite element analysis (FEA) of structures, they lead to powerful tools for the study of existing instrumented structures. In this chapter, we aim to provide an overview of formalisms and tools available to tackle these problems and illustrate them with a few numerical examples.

Markov chain Monte Carlo (MCMC) methods provide powerful computational tools for drawing samples of multi-dimensional random variables whose joint PDF could be known only up to a multiplication constant. By virtue of this capability, these methods become computationally the natural allies to Bayesian approaches to problem formulations. These methods aim to construct Markov chains whose steady state PDFs match the target PDFs and, under the assumption of ergodicity, a single realization of the chain is construed to provide the desired samples of the vector random variable of interest. These methods yield correlated samples and also require numerical diagnostics to determine if the chain has reached steady state. Following seminal contributions by Metropolis et al., (1953) and Hastings (1970), the literature on MCMC has grown extensively (Brooks *et al.*, 2011; Gilks *et al.*, 1996; Kroese *et al.*, 2011; Liu, 2001; Robert and Casella, 2004). When dealing with real time applications, in which measurement data arrive sequentially in time and inferences need to be made in real time, or, in dealing with complicated time evolutions of system states, the sequential Monte Carlo (SMC) simulation methods become more useful. For this class of problems, the linear Gaussian state space models admit closed-form solutions as was propounded in the celebrated work of Kalman (1960). This led to the development of approximate analytical methods for treating nonlinear systems via linearization/ Gaussianization (Evensen, 2006; Jazwinski, 1970; Maybeck, 1982). Gordon *et al.*, (1993) introduced recursive Monte Carlo simulation-based methods for treating nonlinear systems and non-Gaussian noise models, and following this, the simulation-based methods have been widely researched (Bocquet *et al.*, 2010; Cappé *et al.*, 2007; Doucet *et al.*, 2001; Doucet and Johansen, 2008; Kantas *et al.*, 2011; Ristic *et al.*, 2004; Tanizaki, 1996).

The growth of the subject has been due to contributions from diverse fields such as computational statistics, climate modeling, robotics, economics, signal processing, and control engineering. In the area of structural engineering some of the early studies in this area of research are found in (Beck, 1978; Hoshiya and Saito, 1984; Imai *et al.*, 1989; Shinozuka *et al.*, 1982) and a representative list of later publications are contained in (Beck and Au, 2002; Ching *et al.*, 2006; Manohar and Roy, 2006; Wang and Haldar, 1997; Yuen and Kuok, 2011). The recent book by Yuen (2010) (with 303 references) contains a comprehensive account of application of Bayesian methods in civil engineering.

In this work, we will not delve into algorithmic details of MCMC and SMC methods; instead we will demonstrate how the methods can be used to solve problems of interest in structural engineering. It may also be noted that the general class of methods for nonlinear system identification includes a diversity of tools (Bendat, 1998; Imregun, 1998; Kerschen *et al.*, 2006; Nelles, 2001; Worden and Tomlinson, 2001). We focus our attention in this Chapter on simulation-based methods which treat the problem of model updating within the probabilistic framework. The material presented herein is biased by our recent studies (Ghosh *et al.*, 2007; Ghosh *et al.*, 2008; Namdeo and Manohar, 2007; Namdeo, 2007; Nasrellah, 2009; Nasrellah and Manohar, 2010; Nasrellah and Manohar, 2011; Nasrellah and Manohar, 2011; Radhika and Manohar, 2010; Radhika and Manohar, 2011; Radhika and Manohar,2012; Radhika, 2012; Sajeeb *et al.*, 2007; Sajeeb *et al.*, 2009; Sajeeb *et al.*, 2010; Sundar and Manohar, 2013; Tipireddy *et al.*, 2009). These studies have focused on algorithmic issues and on treating measurement data emanating from laboratory and field investigations.

2. Statically Loaded Structures: MCMC-Based Methods

Consider a statically loaded structure governed by the equation

$$K(\theta)x+Q(x,\theta)=\Gamma(\theta) \tag{1}$$

Here $x=n\times 1$ vector of unknown displacements, $\theta=p\times 1$ vector of random variables which collectively represent all the uncertainties in the problem (including those associated with structural properties and load characteristics) with joint PDF $p(\theta)$, $K=n\times n$ structure elastic stiffness matrix, $Q(x,\theta)=n\times 1$ vector of functions representing the nonlinear behavior of the system, and $\Gamma(\theta)=n\times 1$ vector of equivalent nodal forces. Furthermore, we assume that the structure is instrumented with a set of s sensors and a set of measurements from these sensors is available for N episodes of loading conditions. These measurements could be on structural strains, displacements, or reactions transferred to the supports and the model for measurements is expressed as:

$$y_j=h_j\left[\theta,x_j(\theta)\right]+v_j; j=1,2,\cdots,N \tag{2}$$

Here x_j is such that $K(\theta)x_j+Q(x_j,\theta)=\Gamma_j(\theta)$; y_j is a $s\times 1$ vector of measurements from the s sensors and for the j-th episode of loading; v_j is a $s\times 1$ vector of random variables which denote the collective effect of measurement noise in the j-th loading episode and errors in relating measured y_j with the system state vector x_j. It is assumed that v_j is an independent sequence of Gaussian vector random variables with zero mean and covariance $\langle v_i v_j^t\rangle=\Sigma_{v_i}\delta_{ij}$; here δ_{ij} is the Kronecker delta function, $\langle\bullet\rangle$ is the expectation operator, and the superscript t denotes matrix transposition. We also take that v_j and θ are independent. In further work, for the sake of simplicity, we write $h_j\left[\theta,x_j(\theta)\right]$ simply as $h_j(\theta)$. We introduce the notation $y_{1:N}=\left[y_1 \quad y_2 \quad \cdots \quad y_N\right]$ to denote all the observations concatenated into a single matrix. The problem of system identification consists of determining the posterior PDF $p(\theta\,|\,y_{1:N})$. By using the definition of conditional PDF and by applying Bayes' theorem we get

$$p(\theta\,|\,y_{1:N})=\frac{p(y_{1:N}\,|\,\theta)\,p(\theta)}{\int p(y_{1:N}\,|\,\theta)\,p(\theta)\,d\theta} \tag{3}$$

By denoting $C=\left[\int p(y_{1:N}\mid\theta)p(\theta)d\theta\right]^{-1}$, the normalization constant, and by using Eq. (2), we get

$$p(\theta\mid y_{1:N})=C\prod_{i=1}^{N}p(y_i\mid\theta)p(\theta) \tag{4}$$

It follows from Eq. (2) that $p(y_i\mid\theta)$ is a Gaussian PDF with mean $h_j(\theta)$ and covariance Σ_{v_j}. The quantity $\prod_{i=1}^{N}p(y_i\mid\theta)$ is called the likelihood function.

The above solution remains formal since the normalization constant C is not easy to evaluate and hence it would not be possible to evaluate probabilities of specific events nor evaluate moments of θ conditioned on measurements. However, as has already been noted, the MCMC tools enable us to draw samples from $p(\theta\mid y_{1:N})$, even when C is not known, based on which inferences on θ could be made. Besides, these tools can easily be combined with readily available professional FEA codes while computing the likelihood function. It must be noted that for each draw of θ, the evaluation of $p(y_i\mid\theta)$ requires a call to the code that analyses Eq. (1) and it is in implementing this step that most computational effort would be spent.

The steps involved in simulating N_0 number of samples from $p(\theta\mid y_{1:N})$ using Metropolis-Hastings algorithm are summarized as follows:

1. Set length of chain M (user defined scalar, say $\approx 10N_0$)
2. Set $r=1, l=1$, $N_b=M/10$, burn=1, $n_c=$ some scalar $>N_0$. Initialize $\theta(=\theta_1)$.
3. Assume a proposal PDF $q(\psi\mid\theta_r)$ (for example, normal PDF with mean= θ_r and covariance= $\sigma^2 I$).
4. Evaluate $h_j(\theta_r), j=1,2,\ldots N$ by analyzing the structure using FEA code.
5. Evaluate $p(\theta_r\mid y_{1:N})$ using Eq. (4) (in terms of C)
6. Draw $\psi\sim q(\psi\mid\theta_r)$. Determine $h_j(\theta), j=1,2,\ldots N$ for $\theta=\psi$ through the FEA code. Evaluate $p(\psi\mid y_{1:N})$ using Eq. (4) (in terms of C).
7. Simulate $u\sim U[0,1]$.
8. Define $\alpha(\theta_r,\psi)=\min\left[1,\dfrac{p(\psi\mid y_{1:N})q(\theta_r\mid\psi)}{p(\theta_r\mid y_{1:N})q(\psi\mid\theta_r)}\right]$
9. Set $\theta_{r+1}=\begin{cases}\psi & \text{if } \alpha(\theta_r,\psi)\ge u\\ \theta_r & \text{otherwise}\end{cases}$ and

$$p(\theta_{r+1} \mid y_{1:N}) = \begin{cases} p(\psi \mid y_{1:N}) \text{ if } \alpha(\theta_r, \psi) \geq u \\ p(\theta_r \mid y_{1:N}) \text{ otherwise} \end{cases}$$

10. If burn = 0, go to step 11 (Determining the burn-in period, N_b)

    Else

    If $r = n_c l$

    a. Set $X = \theta_{1:r}$. For $i = 1:r$ do the following
    - Set $Y = \theta_{i:(r+i-1)}$
    - Calculate the autocorrelation coefficient, $\tau_i = \dfrac{\sigma_{XY}}{\sigma_X \sigma_Y}$

    b. If $|\tau_i - \tau_{i-1}| < \varepsilon$, set $N_b = r$ and burn = 0

       (Here $\varepsilon(=10^{-2})$ is a user defined tolerance limit)

       Else set $l = l + 1$

    c. If $N_b > \dfrac{M}{2}$, set $M = 2M$, go to step 2.

       Else go to step 11.

       Else go to step 12

11. If $r - N_b \geq N_0$, required samples= $\theta_{(N_b+1):(N_b+N_0)}$ (Check for the required number of samples). Else go to step 12
12. Set $r = r + 1$.
13. If $r > M$, required samples= $\theta_{1:M}$, end. Else go to step 6.

We illustrate this approach by considering the nonlinear instrumented frame shown in Fig. 1. The frame is modeled on the Abaqus software using beam elements (B21), resulting in a model with 252 degrees of freedom (DOFs). The material of the beam is taken to display bilinear stress-strain behavior. The parameters to be identified are taken to be the loads $P_1, P_2, \cdots P_5$, Young's modulus (E), yield stress (f_y), ultimate stress (f_u) and ultimate strains (ε_u) for the three columns and two beams. In all, there are 25 parameters to be identified. The multi-dimensional prior PDF for load and material characteristics is constructed based on the models that are typically used in reliability analysis at the design stage. Thus, material properties for steel are taken to be log-normally distributed with coefficient of variation and correlation coefficient (ρ) as given in the JCSS report.

The load parameters P_1-P_5 are modeled as follows: P_1~Gumbel (25.0 kN/m, 4.05 kN/m); P_2~Normal (35.0kN/m, 8.75kN/m); P_3~Normal (30.0kN/m, 7.5kN/m); P_4~Normal (28.0 kN, 7.0 kN) and P_5~Normal (25.0 kN, 6.25 kN); Furthermore, P_2 and P_3 are taken to be correlated with

$\rho = 0.3$ and similarly, P_4 and P_5 have a correlation of $\rho = 0.2$. The loads and material properties are taken to be independent. Based on this partial information, a Nataf's model for the multi-dimensional prior PDF is constructed. The system identification using 15000 MCMC samples (with initial 20000 realizations discarded to achieve stationarity) with three episodes of measurements is performed. The measurements were synthetically generated with $f_y = 248.5\text{MPa}$, $f_u = 398.0\text{MPa}$, $E = 199.0\text{GPa}$ and $\varepsilon_u = 0.3475$. Based on the MCMC samples drawn from $p(\theta \mid y_{1:N})$, the mean and standard deviation of Young's modulus of the material in the three columns and two beams are estimated respectively as, mean (GPa): 199.65, 199.22, 209.18, 210.65, 193.75 and standard deviation (GPa): 6.1193, 4.0715, 7.4161, 5.5566, 5.1906. Similarly, for the yield stress the estimates are, mean (MPa): 233.85, 270.47, 245.55, 253.22, 252.57 and standard deviation (MPa): 18.488, 14.277, 11.546, 15.671, 15.891. These estimates compare well with corresponding values assumed in simulating the measurements. It must be noted that once $p(\theta \mid y_{1:N})$ is estimated, it could further be used to update structural reliability models. Sundar and Manohar (2013) have explored the procedure for extending first order reliability methods in reliability model updating and have also discussed the solutions for an exactly solvable case.

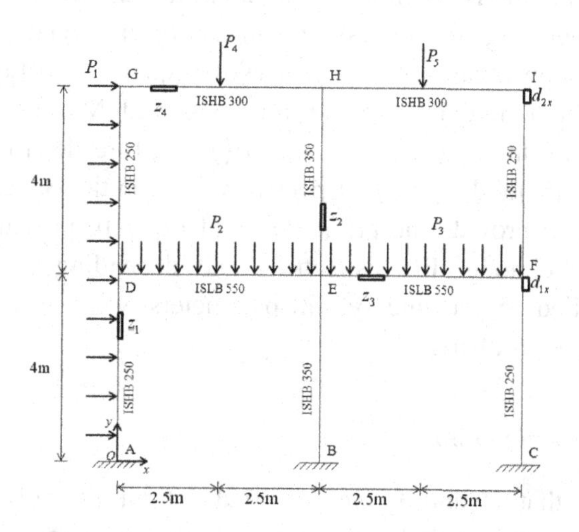

Fig. 1. Statically loaded frame undergoing large displacements; the material is taken to possess bilnear stress-strain characteristics; the nomenclature for rolled sections (ISHB) follows the Bureau of Indian Standards specification (IS 808:1989).

3. Dynamically Loaded Structures: Sequential Monte Carlo Approach

Here we consider the system equations in the state space form given by

$$\begin{aligned} x_{k+1} &= h_k\left(x_k, \theta, u_k, w_k\right); k = 0,1,2,\cdots \\ y_k &= q_k\left(x_k, \theta, u_k, v_k\right); k = 1,2,\cdots \end{aligned} \tag{5a, b}$$

The first of these equations is called the process equation and the second, the measurement equation. Table 1 describes the various variables involved in the above equation. For details regarding the deduction of above equations, from equations formulated in configuration space, refer to Kloeden *et al*., (1999). The noise term w_k in Eq. (5a) represents the modeling error in deriving the process equation. The sources of this error could include idealization of constitutive laws, damping models, displacement fields, boundary conditions, joint flexibility, and numerical errors resulting from spatio-temporal discretization. Similarly, v_k in Eq. (5b) accounts for errors in mathematical models that relate measured quantity to system states and also experimental errors due to electronic noise, inaccurate calibration, digitization during data acquisition, and sensor-structure interactions. The assumption that w_k and v_k are taken to be white sequences simplifies the mathematical formulations and implies that all systematic errors have been taken into account in deriving the process and measurement equations. Under these assumptions, it follows that x_k forms a Markov sequence and Eqs. (5a,b) can be taken to represent a model for a partially observed Markov process. In the analysis we assume that $p(x_{k+1} \mid x_k)$ and $p(y_k \mid x_k)$ are determinable from the knowledge of $p_{w_k}(w)$ and $p_{v_k}(v)$ respectively. Within the Bayesian framework, Eq. (5a) is taken to provide the prior PDF and Eq. (5b) provides the basis for computing the likelihood. In this setting and depending on the knowledge available on applied actions and system parameters, we now consider different classes of inference problems.

3.1. *Hidden state estimation*

Here we assume that θ and u_k are specified deterministically and we aim to estimate the joint posterior PDF $p(x_{0:k} \mid y_{1:k})$ (known as smoothing PDF) and its marginal $p(x_k \mid y_{1:k})$ (known as the filtering PDF) or any of the associated marginals/moments. Attention could also be focused on characterizing the posterior PDF of some functions of x_k as well. A complete solution to the characterization of $x_{0:k} = [x_0 \ \ x_1 \ \ \cdots \ \ x_k]$ and $y_{1:k} = [y_1 \ \ y_2 \ \ \cdots \ \ y_k]$ consists of determination of the joint PDF $p(x_{0:k}, y_{1:k})$. It can be easily shown that

$$p(x_{0:k}, y_{1:k}) = p(y_{1:k} \mid x_{0:k}) p(x_{0:k}) = p(x_0) \prod_{i=1}^{k} p(y_i \mid x_i) p(x_i \mid x_{i-1}) \tag{6}$$

Table 1. Description of quantities appearing in Eqs. (5a,b).

No	Quantity	Size	Description
1	k	scalar	Discretized time variable
2	x_k	$n_x \times 1$	State vector; x_0 is a random vector with specified PDF $p(x_0)$.
3	h_k	$n_x \times 1$	Nonlinear function
4	θ	$p \times 1$	Time invariant system parameters
5	u_k	$n_u \times 1$	External time varying actions.
6	y_k	$n_y \times 1$	Measurement vector.
7	q_k	$n_y \times 1$	Nonlinear function which relates measured quantities to system states and parameters
8	w_k	$n_w \times 1$	Process noise; modeled as a discrete parameter continuous state random process; w_k and w_j are taken to be independent for all $k \neq j = 1,2,\cdots$; the PDF $p_{w_k}(w)$ is specified.
9	v_k	$n_v \times 1$	Measurement noise; modeled as a discrete parameter continuous state random process; v_k and v_j are taken to be independent for all $k \neq j = 1,2,\cdots$; x_0, w_k, and v_j are taken to be independent for all $k \neq j = 1,2,\cdots$; the PDF $p_{v_k}(v)$ is specified.

If one takes the process and measurements equations to be of the form $x_{k+1} = h_k(x_k, \theta, u_k) + w_k;\ k = 0,1,2,\cdots$ and $y_k = q_k(x_k, \theta, u_k) + v_k;\ k = 1,2,\cdots$, the above equation can be simplified to get

$$p(x_{0:k}, y_{1:k}) = p(x_0) \prod_{i=1}^{k} p_{v_i}\left[y_i - q_i(x_i, \theta, u_i)\right] p_{w_i}\left[x_i - h_{i-1}(x_{i-1}, \theta, u_{i-1})\right] \tag{7}$$

Analytical determination of $p(x_{0:k}, y_{1:k})$ or $p(x_{0:k} \mid y_{1:k})$ based on above formulation is seldom possible and one resorts to numerical solutions based on Monte Carlo simulation techniques. In principle, one can use MCMC methods and make inferences based on random draws from the above PDF-s. However, in practice difficulties will be encountered in constructing suitable multi-dimensional proposal PDF-s. The dimension of the PDF-s $p(x_{0:k}, y_{1:k})$ or $p(x_{0:k} \mid y_{1:k})$ increase as and how more measurement become available. Moreover, the standard form of MCMC does not possess a recursive format. If data arrive sequentially in time or it is essential to make inferences in real time, it becomes desirable to assimilate the data in sequence. The basic idea here is to commence from $k = 0$ and propagate the posterior PDF-s, $p(x_{0:k} \mid y_{1:k})$ and

$p(x_k \mid y_{1:k})$, in time, using Markov property of x_k (that is, by using the governing Chapman-Kolmogorov equation) and update these PDFs using Bayes' theorem whenever a measurement becomes available (see, Fig. 2 for a schematic).

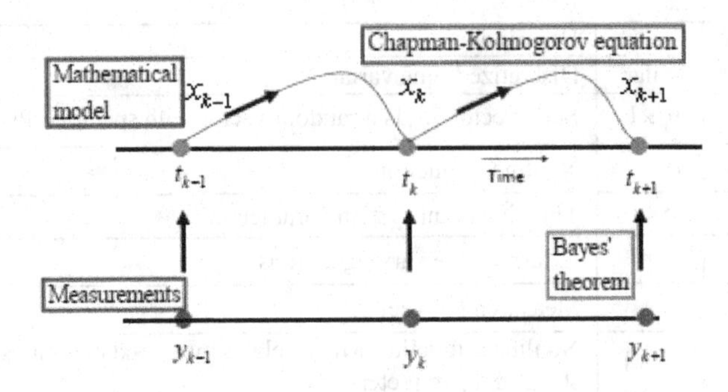

Fig. 2. Schematic of prediction using Markov property and updating by Bayes' theorem.

This idea leads to a pair of functional recursive relations for the filtering PDF $p(x_k \mid y_{1:k})$ given by (Gordon *et al.*, 1993; Ristic *et al.*, 2004):

$$\begin{aligned} p(x_k \mid y_{1:k-1}) &= \int p(x_k \mid x_{k-1})\, p(x_{k-1} \mid y_{1:k-1})\, dx_{k-1} \\ p(x_k \mid y_{1:k}) &= \frac{p(y_k \mid x_k)\, p(x_k \mid y_{1:k-1})}{\int p(y_k \mid x_k)\, p(x_k \mid y_{1:k-1})\, dx_k} \end{aligned} \tag{8}$$

The above equations are, respectively, the equations of prediction and updating. Similarly, for the multi-dimensional posterior PDF, one obtains

$$p(x_{0:k+1} \mid y_{1:k+1}) = p(x_{0:k} \mid y_{1:k}) \frac{p(y_{k+1} \mid x_{k+1})\, p(x_{k+1} \mid x_k)}{p(y_{k+1} \mid y_{1:k})} \tag{9}$$

Analytical solutions to the above equations is, in general, not possible, and, in this sense, these equations remain formal in nature. For linear state space models with additive Gaussian noises, Eq. (8) can be shown to lead to the well known Kalman filter. For more general class of models one has to resort to analytical or numerical approximations (Jazwinski, 1970; Maybeck, 1982) or employ full fledge Monte Carlo simulations. The basic idea in the latter class of approaches is to represent the posterior PDF in terms of a set of Monte Carlo samples with associated weights and propagate them in time using recursive computations (Cappé et al., 2007; Doucet et al., 2001; Doucet and Johansen, 2008; Gordon et al., 1993; Ristic et al., 2004; Tanizaki, 1996). In several applications in structural engineering it becomes possible to partition the state vector x_k as $x_k = \begin{bmatrix} x_k^l & x_k^n \end{bmatrix}^t$ so that the state space model can be recast to read as

$$\begin{aligned} x_{k+1}^l &= f_k^l\left(x_k^n\right) + A_k^l\left(x_k^n\right)x_k^l + F_k^l + G_k^l\left(x_k^n\right)w_k^l \\ x_{k+1}^n &= f_k^n\left(x_k^n\right) + A_k^n\left(x_k^n\right)x_k^l + F_k^n + G_k^n\left(x_k^n\right)w_k^n; k = 0,1,2\cdots \\ y_k &= h_k\left(x_k^n\right) + C_k\left(x_k^n\right)x_k^l + v_k; k = 1,2,\cdots \end{aligned} \tag{10}$$

Here the superscripts l and n, respectively, stand for linear and nonlinear states. It can be discerned from the above equation that, conditioned on x_k^n, the state space equation governing x_k^l is linear and can consequently be analyzed using the Kalman filter. This would mean that a part of the problem can be handled exactly using Kalman filter and the more computationally intensive Monte Carlo simulation methods need to be used only for treating the remaining part of the problem. Clearly, the sampling variance here can be expected to reduce when compared with the variance if entire problem were to be tackled using only the simulation approach. This way of achieving reduction in sampling variance is referred to as Rao-Blackwellization (Cappé *et al.*, 2007; Doucet and Johansen, 2008; Ristic *et al.*, 2004; Schön and Gustafsson, 2005). The recent study by Radhika and Manohar (2012) explores this approach for solving a few problems in structural dynamics.

3.2. *Combined state and force identification*

Here we take the vector of applied actions u_k to be partially or completely unknown. The vector is treated as being random in nature with a postulated prior PDF. The system parameter vector θ is taken to be known. The problem on hand consist of determining the posterior PDF $p\left(x_k, u_k \mid y_{1:k}\right)$. It may be noted that the vector u_k could also include time varying system parameters or parametric excitations. We tackle this problem by augmenting Eq. (5a) with an additional process equation for u_k of the form $u_{k+1} = u_k + \overline{w}_k; k = 0,1,2,\cdots$ with $\overline{w}_k$ being a sequence of independent Gaussian random variables. We consider the extended vector$\left(x_k \quad u_k\right)^t$ and determine $p\left(x_k, u_k \mid y_{1:k}\right)$ by considering the problem as a problem of hidden state estimation. To illustrate this we consider an instrumented building idealized as a single degree of freedom (SDOF) system (Fig. 3a) and subjected to El Centro earthquake support motions. It is assumed that the displacement of the mass is measured and it is required to estimate the applied support displacement and velocity. The governing equation of motion and the measurement equation are obtained as:

$$dx_1 = x_2 dt; dx_2 = (-2\eta\omega x_2 - \omega^2 x_1 + 2\eta\omega x_4 + \omega^2 x_3)dt + dW_1(t)$$
$$dx_3 = x_4 dt; dx_4 = dW_2(t); x_1(0) = 0; x_2(0) = 0; x_3(0) = x_{30}; x_4(0) = x_{40} \tag{11}$$
$$y_k = x_k^1 + \nu_k; k = 1, 2, \cdots$$

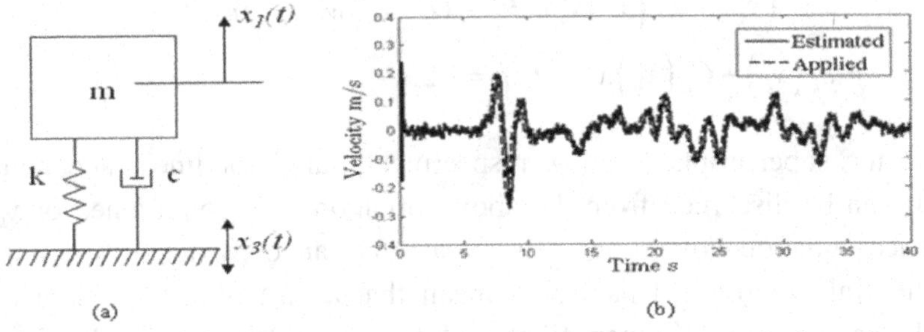

Fig. 3. Identification of transient support motions based on measured displacement response; $\eta = 0.05$; $\omega = 2\pi$ rad/s; $\Sigma_{W_1} = 0.01\,(\text{m/s/s})^2$ and $\Sigma_{W_2} = 0.05(\text{m/s/s})^2$; (a) sDOF system under dynamic support motion; (b) expected support velocity conditioned on measured displacement $x_1(t)$.

It may be noted that the applied support displacement and velocity, $x_3(t)$ and $x_4(t)$ are now treated as additional states and $W_i(t); i = 1, 2$ are the artificial process noises which are taken to be zero mean Gaussian white noise processes with $\langle W_i(t) W_i(t+\tau) \rangle = \Sigma_{W_i} \delta(\tau)$. Here $\delta(\bullet)$ is the Dirac delta function. After discretization (Kloeden *et al.*, 1999), one obtains a linear state space model with additive Gaussian noises. Consequently, the resulting problem of state estimation can be tackled using the Kalman filter. Figure 3b compares the estimated support velocity with the applied velocity and a satisfactory mutual agreement is revealed. It was observed that the choice of Σ_{W_2} played a crucial role in the successful implementation of the method: too small or too large a noise was found to lead to unsatisfactory performance; a satisfactory choice, however, could be established by tuning the noise and comparing the posterior mean with the noisy measurement. The approach presented here can be extended to nonlinear systems where the resulting state space model would turn out to be nonlinear thereby necessitating the use of sequential Monte Carlo filters to tackle the problem (Radhika, 2012).

3.3. *Combined state and parameter estimation*

This class of problems is perhaps the most challenging amongst the model updating problems considered in this work. These problems can be tackled entirely within the Bayesian framework or by combining Bayesian methods for state estimation with maximum likelihood methods for parameter estimation. In the first option, we model the unknown system parameter vector θ as a vector of random variables with a prior PDF $p_0(\theta)$ and the forcing function u_k is taken to be specified deterministically. The problem on hand consists of determining $p(x_k,\theta \mid y_{1:k})$. In the approach based on maximization of the likelihood function, the vector θ is typically treated as being deterministic and the optimization algorithm needs to be incorporated into the recursive formulae for state estimation. In many applications, the parameters of process and measurement noises, w_k and v_k, could be unknowns. Then, these parameters can also be estimated as a part of the system identification procedure, in which case, the parameter vector θ can be extended to include the noise parameters as well.

3.3.1. *Method of augmented states and global iterations*

Here we declare the vector of system parameters θ to be identified as additional state variables and augment the original state vector by these artificial state variables. Since the system parameters often multiply the system states, the problem of state estimation here becomes nonlinear in nature even when the original process and measurement equations are linear. Consequently, the problem here could be solved using extended Kalman filtering (Corigliano and Mariani, 2004; Ghosh *et al.*, 2007; Hoshiya and Saito, 1984; Imai *et al.*, 1989; Yun and Shinozuka, 1980) or particle filtering strategies (Ching *et al.*, 2006; Ghosh *et al.*, 2008; Manohar and Roy, 2006; Sajeeb *et al.*, 2009). Here, for the sake of computational expediency, two additional features are introduced into the solution of the state estimation problem. Firstly, we assume that parameter vector θ evolves artificially in time. More specifically, we assume that θ_k constitutes a simple random walk so that it obeys the equation $\theta_{k+1} = \theta_k + r_k; k = 0,1,2,\cdots$, where r_k is a sequence of zero mean, independent random variables (which typically are taken to be Gaussian) and $\theta_0 \sim p(\theta_0)$. This facilitates exploration of the parameter space beyond what is initially assumed at $k = 0$ (Ghanem and Shinozuka, 1995; Imai *et al.*, 1989; Storvik, 2002). It may be noted that the representation for θ_k is asymptotically exact as $\langle r_k^2 \rangle \to 0$. The second aspect has to do with choice of prior for θ. The initiation of the filtering algorithm, as has been indicated already, requires an assumption to be made on the initial form of the PDF $p(\theta_0)$. An intuitive choice would be to take θ_0 to be uniformly distributed over a plausible range of values of its components and the components themselves to be independent. To enhance the performance of the identification procedure, an additional step involving a global iteration can also be employed. Here, the guess on PDF of θ_0 is updated at the end of a given cycle of filtering by the final $p(\theta_N \mid y_{1:N})$ and is used as the starting guess for the next cycle of filtering. This global iteration loop is repeated till a satisfactory convergence on the expected value of θ conditioned on measurements is obtained. It may be noted that the idea of this step was proposed by Hoshiya and Saito (1984). For the purpose of illustration, we consider the problem of identification of Coulomb friction parameter in a nonlinear oscillator (Manohar and Roy, 2006). The system under study is given by

$$\begin{aligned} dx_1 &= x_2 dt; dx_2 = (-x_3 \operatorname{sgn}(x_2) - \omega^2 x_1 + P\cos\lambda t)dt + dW_1(t) \\ dx_3 &= dW_2(t); x_1(0) = x_{10}; x_2(0) = x_{20}; x_3(0) = 0 \end{aligned} \tag{12}$$

Here x_3 is the Coulomb friction parameter to be identified. $W_i(t); i=1,2$ are the artificial process noises which are taken to be zero mean Gaussian white noise processes with $\langle W_i(t)W_i(t+\tau)\rangle = \Sigma_{W_i}\delta(\tau)$. After discretizing the above equation, a map of the form Eq. (5a) is obtained. In the numerical work we take $\omega = 4\pi$ rad/s, step size $\Delta = 1.5625\times10^{-3}$ s, $\lambda = 0.8\omega$, $\Sigma_{W_1} = P/10$N, $P = 10$N and 1000 number of particles. The initial displacement and velocity are taken to be zero mean, mutually independent Gaussian random variables with standard deviations $\sigma_{10} = 0.02$ m and $\sigma_{20} = \lambda\sigma_{10}$ respectively. The measurement equation is taken to be of the form

$$y_k = -x_k^3 \operatorname{sgn}(x_k^2) + \omega^2 x_k^1 + v_k; v_k \sim N(0, 3.678); k = 1, 2, \cdots \tag{13}$$

Synthetic data on the reaction transferred to the supports is generated taking the friction parameter value to be 0.05. In the filter computations, it is assumed that $\Sigma_{W_2} = 0.0025$. At k = 0, x_3 is taken to be distributed uniformly in the range $[0.001, 0.004]$ and independent of x_1 and x_2. Figure 4a compares the estimate of reaction transferred to the support with the measurement and Fig. 4b shows results of identification of the Coulomb friction parameter. The estimate of the parameter was found to be in the range of 0.055 to 0.068: this can be considered as being reasonable approximation to the true value of 0.05.It may be noted that the partitioning of states, as in Eq. (10), can also be implemented in the context of system parameter identification (Sajeeb *et al.*, 2010).

3.3.2. *Method of maximum likelihood*

In a commonly used form of this method, the unknown parameter vector θ is treated as being deterministic and is determined by maximizing a likelihood function. This maximization has to be now carried out in conjunction within the framework of recursive calculations for the state estimation. Maybeck (1982) (Chapter 10, Vol II) has discussed several candidate likelihood functions and examined their relative merits. Specifically, the functions considered include the logarithm of the following PDFs: $p(x_k,\theta|y_{1:k})$, $p(y_{1:k}|x_k,\theta)$, $p(x_k|y_{1:k},\theta)$, $p(x_k,y_{1:k}|\theta)$, $p(x_k,\theta|y_{k-J+1:k})$, $p(y_{k-J+1:k}|x_k,\theta)$, $p(y_{k-J+1:k}|y_{1:k-J},x_k,\theta)$, $p(x_k|y_{k-J+1:k},\theta)$, $p(x_k,y_{k-J+1,k}|\theta)$ and $p(x_k,y_{k-J+1,k}|y_{1:k-J},\theta)$. The first four of these functions yield growing-memory algorithms and the remaining provide fixed-length memory parameter estimators with either growing-length or fixed length state estimators. When $p(x_k,\theta|y_{1:k})$ or $p(x_k,\theta|y_{k-J+1:k})$ is used as the likelihood function, θ is treated as a random variable and maximization of the likelihood function is carried out with respect to the parameters of the assumed PDF for θ. For all other choices, θ is considered to be deterministic. Maybeck recommends the use of $p(x_k,y_{1:k}|\theta)$ and has presented formulary based on gradient based approaches for the solution of the problem of combined state and parameter estimation in linear-Gaussian state space models. Extensions to general nonlinear state space models are available in (Andrieu and Doucet, 2003; Ghanem and Shinozuka, 1995; Ionides *et al.*, 2006; Kitigawa, 1998). Kitigawa (1998) has noted the following two difficulties while applying maximum likelihood to practical problems: (a) repeated application of a numerical optimization procedure for evaluating the likelihood function results in significant computational demands, and (b) the log-likelihood computed by the Monte Carlo filter is subject to sampling error; therefore, precise maximum likelihood parameter estimates can be obtained only by using a very large number of particles or by parallel application of many Monte Carlo filters.

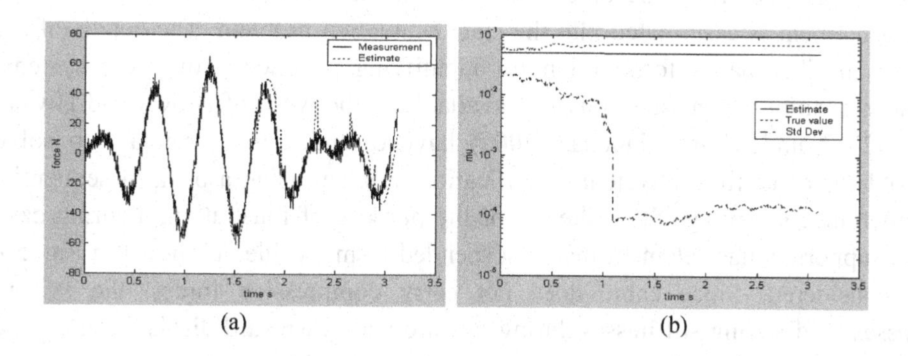

Fig. 4. Studies on the Coulomb oscillator (Manohar and Roy, 2006); (a) measured and estimated system state; (b) estimation of the nonlinearity parameter; results on the estimate of the mean and standard deviation; time average of the mean = 0.0682.

3.3.3. *Bank of filter approach*

Here the system parameters to be identified are treated as a vector of discrete random variables and a bank of filters is used to perform state estimation with each member of the bank corresponding to a possible state of the system parameters (Brown and Hwang, 1992; Maybeck, 1982; Namdeo and Manohar, 2007). No artificial time evolution to the elements of θ is ascribed. For the purpose of illustration consider the evaluation of the expectation

$$a_{k|k} = \langle x_k \mid y_{1:k} \rangle = \int x_k \int p(x_k, \theta \mid y_{1:k}) d\theta dx_k = \int p(\theta \mid y_{1:k}) \hat{x}_k(\theta) d\theta \tag{14}$$

If we postulate that the parameter θ to be a vector of discrete random variables with L number of states, then the above equation can be re-written as (also see Fig. 5a)

$$a_{k|k} = \sum_{i=1}^{L} \hat{x}_k(\theta_i) P[\theta = \theta_i | y_{1:k}] \tag{15}$$

The problem at hand consists of determining the weights $P[\theta = \theta_i \mid y_{1:k}]$ and the estimate $\hat{x}_k(\theta_i)$. Clearly, the value of the weights $P[\theta = \theta_i \mid y_{1:k}]$ would be influenced by the measurement $y_{1:k}$ and, as the measurements are assimilated, the scheme learns which of the filters are the correct ones, and its weight factor $P[\theta = \theta_i \mid y_{1:k}]$ approaches unity while the others are going to zero. This would mean that the filter becomes adaptive or self-learning in nature as the measurement is assimilated into the state estimation problem. The details of the Kalman filter based formulation for identifying parameters of linear systems under additional Gaussian noises is available in the work of Brown and Hwang (1992). Namdeo and Manohar (2007) have extended the formulary to solve problems of nonlinear system identification problem using a bank of sequential importance sampling (SIS) filters. For the purpose of illustration, a spring-dash pot supported mass element that is suspended from a cable, as shown in Fig. 5b, is considered. Since cable does not carry compressive forces, the system possesses differing stiffnesses during upward and downward displacements. The governing equation of motion for this system can be shown to be given by

$$dx_1 = x_2 dt; dx_2 = \left(-\frac{c}{m} x_2 - \frac{k_o}{m} x_1 - \frac{k_s}{m} x_1 U[x_1 + x_s] - \frac{f(t)}{m} \right) dt + dW(t) \tag{16}$$

with $f(t) = m\ddot{x}_s(t) + k_s x_s U[x_1 + x_s]$. Here $U[\bullet]$ is the Heaveside step function. $W(t)$ is the process noise which is taken to be zero mean Gaussian white noise with $\langle W(t)W(t+\tau)\rangle = \Sigma_w \delta(\tau)$. The measurement here is taken to be made on the velocity response, and, accordingly the measurement equation reads $y_k = x_k^2 + v_k$; $k = 1,2,\cdots$. In the numerical work we take m = 1.0 kg, η = 0.04, k_0= $16\pi^2 m$ N/m, $\omega = 4\pi$ rad/s, Σ_W = 0.0051 N, and Δ = 0.0163 s. The system is assumed to start from rest. We consider the measurement noise having zero mean and standard deviation, Σ_v = 0.0013 m/s. Synthetic data are generated with a reference value of k_s = 245.0 N/m. The support displacement to be a sample of a stationary random process generated using the Fourier representation $x_s(t) = \sum_{i=1}^{n} \sigma_i (\bar{v}_i \cos\omega_i t + \bar{w}_i \sin\omega_i t)$. Here $(\bar{v}_i, \bar{w}_i), i = 1,2,\cdots,n$ are taken to be identical, independently distributed Gaussian random variables with zero mean unit standard deviations. It is also assumed that $\bar{v}_i$ and $\bar{w}_k$ are independent for all i and k. The constants $\{\sigma_i\}_{i=1}^{n}$ and $\omega_i = i\omega_0$ are taken to be specified and deterministic with $\omega_0 = 2\pi$ rad/s, n = 5, and amplitude of support displacement is σ_j = 0.001m for all $j = 1,2,\cdots,5$. Figures 6a,b show the result on estimation of cable stiffness and system velocity. We get the estimate for the expected value of cable stiffness to be about 244.0 N/m which compares very well with the reference value of k_s = 245.0 N/m. Similarly, the estimated expected value of system velocity shows good comparison with the measured velocity.

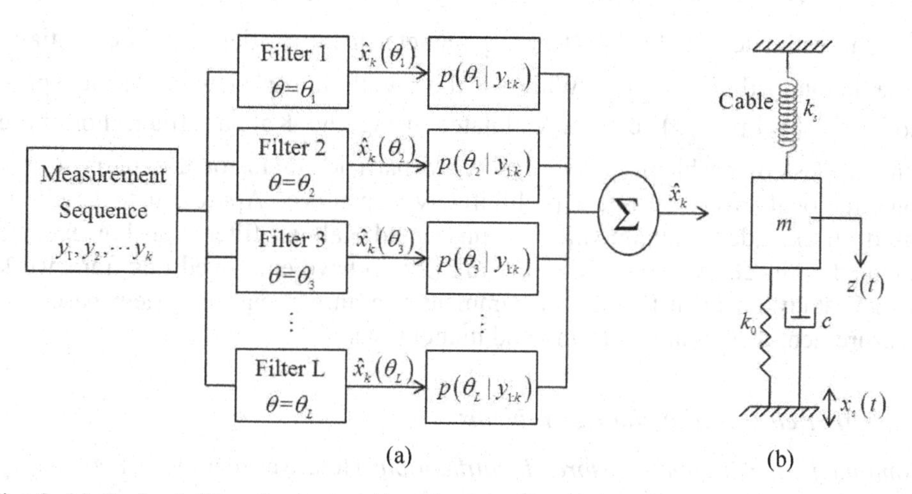

Fig. 5. (a) Bank of filters for system identification; (b) Nonlinear sDOF system with bilinear stiffness characteristics (Namdeo, 2007).

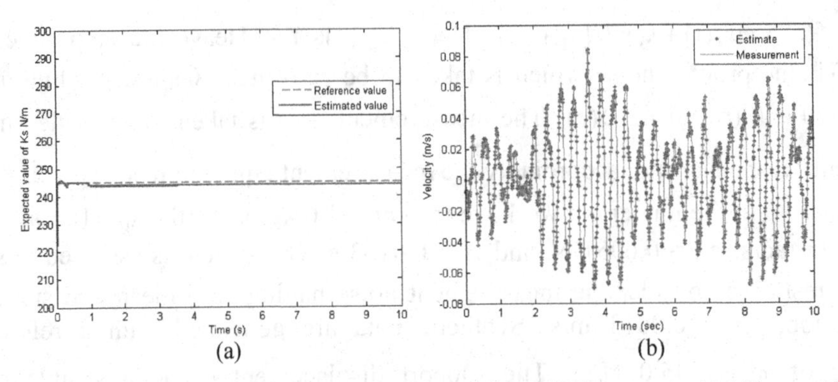

(a) (b)

Fig. 6. Estimation of cable stiffness parameter of the nonlinear sDOF system with bilinear stiffness characteristics using bank of SIS filters; (a) expected value of ; (b) estimation of the velocity state of the system using bank of SIS filters (Namdeo, 2007).

3.3.4. *Combined MCMC and Bayesian filters*

As in the previous section, in this method also, no artificial time evolution is ascribed to θ. Instead of treating θ as a discrete random vector variable, we take θ now to possess continuous states. Based on the application of Bayes' theorem we write (Khalil *et al.*, 2010; Kitigawa, 1998):

$$p(\theta \mid y_{1:N}) \propto p(y_{1:N} \mid \theta) p(\theta) = p(\theta) \prod_{k=1}^{N} \int p(y_k \mid x_k, \theta) p(x_k \mid y_{1:k-1}, \theta) dx_k \tag{17}$$

where $y_{1:N}$ denotes the complete set of measurements available. The idea here is to use MCMC to draw samples from $p(\theta \mid y_{1:N})$ and, while doing so, use Bayesian filters to estimate $p(x_k \mid y_{1:k-1}, \theta)$. The parameter identification here is essentially done in an offline manner. While dealing with linear Gaussian state space models, $p(x_k \mid y_{1:k-1}, \theta)$ can be estimated using the Kalman filter. For more general class of problems, if one resorts to particle filters for this purpose, the computational effort becomes prohibitively expensive. Approximate solutions based on extended Kalman filters or unscented Kalman filters can however be obtained with relative ease. Khalil *et al.*, (2010) have employed this framework in their study on identification of parameters in an aeroelastic system based on measurement data obtained from wind tunnel tests.

3.4. *Other classes of updating problems*

Combined state, system and force identification: Here we treat the unknowns u_k and θ as being random and aim to determine $p(x_k, u_k, \theta \mid y_{1:k})$ from which one obtains $p(x_k \mid y_{1:k})$ (state estimation), $p(\theta \mid y_{1:k})$ (system identification), and $p(u_k \mid y_{1:k})$ (force identification). This class of system identification problems, in which applied actions are not known or partially known, is referred to as blind system identification problems (Manohar and Roy, 2006).

Sensitivity model updating: Here we consider the evaluation of the posterior PDF of the gradients $\gamma_{kj} = \frac{dx_k}{d\theta_j} (j = 1,2,\cdots p)$ evaluated at a reference $\theta = \theta^*$ and for $k = 1,2,\cdots$. This problem can be tackled as a problem of state estimation by augmenting the state vector x_k with the gradients γ_{kj}. In this problem θ is fixed at θ^* and the forcing u_k is taken to be known (Radhika and Manohar, 2011; Radhika and Manohar, 2012).

Reliability model updating: Here we consider a performance function $\chi(x_k)$ and aim to evaluate the posterior probability $\mathrm{P}[\chi(x_k) \le \chi^* \forall k \in [1,k^*] | y_{1:k}]$ where χ^* is specified limit on acceptable value of χ. This probability, in turn, can be evaluated in terms of extreme of $\chi(x_k)$ over $k \in [1,k^*]$. That is, the required probability is $\mathrm{P}[\max_{k\in(1,k^*)} \{\chi(x_k)\} \le \chi^* \mid y_{1:k}]$. It may be noted that the probability here is being evaluated across the ensemble excitation u_k and θ can be assumed to be known. The measurements may or may not include data on u_k (Ching and Beck, 2007; Ionides *et al.*, 2006).

4. Finite Element Model Updating with Combined Static and Dynamic Measurements

When Bayesian framework is employed in dealing with condition assessment of large scale engineering structures, such as bridges and multi-storied buildings, it is most advantageous to employ professional FEA packages in structural modeling. In such situations, it becomes infeasible to re-cast the governing equations in the standard form as given in Eq. (5a). Moreover, in structural condition assessment studies, measurement data could be available from diverse testing conditions, some of which could be under static loads and some under dynamic excitations. The use of such data in system identification studies poses interesting questions and some of the relevant issues have been examined in (Nasrellah and Manohar, 2011; Nasrellah and Manohar, 2011; Tipireddy *et al.*, 2009). In this section we provide a brief account of these investigations. We consider the structural model with parameter vector θ.The structural behavior could be linear or nonlinear and static or dynamic. We consider that a set of N_T different tests have been conducted on the structure. These tests could be static and (or) dynamic in nature (see Table 2). Associated with each of the test scenarios, we introduce independent variables $\{a_i\}_{i=1}^{N_T}$ such that $a_{il} \le a_i \le a_{iu}$ and consider the system response in *i*-th test to evolve in the variable a_i.

While for dynamic tests the time variable t is the obvious choice for the independent variable and system response indeed 'evolves' in t, for other tests, the notion of independent variable and evolution along this variable, is introduced deliberately to facilitate data assimilation via state estimation techniques. When measurement data originate from the alternative tests listed in Table 2, we introduce a common independent variable τ given by

$$\tau = \frac{\alpha - \alpha_{\min}}{\alpha_{\max} - \alpha_{\min}} = \frac{\beta - \beta_{\min}}{\beta_{\max} - \beta_{\min}} = \frac{\omega - \omega_{\min}}{\omega_{\max} - \omega_{\min}} = \frac{t - t_{\min}}{t_{\max} - t_{\min}} \tag{18}$$

Clearly, as the variables α, β, ω and t take values over their respective ranges, the variable τ takes values from 0 to 1. The system parameters of interest $\theta_i, i = 1,2,\cdots,p$ are independent of the variables α, β, ω and t and hence independent of τ. If the variable τ is discretized such that $\tau_k = \tau / N_\tau$, and, by using the notation $\theta_{ik} = \theta_i(\tau = \tau_k)$, the above statement translates into the equation

$$\theta_{ik+1} = \theta_{ik}; i = 1,2,\cdots p; k = 1,2,\cdots,N_\tau \tag{19}$$

We interpret this equation to constitute the process equation. Suppose that, for each of the tests mentioned in Table 2, we use a set of N_s sensors, we could acquire a suite of measurements and we denote this by a $N_s \times N_\tau$ matrix $y_{1:N_\tau}$. Clearly, this matrix depends upon the system parameter vector θ and we need to postulate a mathematical model to capture the dependence. To achieve this, we postulate that a set of finite element (FE) models for the structure under study can be formulated so that each of this model correspond to one testing scenario listed in Table 2. Each of the FE models would have θ as the common set of model parameters.

In the present study we assume that these FE models reside in professional FE softwares that are commercially available and, in our work, we have utilized the NISA family of FE softwares. We represent the model for measurements using the equation

$$y_k = h_k(\theta_{1k}, \theta_{2k}, \cdots \theta_{pk}) + \xi_k; k = 1,2,\cdots,N_\tau \tag{20}$$

Table 2. The artificial independent variable a_i for different types of tests.

No.	Test	Independent variable
1	Incremental load-displacement test in which externally applied loads are incremented as $P_j = \alpha_j P_0$, $j = 1,2,\cdots N_1$ where P_0 is a reference value.	$a = \alpha$ with $a_l = 1$ and $a_u = N_1$
2	Quasi-static moving load test in which a set of forces move quasi-statically and the position of the leading load from one of the ends of the structure is denoted by β.	$a = \beta$ with $a_l = \beta_{\min}$ and $a_u = \beta_{\max}$
3	Measurement of a set of frequency response functions of the structure using either impulse hammer test or modal shaker test over the frequency range $\omega_{\min} \le \omega \le \omega_{\max}$.	$a = \omega$ with $a_l = \omega_{\min}$ and $a_u = \omega_{\max}$
4	Measurement of response time histories over the time duration $t_{\min} \le t \le t_{\max}$.	$a = t$ with $a_l = t_{\min}$ and $a_u = t_{\max}$

Here $h_k(\theta_k)$ denotes the set of outputs from the suite of FE models with input parameters θ_k and serves as an approximation to the measured quantities $y_{1:N_\tau}$. The quantity $\xi_k, k = 1,2,\cdots N_\tau$ represents a sequence of $N_s \times 1$ vector of identical and independent random variables with a specified PDF given by $p(\xi_k)$. This random quantity models the errors made in relating the measured quantities $y_{1:N_\tau}$ with the 'system states' θ through the postulated FE model and also the effect of measurement noise. The problem on hand consists of determining the posterior PDF $p(\Theta_k \mid y_{1:k})$ of the system parameters $\Theta_k = \{\theta_{1k} \ \theta_{2k} \ \cdots \ \theta_{pk}\}^t$ conditioned on the measurements $y_{1:N_\tau}$. It is found computationally expedient to add a small artificial noise to the process equation so that Eq. (19) is replaced by

$$\theta_{ik+1} = \theta_{ik} + w_{ik}; i = 1,2,\cdots p; k = 1,2,\cdots,N_\tau \tag{21}$$

Here $w_k = \{w_{ik}\}_{i=1}^p$ constitute a vector of random variables with joint PDF $p(w_k)$. It is also assumed that for, $k = 1,2,\cdots,N_\tau$, w_k form a sequence of independent random vectors and this sequence is independent of $\xi_k, k = 1,2,\cdots,N_\tau$. The number of calls to the FE codes depends upon the number of observation points N_τ and the major computational effort in implementing the algorithm is spent in this step. Bearing in mind that system parameters are intrinsically time invariant, and, also, with a view to lessen the number of calls to the FE codes, we group the N_τ discrete values of τ into R groups with N_i number of discrete values of τ in the i-th group with $\sum_{i=1}^{R} N_i = N_\tau$. We assume that Θ_i remains constant for all values of τ lying within the i-th group. The steps involved in characterizing

$p(\Theta_k \mid y_{1:k})$ based on bootstrap filter (Gordon *et al.*, 1993) are as follows (Nasrellah and Manohar, 2011):

1. Set $r = 0$. Generate N samples $\{\Theta^*_{i,0}\}_{i=1}^N$ from the initial PDF $p(\Theta^*_0)$ and $\{w_{i,0}\}_{i=1}^N$ from $p(w)$.
2. Evaluate $\{\Theta^*_{i,r+1}\}_{i=1}^N$ using $\Theta_{r+1} = \Theta_r + w_r$. Set $r = r+1$.
3. Run FE codes and generate $h_k\left[\Theta^*_{i,r+1}\right]$ and evaluate $p(y_k \mid \Theta^*_{i,r})$ for $(r-1)n_\tau \le k \le rn_\tau$ where $n_\tau = N_\tau / R$.
4. Consider the k-th measurement y_k. Define
$$q^*_{k,j} = p\left(y_k \mid h_k\left[\Theta^*_{j,r}\right]\right) \Big/ \sum_{i=1}^{N} p\left(y_k \mid h_k\left[\Theta^*_{j,r}\right]\right)$$
5. Evaluate $q_i = \frac{1}{n_\tau} \sum_{k=(r-1)n_\tau}^{rn_\tau} q^*_{k,i}$.
6. Define the probability mass function $P\left[\Theta_r(j) = \Theta^*_{i,r}\right] = q_i$. Generate N samples $(\Theta^*_{i,r})_{i=1}^N$ from this discrete distribution.
7. Evaluate $a_{r|r} = \frac{1}{N}\sum_{i=1}^{N} \Theta_{i,r}$ and $\Sigma_{r|r} = \frac{1}{N-1}\sum_{i=1}^{N} (\Theta_{i,r} - a_{r|r})^t (\Theta_{i,r} - a_{r|r})$.
8. Go to step 2 if $r < R$; otherwise, stop.

In the numerical work, the above steps are combined with the global iteration step mentioned in Sec. 3.3.1.

For the purpose of illustration, we summarize a few results from our studies on condition assessment of existing railway bridges in India (Chandra Kishen *et al.*, 2011; Nasrellah and Manohar, 2011). The goal of these studies has been to assess the capability of existing bridges to carry enhanced freight loads. As a part of these studies, five existing bridges in South India have been extensively instrumented and bridge response to a suite of static/dynamic loads resulting from moving train formations have been measured. The loads here are either diagnostic in nature (their spatial distribution, speeds and magnitudes are measured) or due to ambient train traffic (when load characteristics are not precisely known). For the purpose of illustration, we consider 120-years old, five span, masonry arch bridge with stone masonry piers and abutments, brick masonry arch barrels and filler material made up of graded soil. The bridge spans are 7.70, 7.70, 7.70, 17.71, and 17.30 m; the first three smaller spans have been filled up and closed and only two main spans (17.71 m and 17.30 m) remain operational. These two spans were instrumented with strain gauges (20 numbers),

LVDTs (6 numbers), tilt meter (1 number) and uniaxial/multiaxial accelerometers (10 number of channels). In one of the tests that was conducted, a formation with 40 axles (with known weight and geometry) moved quasi-statically on the bridge so that it was made to halt on the bridge whenever an axle crossed a reference point ('rp') marked on the rail at the left springing level (Fig. 8) and the static response of the bridge was measured. In the identification study the position of leading axle was treated as the artificial independent variable (parameter a in Table 2) and $\tau = a/L_0$ where L_0 is the length of the formation. A 2-dimensional FE model based on the use of 4-noded quadrilateral elements with 2-DOFs per node, and with 2230 number of DOFs was used. To include the effects that the inactive arch spans would have on the active spans, the boundary conditions at the interface were represented through a set of discrete two-dimensional springs (16 numbers). The FE model thus consisted broadly of five major entities: two arch barrels, abutments, pier, discrete springs at one end, outer filler material (consisting of rails, ballast, sleepers and sandy soil) and inner soil layer (consisting of sandy soil). For the purpose of parameter identification, the bridge structure was divided into 42 zones such that within each zone the structural parameters were taken to be constant The Young's modulus and Poisson's ratio within each zone in the arch barrel, fillers, abutment and piers and the discrete springs at the end were taken to be the system parameters to be identified. Thus the total number of parameters to be identified was 76.

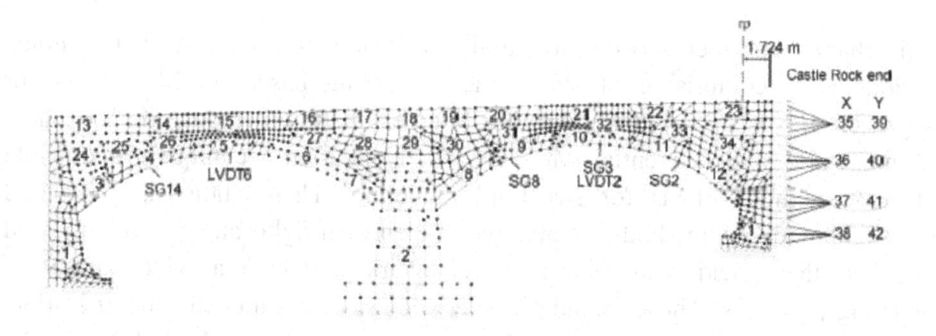

Fig. 8. Plane stress FE model for the bridge structure; the numbers in circles correspond to zones used in identification process; five strain gauges and two LVDT-s used in the present study are also shown; the point 'rp' shown on the figure corresponds to the reference point used in the quasistatic moving load test (Nasrellah, 2009).

Figure 8 also shows the various sensors that were used in the present study. This consisted of 4 strain gauges and two LVDTs. Figure 9a shows the evolution of one of the system parameters (Young's modulus of the arch barrel) as a function of the global iterations. A set of 100 particles were employed and calculations were performed for 50 global iterations. The prediction of mid-span transverse displacement from the identified model and the initial model are compared with field measurements in Fig. 9b. Clearly, at the end of data assimilation, the prediction from the initial model gradually moves towards the measured values thereby indicating the satisfactory performance of the identification procedure.

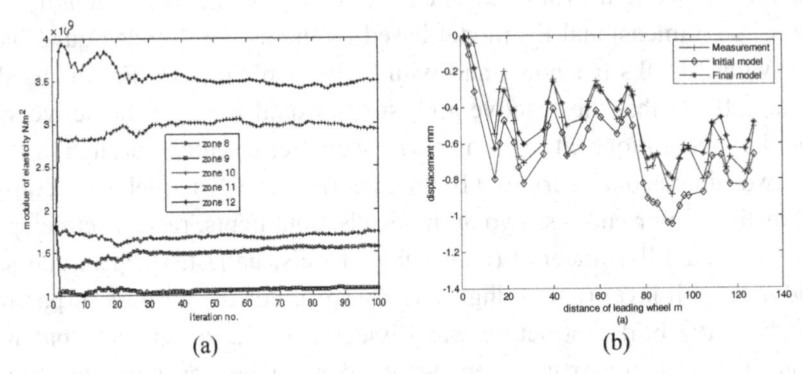

Fig. 9. (a) Estimation of system paramters for the arch bridge problem; results for Young's modulus of the arch barrel; (b) Comparison of predictions on system response from identified and initial models with corresponding measurements (LVDT-2) (Nasrellah, 2009).

5. Closing Remarks

This chapter has focussed on structural model updating methods that combine various tools comprising of state space modeling based on Markov vector approach, Bayesian and maximum likelihood estimation methods, Markov chain Monte Carlo and sequential Monte Carlo simulation techniques, and finite element method (FEM) for structural modeling. These statistical tools and structural modeling methods are matured in their own rights and, when harnessed together, they yield comprehensive framework to tackle a wide variety of updating problems. These include problems of system parameter identification, hidden state estimation, force identification, and sensitivity and reliability model updating. Brief illustrations demonstrating some of these capabilities have been presented. In closing, we draw attention to a few issues that need better resolution and also suggest a few avenues for further research.

The state space representation treats modeling errors as white noise processes. In the context of FEA of structures, the modeling errors arise out of chielfly two sources: (a) simplifying assumptions made on constitutive laws and strain-displacement relations, boundary condtions, joint flexibilities, energy disspation models, interactions with soil and environment, and assumed spatial variations in deformations, and (b) computational errors resulting from spatio-temproal discretization of field variables. In any given situation, these assumptions are made consciously to achieve a desired level of engineering sophistication in modeling. Added to these errors, of course, are the other issues related to limitation of knowledge on underlying processes (such as, for example, deterioration mechansims, details of constitutive laws, and accumulation of fatigue damage). If the question of characterizing the process and measurement noises is examined in this backdrop, the adoption of white noise models seems not well justified. Procedures to tune these noise parameters seem all the more difficult to establish. The issues become further complicated when artificial dynamics are ascribed to either time varying quantitites (Sec. 3.2) or to static parameters (Sec. 3.3.1). Thus, it would seem that the use of white noise models in this context must be viewed more within the philosphical framework of enabling powerful tools of Markov process theory to problem on hand than as means for satisfactory representation of modeling errors. In the same vein, the noise parameters perhaps need to be adjusted to obtain stable numerical results.

Imbedding the updating tools within the framework of FEM, although, in principle, poses no fundamental obstacles, from the point of view of efforts already expended in development of professional FEA codes, however, it could pose several difficulties. For instance, the deduction of discrete state space models by treating the semi-discretized equations of motions as a set of SDE-s and subsequent discretization in time using Ito-Taylor expansions, is quite alein to practices followed in FEA of dynamical systems. Similarly, the implicit integration schemes often used in FEA of dynamical systems are not well suited for filtering applications. If one altogether ignores the modeling errors incurred in arriving at governing equillibrium equations, and restrict the noise models only to meaurement equations, one could readiliy utilize the currently available professional FEA packages, at least for problems of parameter identification, by externally interfacing them with updating tools developed on independent platforms (Sec. 4). How far such approaches would yield satisfactory reuslts, is a question that requires further explorations.

In implementing updating tools for large FEA models, most computational effort gets invariably spent in performing sample response calculations. Strategies such as parallelizing the codes or use of powerful computational hardware can obviosuly offset the difficulties. Another alternative would be to replace judiciously the FEA codes by simpler surrogate models (as is commonly done in reliability calculations using response surfaces). Such replacements seem to be well suited, for instance, in MCMC based approaches requiring initial calculations to traverse the burn-in period.

The idea of using global iterations in parameter identification (Sec. 3), although is intutively appealing, is not free of pitfalls. The global iteration step essentially helps one to construct improved prior models by using measured data. This essentially contradicts the Bayseian outlook which expects measured data to speak for itself and the choice of prior is based on engineering judgement to be excercised before measurements are made. The danger of global iteration is that we may end up fitting data by adjusting parameters of what in reality could be a poor mathematical model. A way to resolve this issue, at least partially, would be to restrict global iterations only to a segment of measured data and assimilate the remaing data without global iterations. Also, reserve some of the measurements for validation purposes and not to use them in parameter identification process.

In the identification procedures discussed in this Chapter, the parameters are treated as random variables or as being determinsitic. It would be of interest to explore methods that can treat alternative models for paramter uncertainty such as interval, convex, fuzzy and hybrid models (Ben-Haim, 1996).

Updating problems, beyond problems of system identification, can be handled within the framework of methods discussed in this Chapter. While problems of sensitivity and relaibility model updating have been briefly mentioned herein, their applications to practical problems of developing repair, retrofit, and (or) maintenance strategies and also for updating fragility models (Der Kiureghian, 1999) require further efforts. While we have addressed questions of local sensitivity model updating, similar questions on global sensitivity model updating require newer efforts.

The measurment data considered in this chapter are taken to emanate from traditional sensors such as strain gauges, LVDTs, and accelerometers. Assimilating data from moderen sensing devices such as laser vibrometers, fibre optics for distributed sensing, acoustic emission sensors, and digital cameras require special efforts espcially in handling large amount of data. Assessing the relative roles of physical models and measurement data, especailly when extensive data become regularly available from large infrastructures, require further studies.

The structural model updating methods discussed in this Chapter, by definition, are model dependent. It is important to note that, in a given context, several alternative FEA models can be conceived. There exisits a need for developing numerical procedures and codes to assign figures of merit to the competing models based on their ability to explain observed experimental data and yet the same time remain robust. Frameworks for achieving this have been outlined, for example, in the works of Burnham and Anderson (2002) and Yuen (2010). The development of model selection approaches is particularly needed in the context of application of FEM to complicated problems involving nonlinear material modeling, interacting systems, or multi-physics problems.

The process noise in our study is introduced in the eqaution of motion in the configuration space and the state space equations are derived using theory of stochastic differential equations (SDEs). This helps to represent and treat the process noise in a systematic and parsimonious way. A higher order discretization scheme permits larger step size to be used but would introduce non-Gaussian terms into the noise vector. Several implicit schemes for discretizing SDEs exist (Kloeden *et al.*, 1999) but they seem to be not suited for filtering applications. It is to be noted that the discretization schemes used in integrating dynamic equilibrium equations in standard FEM have different moorings and their use in filtering problems and several issues related to accuracy and stability need resolution. It may also be noted that if modeling errors are introduced at the level of partial differential equations, additional efforts are needed to spatially discretize these errors. The effort required to deal with this would go beyond the ambit of theory of SDEs used in the present study.

Acknowledgements

A part of this work has been funded by the BRNS (DAE), Government of India. CSM had useful discussions with Professor Abhijit Sarkar of Carleton University on the topics of this paper.

References

Andrieu, C. and Doucet, A. (2003). Online expectation maximization type algorithms for parameter estimation in general state space models, *Proc. Int. Conf. 6-10 April 2003*, VI - pp. 69-72.

Beck, J. L. (1978). Determining models of structures from earthquake records. (Report No. EERL 78-01, California Institute of Technology, Pasadena).

Beck, J. L. and Au, S. K. (2002). Bayesian updating of structural models and reliability using Markov Chain Monte Carlo simulation, J. Eng. Mech., 128 , pp. 380-391.

Ben-Haim, Y. (1996). *Robust reliability in the mechanical sciences*. (Springer Verlag, Berlin).

Bendat, J. S. (1998). *Nonlinear System Techniques and Applications*. (Wiley, New York).

Bocquet, M., Pires, C. A. and Wu, L. (2010). Beyond Gaussian statistical modeling in geophysical data assimilation, Monthly weather review, 138, pp. 2297-3023.

Brooks, S., Gelman, A., Jones, G. L. and Meng, X. L. (2011). *Handbook of Markov chain Monte Carlo*. (Chapman and Hall/CRC, Boca Raton).

Brown, R. G. and Hwang, P. Y. C. (1992). *Introduction to random signals and applied Kalman filtering*, 2nd Ed. (John Wiley and Sons, Inc., New York).

Burnham, K. P. and Andersen, D. R. (2002). *Model selection and multimodel inference*. (Springer Verlag New York).

Cappé, O., Godsill, S. J. and Moulines, E. (2007). An overview of existing methods and recent advances in sequential Monte Carlo, Proceeding of the IEEE, 95, pp. 899-924.

Chandra Kishen, J. M., Ananth Ramaswamy. and Manohar, C. S. (2011). Safety assessment of a brick masonry arch bridge: field testing and simulations, Accepted for publication, J. Bridge Eng., ASCE.

Ching, J., Beck, J. L. and Porter, K. A. (2006). Bayesian state and parameter estimation of uncertain dynamical systems, Probab. Eng. Mech., 21, pp. 81–96.

Ching, J. and Beck, J. L. (2007). Real-time reliability estimation for serviceability limit states in structures with uncertain dynamic excitation and incomplete output data. Probab. Eng. Mech., 22, pp. 50-62.

Corigliano, A. and Mariani, S. (2004). Parameter identification in explicit structural dynamics: performance of the extended Kalman filter, Comput. Meth. Appl. Mech. Eng., 193(36-38), pp. 3807–3835.

Der Kiureghian, A. (1999). Seismic fragility assessment of structural system: towards a synthesis approach, *Transactions of the 15th International Conference on Structural Mechanics in Reactor Technology*, Seoul, Korea, I-pp. 75-92.

Doucet, A., Freitas, N. de. and Gordon, N. (2001). *Sequential Monte Carlo methods in practice*. (Springer, New York).

Doucet, A. and Johansen, A. M. (2008). A tutorial on particle filtering and smoothing: fifteen years later. (http://www.cs.ubc.ca/ ~arnaud/ doucet_ johansen_tutorialPF.PDF accessed on 14th March 2009).

Evensen, G. (2006). *Data assimilation: Ensemble Kalman filter*. (Springer, Berlin).

Ghanem, R. and Shinozuka, M. (1995). Structural system identification I: Theory, J. Eng. Mech., ASCE, 121, pp. 255-264.

Ghosh, S., Roy, D. and Manohar, C. S. (2007). New forms of extended Kalman filter via transversal linearization and applications to structural system identification, Comput. Meth. Appl. Mech. Eng., 196, pp.5063-5083.

Ghosh, S., Manohar, C. S. and Roy, D. (2008). Sequential importance sampling filters with a new proposal distribution for parameter identification of structural systems, Proc. R Soc. London, A, 464, pp. 25-47.

Gilks, W. R., Richardson, S. and Spiegelhalter, D. J. (1996). *Markov Monte Carlo in practice*. (Chapman and Hall, Boca Raton).

Gordon, N. J., Salmond, D. J. and Smith, A. F. M. (1993). Novel approach to nonlinear/non-Gaussian Bayesian state estimation, IEE Proceedings-F, 140(2), pp. 107-113.

Hastings, W. K. (1970). Monte Carlo sampling methods using Markov chains and their applications, Biometrika, 57(1), pp. 97-109.

Hoshiya, M. and Saito, E. (1984). Structural identification by extended Kalman filter, J. Eng. Mech., ASCE, 110, pp. 1757-1772.

Imai, H., Yun, C. B., Maruyama, O. and Shinozuka, M. (1989). Fundamentals of system identification in structural dynamics, Probab. Eng. Mech., 4, pp. 162-173.

Imregun, M. (1998). A survey of nonlinear analysis tools for structural systems, Shock Vib. Dig., 30, pp.363-369.

Ionides, E. L., Breto, C. and King, A. A. (2006). Inference for nonlinear dynamical systems, Proceedings of National Academy of Sciences, 103(49), pp. 18438-18443.

Jazwinski, A. H. (1970). *Stochastic processes and filtering theory*. (Dover, Mineola).

JCSS. (2001). Probabilistic model code, Joint committee on structural safety. (http://www.jcss.byg.dtu.dk/Publications/Probabilistic_Model_Code.aspx,).

Kalman, R. E. (1960). A new approach to linear filtering and prediction problems, Transactions of ASME, J. Basic Eng., 82 (Series D), pp. 35-45.

Kantas, N., Doucet, A., Singh, S. S., Maciejowski, J. M. An overview of sequential Monte Carlo methods for parameter estimation in general state space models. (www.cs.ubc.ca/, accessed on 1st December 2011).

Kerschen, G., Worden, K., Vakakis, A. F. and Golinval, J. C. (2006). Past, present and future of nonlinear system identification in structural dynamics, Mech. Syst. Signal Process, 20, pp. 505-592.

Khalil, M., Poirel, D. and Sarkar, A. (2010). Parameter estimation of a fluttering aeroelastic system in the transitional Reynolds number regime, *Proc. ASME 3rd US Eur. Fluids Eng. Summer Meeting*, Montreal, Canada.

Kitigawa, G. (1998). A self-organizing state-space model, J. Am. Stat. Assoc., 93(443), pp. 1203-1215.

Kloeden, P. E. and Platen, E. (1999). *Numerical Solution of Stochastic Differential Equations*. (Springer, Berlin).

Kroese, D. P., Taimre, T. and Botev, Z. I. (2011). *Handbook of Monte Carlo methods*. (Wiley, New Jersey).

Liu, J. S. (2001). Monte Carlo strategies in scientific computing. (Springer, New York).

Manohar, C. S. and Roy, D. (2006). Nonlinear structure system identification using Monte Carlo filters, Sadhana, Academy Proceedings in Engineering, Indian Academy of Science, 31(4), pp. 399-427.

Maybeck, P. S. (1982). Stochastic models, estimation and control, Volumes I-III. (Academic press, New York).

Metropolis, N., Rosenblueth, A. W., Rosenblueth, M. N., Teller, A. H., and Teller, E. (1953). Equations of state calculations by fast computing machines, Journal of Chemical Physics, 21(6), pp. 1087-1091.

Namdeo, V. and Manohar, C. S. (2007). Nonlinear structural dynamical system identification using adaptive particle filters, J. Sound Vib., 306, pp. 524-563.

Namdeo, V. (2007). Novel strategies for nonlinear structural system identification using particle filtering and force state map construction, MSc(Engg) thesis, Department of Civil Engineering, Indian Institute of Science, Bangalore.

Nasrellah, H. A (2009). Dynamic state estimation techniques for identification of parameters of finite element structural models. PhD thesis, Department of Civil Engineering, Indian Institute of Science, Bangalore.

Nasrellah, H. A. and Manohar, C. S. (2010). A particle filtering approach for structural system identification in vehicle-structure interaction problems, J. Sound Vib., 329(9), pp. 1289-1309.

Nasrellah, H. A. and Manohar, C. S. (2011). Finite element method based Monte Carlo filters for structural system identification, Probab. Eng. Mech., 26 (2011) pp. 294–307.

Nasrellah, H. A. and Manohar, C. S. (2011). Particle filters for structural system identification using multiple test and sensor data: a combined computational and experimental study, Struct. Control Health Monit., 18, pp. 99–120.

Nelles, O. (2001). *Nonlinear System Identification*. (Springer-Verlag Berlin Heidelberg, N.Y.).

Radhika, B. and Manohar, C. S. (2010). Reliability models for existing structures based on dynamic state estimation and data based asymptotic extreme value analysis, Probab. Eng. Mech., 25, pp. 393-405.

Radhika, B. and Manohar, C. S. (2011). Updating response sensitivity models of nonlinear vibrating structures using particle filters, Comput. Struct., 89(11-12), pp. 901-911.

Radhika, B. and Manohar, C. S. (2012). Nonlinear dynamic state estimation in instrumented structures with conditionally linear Gaussian substructures, Probab. Eng. Mech., 30, pp. 89-103.

Radhika, B. (2012). Monte Carlo simulation based response estimation and model updating in nonlinear random vibrations, PhD thesis, Department of Civil Engineering, Indian Institute of Science.

Ristic, B., Arulampalam, S. and Gordon, N. (2004). *Beyond the Kalman filter*. (Artech House, Boston).

Robert, C. P. and Casella, G. (2004). *Monte Carlo statistical methods*. (Springer, New York).

Sajeeb, R., Manohar, C. S. and Roy, D. (2007). Control of Nonlinear Structural Dynamical Systems with Noise Using Particle Filters, J. Sound Vib., 306(25), pp. 111-135.

Sajeeb, R., Manohar, C. S. and Roy, D. (2009). A Conditionally linearized Monte Carlo filter in nonlinear structural dynamics, Int. J. Nonlinear Mech., 44(7), pp. 776-790.

Sajeeb, R., Manohar, C. S. and Roy, D. (2010). A semi-analytical particle filter for identification of nonlinear oscillators, Probab. Eng. Mech., 25, pp. 35-48.

Schön, T. and Gustafsson, F. (2005). Marginal particle filters for mixed linear/nonlinear state space models, IEEE Trans. Signal Process., 53(7), pp. 2279-2288.

Shinozuka, M., Yun, C. B. and Imai, H. (1982). Identification of linear structural dynamic systems, J. Eng. Mech., ASCE, 108 (6), pp. 1371-1390.

Storvik, G. (2002). Particle filters for state-space models with the presence of unknown static parameters, IEEE Trans. Signal Process., 50(2), pp. 281-289.

Sundar, V. S. and Manohar, C. S. (2013). Updating reliability models of statically loaded instrumented structures, Struct. Saf., pp. 21-30.

Tanizaki, H. (1996). *Nonlinear filters: estimation and applications*, 2nd Ed. (Springer Verlag, Berlin).

Tipireddy, R., Nasrellah, H. A. and Manohar, C. S. (2009). A Kalman filter based strategy for linear structural system identification based on multiple static and dynamic test data, Probab. Eng. Mech., 24, pp. 60-74.

Wang, D. and Haldar, A. (1997). System identification with limited observations and without input, J. Eng. Mech., ASCE, 123(5), pp. 504-511.

Worden, K. and Tomlinson, G. P. (2001). Nonlinearity in structural dynamics: Detection, identification and Modeling. (IOP Publishing, London).

Yuen, K. V. (2010). Bayesian methods for structural dynamics and civil engineering. (Wiley, Chichester).

Yuen, K. V. and Kuok, S. C. (2011). Bayesian methods for updating dynamic models, Appl. Mech. Rev., 64, pp. 1-18.

Yun, C. B. and Shinozuka, M. (1980). Identification of nonlinear structural dynamic systems, J. Struct. Mech., 8(2), pp. 187-203.

Chapter 5

Stochastic Filtering In Structural Health Assessment: Some Perspectives and Recent Trends

S. Sarkar[1], T. Raveendran[2], D. Roy[1*], and R.M. Vasu[2]

[1]*Computational Mechanics Lab, Department of Civil Engineering*
[2]*Department of Instrumentation and Applied Physics*
Indian Institute of Science, Bangalore 560012, India
**Email: royd@civil.iisc.ernet.in*

1. Introduction

The inverse problem of reconstructing parameters of nonlinear vibrating systems, based on limited response measurements, is important in structural health assessment and vibration control, especially where such structures are prone to extreme loading conditions such as strong motion earthquakes and high winds. The detection and characterization of structural damages, such as occurrence of cracks, using vibration data is also a related topic of great interest. The field of study is of particular relevance in the context of problems of earthquake engineering, fracture mechanics etc. wherein structures are typically designed to possess controlled inelastic behavior. Drawing upon their success in dealing with linear system identification problems, frequency and time domain methods have been extended and explored widely in such nonlinear identification problems as well. Frequency domain methods involve development and analysis of higher order spectra (Worden and Manson, 2001) and are basically extensions of classical modal analysis (Ewins, 2000). On the other hand, time domain methods have been mostly limited to state estimation. To handle nonlinearity under the linear framework nonlinear dynamic state estimation problems (Brown and Hwang, 1992; Ewins, 2000; Yun and Shinozuka, 1980) have been linearized and then solved under traditional Kalman filtering scheme.

Most structural system response problems (also called forward problems) can be treated via solutions of system of partial differential equations (PDEs) with some given boundary and initial conditions. A major difficulty here is the modeling error, i.e. an inaccurate forward model, wherein the computed response could be irreconcilable with the measured response data. Model errors often arise due to the many assumptions in deriving the mathematical models. Difficulties also arise in modeling complex system geometries since domain discretization techniques might only provide a gross approximation to the original geometry. Also, owing to corrosion or other environmental effects that are difficult to account for within a mechanical model, there could be time-dependent uncertain variations in system parameters (stiffness, damping etc.) of existing structures and such modeling errors may not be adequately represented by just adding white noise in the process equations. Nevertheless, in many cases of practical interest wherein model errors are not too pronounced to demand a more elaborate stochastic representation than low-intensity white noise processes, dynamic state and parameter estimation techniques can offer an effective framework for assessing the current functional features (e.g. the load carrying capacity or the residual life) of a structural system. The problem of identification of linear structural systems is widely studied through a number of methods in time and frequency domains (Kotecha and Djuric, 2001; Ewins, 2000; Ljung, 1997; Pintelon and Schoukens, 2001; Peeters and Roeck, 2001; Lieven and Ewins, 2001). Although stochastic filtering is not a new approach to such problems, they are mostly confined to variants of Kalman filtering (KF) (Kalman, 1960). The KF provides the exact posterior probability density function (pdf) of the states of linear dynamical systems (with process noise being additive and Gaussian white) supplemented with linear measurement models, again with the measurement noises also being additive, Gaussian white. Despite its optimality, a major disadvantage of KF is that it cannot handle non-Gaussian noises and structural system nonlinearity, characterized by non-Gaussian response processes. Problems related to the identification of nonlinear structural systems remain an important and largely unexplored area of research. Such studies on nonlinear structural system identification (SSI) assume importance due to the fact that localized forms of nonlinearity are almost invariably associated with the onset of damage. Indeed, in these cases, the measured quantities may also be nonlinearly related to system states. In addition, given the state space representation of the process equation in a stochastic filtering approach, the parameters to be identified are declared as additional states and hence the problem becomes nonlinear even if the process equations are strictly linear to begin with. Applying the KF for such nonlinear SSI problems demands some form of linearization of the process and/or measurement equations. This leads to the sub-optimal extended Kalman filter (EKF). For nonlinear SSI, there are two main classes of methods, viz., those based on the EKF and its variants (Kalman, 1960, Brown and Hwang, 1992), and those based on Monte Carlo simulations, popularly known as particle filters (PFs) (Ristic *et al.*, 2004; Tanizaki, 1996; Tanizaki and Mariano, 1998; Doucet, 1998; Doucet *et al.*, 2000; Ching et al., 2006). The EKF does not account for the non-Gaussian nature of nonlinear system states, and, therefore, can only approximately capture the first two moments of the conditional pdf of the state. Moreover, while EKF requires vector fields in the governing equations to be strictly differentiable, nonlinearity in a damaged structure need not be so. EKF based reconstruction may also be sensitive to process covariance (with updates via Riccatti equation resulting in filter divergence). The need for an elaborate tuning of process noise covariance matrices for higher dimensional systems restricts its practical use.

Some of these problems can be addressed using a PF based strategy within a Monte Carlo (MC) setup. Ensemble Kalman Filter (EnKF) (Lorentzen and Naevdal, 2011), which is one of the basic filters employing MC simulations, preserves the analytical advantages of the KF whilst exploiting the universality of MC simulations.The basic ingredient of the EnKF is the approximation of the error covariance matrix via MC sampling, thereby replacing a Ricatti update characteristic of the KF or the EKF. In ensemble-based data assimilation schemes, the forecast ensemble is generated by evolving an ensemble of initial conditions distributed according to the result of the immediately preceding analysis. Though the reconstructed processes are approximated as Gaussian, filter divergence can be arrested by EnKF. Moreover it requires no derivation of a tangent linear operator or adjoint equations, and no integrations backward in time (Evensen, 2003 and 2004). Since the covariance matrices are generated using an ensemble, the sensitivity to process noise covariance is largely bypassed. Vis-à-vis a PF, sampling variance of the estimate is expected to be lower in the EnKF as it makes use of the analytical KF update formula. But EnKF has hardly been used in SSI despite its potential. PFs, on the other hand, provide a more versatile approach to nonlinear and non-Gaussian estimation problems. The posterior (filtering) distribution is here replaced by an empirical measure through an ensemble of weighted particles (random samples) and is carried forward recursively. As the ensemble size increases, the empirical measure approaches the true filtering PDF. PFs are thus capable of handling system nonlinearity and even a possibly non-Gaussian nature of noises. Different PF schemes and their applications may be found in (Ristic *et al.*, 2004; Tanizaki, 1996; Tanizaki and Mariano, 1998; Doucet, 1998; Doucet *et al.*, 2000; Arulampalam *et al.*, 2002; Gordon *et al.*, 1993; Saha and Roy, 2009). The Sequential Importance Sampling (SIS) scheme forms the basis for many versions of the PF developed and applied over a range of problems over the past few decades (Doucet *et al.*, 2000, Doucet *et al.*, 2001; Arulampalam *et al.*, 2002; Gordon *et al.*, 1993). Most PF strategies involve at least two steps, viz. prediction and updating. The accuracy and efficiency of a PF scheme is dependent on the importance sampling function, typically a Radon-Nikodym derivative of the posterior distribution w.r.t. the prediction distribution, and is used to weigh the predicted particles. The SIR or bootstrap filter (Gordon *et al.*, 1993), a variant of the SIS filter, uses the transitional distribution corresponding to the process equations complied with the latest measurement that is taken care of by the importance sampling function. Here, a third step, called re-sampling, is used to draw (with replacement) uniformly weighted particles from a set of non-uniformly weighted ones. This step tends to discard particles with small weights, replicate those with higher weights and confronts the undesirable prospect of filter degeneracy (Doucet *et al.*, 2000). The Gaussian sum PF (GSPF) (Kotecha and Djuric, 2003; Kotecha and Djuric, 2001), on the other hand, provides a relatively superior alternative wherein the densities (in the prediction and updating steps) are approximated via Gaussian mixtures.

In the context of nonlinear engineering dynamics, process equations are generally written as a system of nonlinear, time-continuous stochastic (ordinary) differential equations (SDEs). Using an integration strategy, the approximate solution of the process SDEs must be written as a map, consistent with the temporally discrete nature of the available observations. However, due to the scarcity of sufficiently accurate schemes for integrating nonlinear SDEs and the general absence of closed-form analytical solutions, errors in the discretization of the process SDEs may result in erroneous estimates (Gobet *et al.*, 2006). In Girsanov linearization method (GLM) based PF (Saha and Roy, 2009), a class of semi-analytical integration schemes based on local linearization (Ching *et al.*, 2006; Evensen, 2003) are used for solving the SDEs and a strategy is suggested to weakly correct for the linearization errors by the Girsanov transformation of measures. This correction (the Radon-Nikodym derivative or the likelihood ratio) is multiplicative and forms a scalar SDE whose formal solution is a stochastic exponential in terms of the linearized trajectories. However, since the stochastic integral in the exponent typically defies an accurate analytical or numerical treatment. Based on the approximation of this term, two different versions of the GLM, viz. GLM-1 and GLM-2, are discussed in (Saha and Roy, 2007). While in GLM-1 a stochastic Taylor expansion is used, GLM-2 employs a Monte-Carlo strategy. Unfortunately, such brute-force computations of the correction terms face serious numerical difficulties manifested through frequent underflows in the computed response statistics.

These limitations can be overcome by a reformulated Girsanov Corrected Linearization Method (GCLM) which is based on rejection sampling and re-sampling (Raveendran *et al.*, 2011). As with the GLM, the time interval of interest is discretized into smaller subintervals and local linearization is performed over each of them. However, unlike the GLM, the correction scheme in the Girsanov transformation is here implemented through a combined rejection sampling and re-sampling over an ensemble of locally linearized trajectories, thereby bypassing a direct computation of the correction term that is fraught with numerical errors and instability. Using integration by parts, the Radon-Nikodym derivative, which is a stochastic exponential with arguments in the form of stochastic integrals of the linearized solution, is first brought to a canonical form involving no stochastic integrals and then split into two factors. While the rejection sampling is applied using the first factor containing non-stochastic integrals, the selected trajectories are then re-sampled based on the second factor, which is free of integrals and hence readily computable. Since no stochastic Taylor expansions are employed at any stage, the scheme is free of the errors arising from the series truncation or evaluation of the MSIs involved and holds the promise of unparalleled numerical accuracy.

Though PFs overcome most of the issues faced by the variants of the KF and can also handle multiple solutions, they do suffer from the curse of dimensionality. Also, MC simulations make them computationally expensive. The filter divergence and tuning are two other factors which are to be taken care of in PFs. While some work on arresting the filter divergence, which arises due to a discrepancy between prediction and measurement, has been reported, they have hardly been explored in the context of SSI.

2. KF, EKF and EnKF

Under the twin constraints of linearity and Gaussianity, KF gives optimal solutions. KF basically addresses the state ($X \in \mathbb{R}^n$) estimation problem of a system that is governed by linear SDEs. A time-continuous SDE can be written in its time discretized form as given below:

$$X_i = A_{i-1}X_{i-1} + w_{i-1} \tag{1}$$

where i stands for i^{th} instant. Measurement $Z_i \in \mathbb{R}^m$ is related to state vector as

$$Z_i = H_i X_i + v_i \tag{2}$$

$w_i \sim N(0, R)$ and $v_i \sim N(0, Q)$ represent process and measurement noises respectively. They are considered as mutually independent and (Gaussian) white.

R and Q are process and measurement noise covariances respectively. A_{i-1} is an $n \times n$ matrix that relates state vector at the previous time step to the current time step. H_i is an $m \times n$ matrix that relates state vector to the measurement vector at the current time instant. KF is performed in two steps. First the process state vector at the ith instant is estimated ($\hat{X}_i^-$) and then this estimated state vector is updated ($\hat{X}_i$) using feedback in the form of (noisy) measurements. The time updated equations carry forward (in time) the current state and error covariance estimates (P_i^-) to obtain the *a priori* estimates for the next time step. The measurement is used as feedback to evaluate an improved *a posteriori* estimate from the *a priori* estimate. In KF, which is basically a minimum mean-square error estimator, the error in the *a posteriori* estimate, $(X_i - \hat{X}_i)$ is minimized in the $L^2(P)$ sense. Such minimization is equivalent to minimizing the trace of the *a posteriori* covariance matrix P_i in $L^2(P)$. Kalman filtering steps can be summarized as follows:

Prediction step:

$$\hat{X}_i^- = A_{i-1}\hat{X}_{i-1} \tag{3}$$

$$P_i^- = A_{i-1}P_{i-1}A_{i-1}^T + Q_{i-1} \tag{4}$$

Updating step:

$$\hat{X}_i = \hat{X}_i^- + K_{g_i}(Z_i - H_i\hat{X}_i^-) \tag{5}$$

$$K_{g_i} = P_i^- H_i^T (H_i P_i^- H_i^T + R_i)^{-1} \tag{6}$$

$$P_i = (I - K_{g_i} H_i)P_i^- \tag{7}$$

K_{g_i} is referred to as the Kalman gain matrix. KF cannot be applied directly once the process equations or measurement equations become nonlinear. This difficulty is overcome in the EKF. Nonlinear process and measurement equations are linearized around the current estimates by Taylor expansion considering only up to the first order derivatives. For further discussion, the nonlinear process and measurement equations (time discretized) are taken in the form given below:

$$X_i = A_{i-1}(X_{i-1}) + w_{i-1} \quad X \in \mathbb{R}^n \tag{8a}$$

$$Z_i = H_i(X_i) + v_i \quad Z \in \mathbb{R}^m \tag{8b}$$

where $X_{i-1} = X_{i-1}(\omega)$, $\omega \in \Omega$ (the sample set) with $(\Omega, \mathcal{F}, P)$ denoting the probability space (triplet). Equations (8a) and (8b) are linearized around the state vector $X_{r_{i-1}} := X_{i-1}(\omega_r)$ as:

$$X_i = A_{i-1}(X_{r_{i-1}}) + \nabla A_{i-1}(X_{r_{i-1}})(X_{i-1} - X_{r_{i-1}}) + w_{i-1} \tag{9a}$$

$$Z_i = H_i(X_{r_i}) + \nabla H_i(X_{r_i})(X_i - X_{r_i}) + v_i \tag{9b}$$

Once equations (8a) and (8b) are linearized, the same procedure is adopted in the EKF as is followed in the KF. KF and EKF have been extensively applied to structural system identification problems. But if nonlinearity is high, then first order linearization may not be sufficient and the filter may diverge. In order to avoid the sensitivity of estimates by the EKF to process noise covariance (Grewal and Andrews, 2001), EnKF can be used. EnKF is essentially an ensemble-based (Monte Carlo) filter. EnKF partially makes use of the analytical formulas provided by the KF which reduce the sampling variance. This is an important feature as the sampling error (sampling standard deviation) reduces proportionally only to the inverse of the square root of the ensemble size. Indeed, the fact that the EnKF combines the closed-form features of the KF-based estimate update with the PF-like simulation of the error covariance makes it insensitive to process noise covariance. In EnKF (Sakov and Oke, 2008), the filtering pdf is represented by an ensemble $\tilde{X}_i = \{X_i^{(u)}\}_{u=1}^{N}$; N is the ensemble size t_i. Mean and covariance of this ensemble are calculated as:

$$E(\tilde{X}_i) = \frac{1}{N}\sum_{u=1}^{N} X_i^{(u)} \;,\; C_i^- = \frac{A_{i-1}{}^T A_{i-1}}{N-1} \tag{10}$$

where

$$A_i = X_i^{(u)} - E(\tilde{X}_i)r \quad ,r = \{1,1,...,1\}_{1\times N} \tag{11}$$

The Kalman gain (K_{g_i}) matrix used in the regular KF is modified by replacing the state covariance matrix (P_i^-) by sample covariance matrix (C_i^-). u^{th} realization is predicted by the process equation (without the noise) and error covariance by sample covariance (C_i^-):

$$\hat{X}_i^{(u)-} = A_{i-1}\hat{X}_{i-1}^{(u)} \tag{12}$$

$$C_i^- = \frac{A_{i-1}{}^T A_{i-1}}{N-1} \tag{13}$$

Then the predicted state vector and sample covariance matrix are updated using the current measurement feedback as:

$$\hat{X}_i^{(u)} = \hat{X}_i^{(u)-} + K_{g_i}(Z_i + d_i^{(u)} - H_i\hat{X}_i^{(u)-}) \tag{14}$$

$d_i^{(u)}$ is a synthetic vector of perturbations of measurement Z_i.

$$K_{g_i} = C_i^- H_i^T (H_i C_i^- H_i^T + R_i)^{-1} \tag{15}$$

$$C_i = (I - K_{g_i} H_i) C_i^- (I - H_i^T K_{g_i}^T) \tag{16}$$

In all the filters mentioned above, process and measurement equations are time dependent. But a large class of inverse problems arising in structural engineering applications is static (with statically acquired data). Recovering the parameters of such forward models from limited static measurements of the system response is always a challenge. These problems are generally solved within the least squared error minimization framework with the forward model (often in the form of PDEs) acting as constraints. Such an approach involves repeated solutions (inversions) of the forward operators and hence faces the prospect of an unacceptable accumulation of error as iterations progress. This leads to another way of treating such ill posed problems which is known as the pseudo dynamic approach.

2.1 *A pseudo- dynamic approach*

A dynamically evolving process response is required to make a static inverse problem within the statistical filtering framework. Therefore, the static equations are converted into a pseudo-dynamical system whose steady-state response is given by $\lim_{t\to\infty} \delta(t) = K^{-1}F$, where K and F denote the stiffness matrix and the forcing vector respectively (Banerjee *et al.*, 2011). Such a pseudo-dynamical system is not unique and from several possibilities, the following is chosen:

$$\dot{\delta}(t) + K\delta(t) = F + \dot{\xi}_\delta(t) \tag{17}$$

where $\dot{\xi}_\delta(t)$ denotes an additive, Gaussian white noise (vector) process, i.e., $\xi_\delta(t)$ is a Wiener (Brownian motion) process with $\dot{\xi}_\delta(t_0) = 0$. Strictly speaking $\xi_\delta(t)$ is nowhere differentiable with respect to t and equation (17) is more appropriately written in the incremental form:

$$d\delta(t) + K\delta(t)dt = Fdt + d\xi_\delta(t) \tag{18}$$

As the solution $\delta(t)$ evolves over artificial time, it coincides with the static solution of the problem $K\delta = F$ as $t \to \infty$, provided the stiffness matrix K is positive definite and $\dot{\xi}_\delta(t) = 0$. The positive definiteness of K is presently assured as it is derived through the symmetric bilinear form with adequate restraints (against rigid body motion) and the positivity constraint of the physical parameters (e.g. the shear modulus). The exact solution $\delta(t)$ is readily obtainable as:

$$\delta(t) = \exp\{-K(t-s)\}\delta(t_0) + \int_{t_0}^{t} \exp\{-K(t-s)\}Fds + \int_{t_0}^{t} \exp\{-K(t-s)\}d\xi_\delta(s) \tag{19}$$

The last term on the RHS of Eq. (19) is an Ito integral (i.e., a martingale) of zero mean (Klebaner, 2001). Within short time intervals (i.e, when $|t-t_0|$ remains small), the above solution is numerically obtainable to any desired order of accuracy. Alternatively, Eq. (18) may also be solved via an approximate numerical integrator, such as stochastic Euler or Milstein methods (Milstein, 1995). In the context of elasticity imaging, $\delta(t)$ is a hidden Markov process, not adapted to the filtration generated by the measurements. It is however adapted to (and even predictable with respect to) $\mathcal{F}_t^{\xi_\delta} := \sigma(\xi_\delta(s) \mid 0 \le s \le t)$, where $\mathcal{F}_t^{\xi_\delta}$ is the continuous σ-algebra generated by the additive noise vector $\xi_\delta(s)$ for $0 \le s \le t$ (Kallianpur, 1980).

2.2 *A pseudo-dynamic EnKF (PD-EnKF)*

The pseudo-dynamic approach is further elucidated by taking an example of estimation of the shear modulus vector (μ) by the EnKF (i.e. the problem of elastography). The aim of a stochastic filtering technique (including the EnKF) is to estimate the states conditioned on the measurements available till the current time instant. Any filtering technique requires a process equation (which governs the evolution of the states to be estimated) and a measurement equation that functionally relates the observed variables to a subset of the states (modulo noise) and makes available the observed variables in continuous or discrete time. In other words, the σ-algebra (filtration) corresponding to the measurement processes increases (in size) with time. Unfortunately this is not consistent with static elastography, wherein one typically has only one set of (statically recorded) measurements of the axial component of the displacement field (δ^e). This can be bypassed by adding a fictitious (Gaussian) white noise of very small intensity to the measurement equation. This artificially induces a dynamical character to the measurement model and thus enables a Bayesian update of states in time. Thus the linear measurement equation becomes:

$$Z(t) = \delta^e + \eta(t) \tag{20}$$

where $\eta(t) = \eta_0 - \tilde{\eta}(t)$ is the noise process with zero mean and known covariance. Here η_0 is the random noise variable (time-invariant, zero mean) corresponding to the experimental measurement and $\tilde{\eta}(t)$ is the added (fictitious) Wiener noise with very small intensity. The key idea behind this artificial measurement generation is to obtain a non-zero innovation process (in time) so that non-trivial measurement-adapted updates of the states are possible. Such updates can be achieved even if the intensity $\tilde{\eta}(t)$ is very small.

For estimating the parameter vector μ, its elements are declared as additional states ($\dot{\mu} = 0 +$ 'white noise') which makes the estimation problem nonlinear. Optimal solutions of these nonlinear filtering problems are possible through particle filters (Doucet *et al.*, 2000) or geometric filters (Brigo et al., 1998). However a sub-optimal EnKF approach is considered more practicable, as it employs a Kalman state estimate after linearizations of the process and/or measurement equations (as applicable) based on the last available estimates (Corigliano and Mariani, 2004). The μ-augmented process equations are given by:

$$\dot{\delta}(t) + K(\mu(t))\delta(t) = F + \dot{\xi}_{\delta}(t) \tag{21a}$$

$$\dot{\mu}(t) = \dot{\xi}_{\mu}(t) \tag{21b}$$

$\xi_{\delta}(t)$ and $\xi_{\mu}(t)$ are independent Wiener noise vectors. The noise vector $\xi_{\delta}(t)$ takes care of the unknown dynamics of equation (21a) and errors associated with the spatial and time discretization. With $\xi_{\mu}(t)$, μ evolves artificially over time. Equations (21a) and (21b) may be combined to obtain:

$$\begin{aligned}\dot{Y}(t) &= g(\delta(t), \mu(t), t) + \bar{F} + \dot{\xi}(t) \\ &= g(Y(t), \mu(t), t) + \bar{F} + \dot{\xi}(t)\end{aligned} \tag{22a}$$

where $Y(t) = \left[\delta(t)^T \ \mu(t)^T\right]^T$ is the augmented state vector, $\bar{F} = \left\{F^T \ 0^T\right\}^T$ is the augmented force vector and $\xi(t) = \left\{\xi_{\delta}(t)^T \ \xi_{\mu}(t)^T\right\}^T$ is the augmented process noise vector. The linear measurement equation (20) now takes the form

$$Z(t) = HY(t) + \eta(t) \tag{22b}$$

where H is a transformation matrix with entries either 0 or 1. Equation (22a) provides a system of nonlinear SDEs. The estimate $\hat{Y}_{t|t}$ (where $(\hat{\bullet})_{t|t}$) represents an estimated quantity at time t given the measurements up to the same time instant), now has to be evaluated, which is defined as the stochastic process (projected) $\hat{Y}_{t|t} := E(Y_t \mid Z(s); 0 \le s \le t)$. $\hat{Y}_{t|t}$ is $\mathcal{F}_t^Z$ adapted, where $\mathcal{F}_t^Z$ denotes the filtration generated by $Z(s)$, $s \le t$. For the elastography problem given by equations (22a) and (22b), the existence of $\hat{Y}_{t|t}$ may be readily proved via the concept of 'dual predictable projection' (Kallianpur, 1980). Following linearization (about $\hat{Y}_{t-|t-}$) of the process equations within the EnKF, the process Y_t is approximated as Gaussian and $\hat{Y}_{t|t}$ may then be thought of as an orthogonal projection of Y_t on the 'space' spanned by Gaussian (measurement) random variables of the form $\sum_i c_i Z(t_i)$ for all $t_i \le t$, where c_i's are bounded, but otherwise arbitrary, deterministic constants. Now the linearized process equation and its solution may be obtained as:

$$\dot{Y}(t) = g(Y(t_0),t_0) + \left.\frac{\partial g(Y(t),t)}{\partial Y(t)}\right|_{Y(t)=Y(t_0)} \left(Y(t) - Y(t_0)\right) + \bar{F} + \dot{\xi}(t)$$

$$= J_0 Y(t) + \tilde{F}_0 + \dot{\xi}(t) \tag{23}$$

with $J_0 = \left[\frac{\partial g(Y,t)}{\partial Y}\right]_{Y=Y(t_0)}$ and $\tilde{F}_0 = g(Y(t_0),t_0) - J_0 Y(t_0) + \bar{F}$

$$Y(t) = \exp\{-J_0(t-t_0)\}Y(t_0) + \int_{t_0}^{t} \exp\{-J_0(t-s)\}\tilde{F}_0 ds + \int_{t_0}^{t} \exp\{-J_0(t-s)\} d\xi_\delta(s) \tag{24}$$

For convenience of exposition, notational distinctions between the linearized and nonlinear flows are not made. In a practical scenario, measurements are typically available in discrete time and it is thus convenient to write the discretized forms of process and measurement equations as:

$$Y(t_{i+1}) = Y_{i+1}$$

$$= \exp\{-J_i(t_{i+1} - t_i)\}Y_i + \int_{t_i}^{t_{i+1}} \exp\{-J_i(t_{i+1} - s)\}\tilde{F}_i ds + \int_{t_i}^{t_{i+1}} \exp\{-J_i(t_{i+1} - s)\} d\xi_\delta(s) \tag{25}$$

$$Z(t_i) = Z_i = HY_i + \eta(t_i) \tag{26}$$

where $t_i = i\Delta t$, Δt is the pseudo time step (presently assumed to be uniform). $\{\xi(t_i)\}$ and $\{\eta(t_i)\}$ are discretized sequences of noise vectors with covariance matrices Q_i and R_i respectively.

A drastic reduction in the size of the process equation is possible by only declaring the vector μ as the process variables (states). In this case, the governing mechanics of the system, as reflected through the functional relationship between the displacements (or other kinematical descriptors) and the material parameters as well as the external forcing, may be incorporated in the measurement equations. Here, the process and measurement equations may be written as:

$$\dot{\mu}(t) = \dot{\xi}(t) \tag{27}$$

$$Z(t) = DK(\mu)^{-1}F + \eta(t) \tag{28}$$

where D is a rectangular matrix with entries as 0 or 1. Note that, under purely static deformations, the response is given by $\delta = K(\mu)^{-1}F$. In this setting, while the process equation remains linear, the nonlinearity is passed on to the measurement equations. Implementation of this strategy via the pseudo-dynamic EnKF is pursued here. The aim is to obtain the probability distribution $\Pi_{i|1:i} := \Pi(\mu_i \mid Z_{1:i} := \{Z_1^T,...,Z_i^T\}^T)$ (or the associated density $p_{i|1:i}$, if it exists) of the states (in this case, $\mu(t)$ only) conditioned on the measurements $Z_{1:i}$ available till the current time instant t_i. The estimate of μ at t_i may then be given by $\hat{\mu}_i = E_{\Pi_{i|1:i}}(\mu_i)$ (where $E_{\Pi_{i|1:i}}$ denotes the expectation operator under the measure corresponding to $\Pi_{i+1|1:i+1}$). The discretized forms of the above process and measurement equations are:

$$\mu(t_{i+1}) = \mu_{i+1} = \mu_i + \xi_i \tag{29a}$$

$$Z(t_{i+1}) = Z_{i+1} = D\delta(\mu_i) + \Delta\eta_{i+1} \tag{29b}$$

where $t_{i+1} = (i+1)h$, h is the time step (assumed to be uniform) and $\Delta\eta_{i+1} := \eta_{i+1} - \eta_i$.

2.3 *The PD-EnKF algorithm*

The filter-sensitivity to process noise covariance is tackled by adopting a Monte Carlo update of the error covariance instead of the Riccati update. While the family of particle filters (PF-s) (Ghosh *et al.*, 2008; Saha and Roy, 2009; Roy and Manohar, 2006) does that admirably, the sub-optimal EnKF (Evensen, 2003 and 1994) is currently used owing to its inherently analytical features. PF-like simulation in the conventional EnKF also requires that multiple measurement data sets are artificially created in order that the error covariance update remains consistent with the KF formula. In the context of higher dimensional inverse problems, this implies a huge computational overhead. A way out is to use the so-called deterministic EnKF (Sakov and Oke, 2008), wherein the requirement of multiple measurement sets is bypassed via an appropriate modification of the gain term. Given this advantage, deterministic EnKF is opted, even as the basic formulation remains applicable within any stochastic filter.

The error covariance P is approximated as:

$$P = \frac{1}{(N-1)} \sum_{u=1}^{N} (\Sigma_u - \bar{\mu})^T (\Sigma_u - \bar{\mu}) = \frac{1}{(N-1)} \mathrm{A}^T \mathrm{A} \tag{30}$$

The ensemble size is denoted by the integer N and $\Sigma = [\mu(1) \quad \mu(2) \quad .... \quad \mu(N)] \in \mathbb{R}^{m \times N}$ is the matrix of realizations $\mu(j) \in \mathbb{R}^p$, $j \in [1, N]$ and $\bar{\mu}$ is the ensemble mean defined as:

$$\bar{\mu} = \frac{1}{N} \sum_{u=1}^{N} \mu(u) \tag{31}$$

$\mathrm{A}(u) = \mu(u) - \bar{\mu}$ denotes the anomalies. Using the discrete forms of the process and measurement equations, the recursive algorithm is given below:

<u>Step (1)</u>: The forecast ensemble is obtained from last analyzed ensemble $\Sigma_{i|i}$ as:

$$\Sigma_{i+1|i} = \Sigma_{i|i} + \Theta_{i+1} \tag{32}$$

The forecast mean ($\bar{\mu}_{i+1|i}$) is computed via equation (31). Here the *j*-th column of the matrix Θ_{i+1} contains the *j*-th realization of Wiener vector increment $\Delta\xi_{i+1}(j) = \xi_{i+1}(j) - \xi_i(j)$. Now forecast anomalies are obtained as:

$$\mathrm{A}_{i+1|i} = \Sigma_{i+1|i} (I - 1_N) \tag{33}$$

$I \in \mathbb{R}^{N \times N}$ is the identity matrix and $1_N \in \mathbb{R}^{N \times N}$ is a matrix with each entry being *1/N*.

<u>Step (2)</u>: Kalman gain is computed as:

$$G_{i+1} = P_{i+1|i} H_{i+1}{}^T \left[H_{i+1} P_{i+1|i} H_{i+1}^T + R_{i+1} \right]^{-1} \tag{34}$$

where $P_{i+1|i} = \frac{1}{(N-1)} \mathrm{A}_{i+1|i}^T \mathrm{A}_{i+1|i}$ is the predicted error covariance.

Step (3): For obtaining the estimate $\hat{\mu}_{i+1}$ (mean of the filter density), the so-called deterministic EnKF strategy (Sakov and Oke, 2008) is used and the fact that $\bar{\mu}_{i+1|i} = \hat{\mu}_i$.

$$\bar{\mu}_{i+1|i+1} := \hat{\mu}_{i+1} = \hat{\mu}_i + G_{i+1}\left(Z_{i+1} - D\delta_{i+1}\right) \tag{35}$$

The filter anomalies, needed for the ensemble spread over pseudo time, are obtained as:

$$\mathrm{A}_{i+1|i+1} = \mathrm{A}_{i+1|i} - \frac{1}{2} G_{i+1} H_{i+1} \mathrm{A}_{i+1|i} \tag{36}$$

The factor 0.5 in the second term of the RHS of equation (36) produces realizations whose error covariance is the same as that via a KF-based Riccati update modulo a linear approximation in h (Sakov and Oke, 2008).Then the realizations from the filter density is obtained as:

$$\Sigma_{i+1|i+1} = \mathrm{A}_{i+1|i+1} + [\hat{\mu}_{i+1} \quad \hat{\mu}_{i+1} \quad \ldots\ldots \quad \hat{\mu}_{i+1}] \tag{37}$$

The essence of deterministic EnKF is in using the correction factor 0.5 in equation (36). This enables the correct ensemble spread without needing multiple measurement data sets.

2.3.1 *Numerical illustrations on elastography using PD-EnKF*

For the numerical experiments, a planar object, a $10cm \times 10cm$ square, with a (quasi-) uniform element distribution is considered (fig. 1). The aim is to demonstrate the efficiency of pseudo-dynamics approach complied with EnKF. The detailed explanation about the numerical experimentation setup (grid points, boundary conditions etc.) is given in (Banerjee et al., 2009). The comparisons are done between pseudo dynamic EKF (PD-EKF) with PD-EnKF. While PD-EnKF solutions (see fig. 2 for one of them) remain largely unaffected by reasonable variations in process noise covariance, this is not the case with PD-EKF and in fig. 3 the PD-EKF estimate is shown with the best tuned noise covariance matrix. In fig. 3, the results are compared via PD-EnKF and a Tikhonov-regularized Quasi-Newton method (QNM). The background value is treated as the lower bound on the reconstructed μ and it appears to play a crucial role in the quantitative accuracy of QNM-based reconstruction. For a given tolerance, drops in the measure of misfit via PD-EKF and PD-EnKF are shown fig. 4 as functions of the number of iterations. Even with an appropriately tuned noise covariance, PD-EKF takes many more iteration vis-à-vis PD-EnKF to reach the same tolerance level. Unlike PD-EKF, Riccati update of covariance is avoided in PD-EnKF and this eliminates the additional storage requirement for the error covariance matrix leading to higher computational efficiency, especially in higher system dimensions.

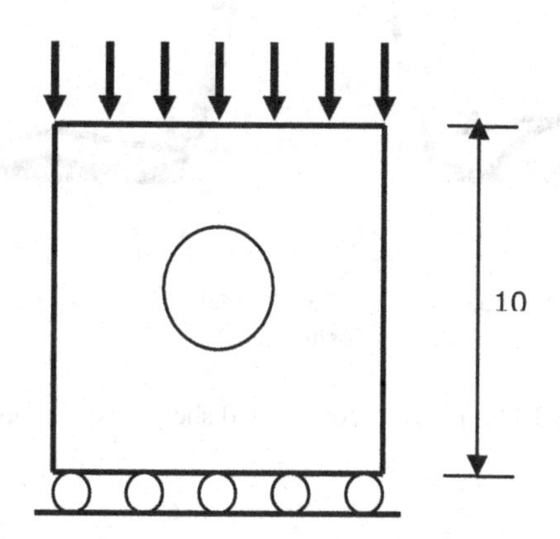

Figure 1 : Square phantom with circular inhomogeneity and loading

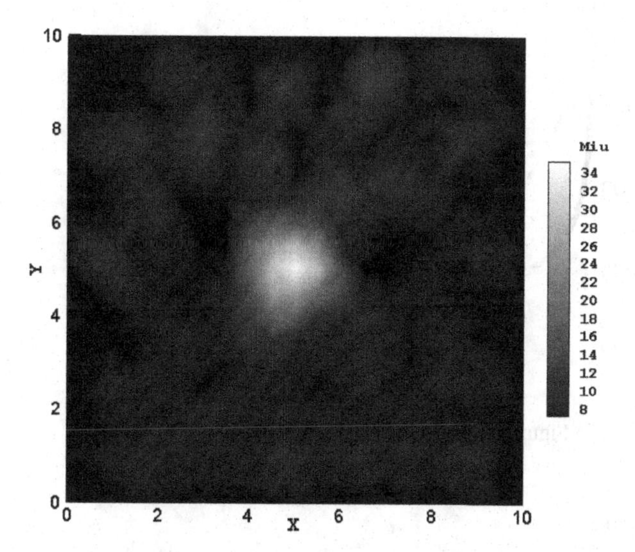

Figure 2: PD-EnKF reconstructed shear modulus from noisy axial measurement; NR = 5%

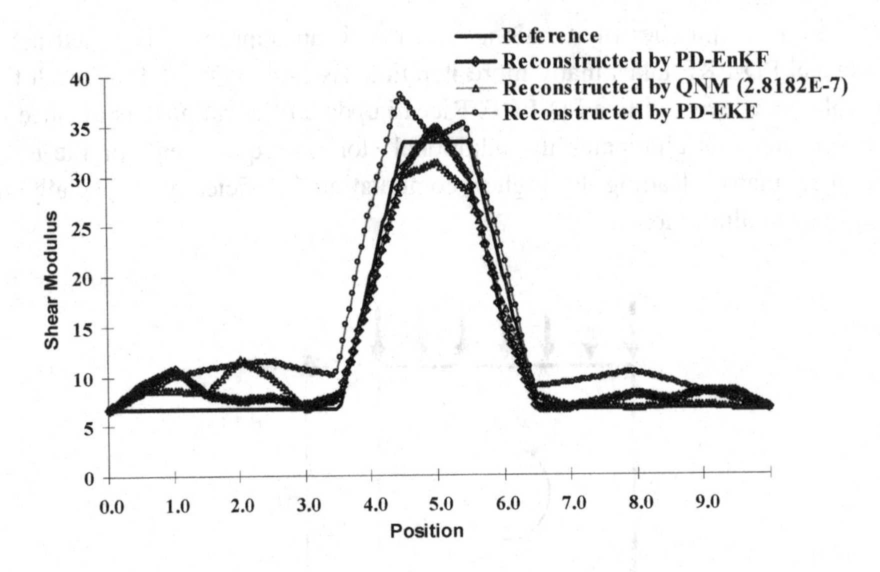

Figure 3: PD-EnKF-reconstructed shear modulus profile

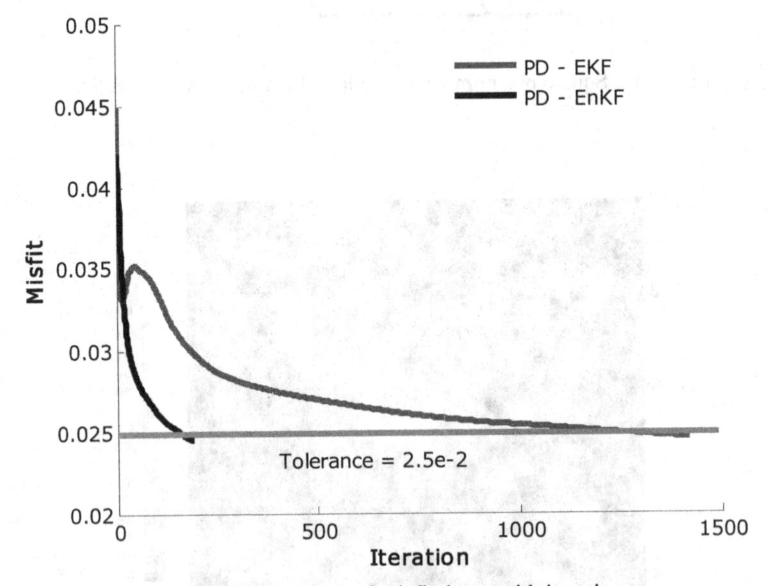

Figure 4: Comparison of misfit drops with iterations

3. Particle Filters

As the PF concept deals with conditional expectations, Bayes' theorem and Ito's integral extensively, in the following sections these topics are briefly revisited.

3.1 *Conditional expectation*

Let $(\Omega, \mathcal{F}_t, P)$ be a complete probability space in which $X:\Omega \rightarrow \mathbb{R}^n$ is a random variable such that expectation of X, $\mathrm{E}[|X|]<\infty$. Now let $\mathcal{H} \subset \mathcal{F}$ be a sub σ-algebra. Then $\mathrm{E}[X \mid \mathcal{H}]$ is defined as the conditional expectation of X under $\mathcal{H}$. $\mathrm{E}[X \mid \mathcal{H}]$ is a function mapping (almost surely) $\Omega \rightarrow \mathbb{R}$ and satisfying the following clauses:

i) $\mathrm{E}[X \mid \mathcal{H}]$ is $\mathcal{H}$-measurable.

ii) $\int_H \mathrm{E}[X \mid \mathcal{H}] dP = \int_H X dP$.

Now let $Y:\Omega \rightarrow \mathbb{R}^n$ be another random variable with $E[|Y|]<\infty$ and let $a,b \in \mathbb{R}$. Some properties of $\mathrm{E}[X \mid \mathcal{H}]$ with respect to Y is given below:

a) $\mathrm{E}[aX + bY \mid \mathcal{H}] = a\mathrm{E}[X \mid \mathcal{H}] + b\mathrm{E}[Y \mid \mathcal{H}]$

b) $\mathrm{E}[\mathrm{E}[X \mid \mathcal{H}]] = \mathrm{E}[X]$

c) If X is $\mathcal{H}$ measurable, then $\mathrm{E}[X \mid \mathcal{H}] = X$

d) If X is independent of $\mathcal{H}$, then $\mathrm{E}[X \mid \mathcal{H}] = \mathrm{E}[X]$

e) If Y is $\mathcal{H}$ measurable, then $\mathrm{E}[Y.X \mid \mathcal{H}] = Y.\mathrm{E}[X \mid \mathcal{H}]$

3.2 *Baye's formula*

As before, let $(\Omega, \mathcal{F}_t, P)$ be a complete probability space and under a new probability measure Q, P is absolutely continuous to Q (i.e. $P \ll Q$). Then for any $\mathcal{G} \subset \mathcal{F}$ and for a random variable X (under the condition that $\mathrm{E}[|X|]<\infty$), one may use Baye's formula to write (almost surely):

$$\mathrm{E}_P(X \mid \mathcal{G}) = \frac{\mathrm{E}_Q(X \frac{dP}{dQ} \mid \mathcal{G})}{\mathrm{E}_Q(\frac{dP}{dQ} \mid \mathcal{G})} \tag{38}$$

The above definition of Baye's theorem is not in a recursive form. One can derive a recursive variant of Baye's theorem which is given below. X is taken as state vector and Z is considered to be measurement vector. Now the recursive formula can be written as:

$$p\left(X_{0:i+1} \mid Z_{1:i+1}\right) = p\left(X_{0:i} \mid Z_{1:i}\right) \frac{p\left(Z_{i+1} \mid X_{i+1}\right) p\left(X_{i+1} \mid X_i\right)}{p\left(Z_{i+1} \mid Z_{1:i}\right)} \tag{39}$$

3.3 *Ito and Stratonovich integrals*

Let $X(t)$ be a diffusion vector process which is solution to an SDE, expressed in its incremental state space form as:

$$dX(t) = a(t, X(t))dt + \sum_{r=1}^{q} \sigma_r(t, X(t))dB_r(t);\ 0 \le t \le T,\ X(0) = X_0 \tag{40}$$

where $\{B_r; r \in [1,q]\}$ is the vector of q scalar Brownian motions (BM-s). The drift vector $a \in \mathbb{R}^n$ and diffusion matrix $\sigma \in \mathbb{R}^n \times \mathbb{R}^n$ must be measurable within time interval T. $\left[\sigma_r^{(j)}(s, X(s)) \mid j \in [1,n], r \in [1,q]\right]$ can be approximated by $\sum_{i=1}^{p} \sigma_r^{(j)}(t_i^*, X(t_i^*))\chi(T_i)$, ($\chi$ is the simple or indicator function over the sub-interval $T_i = (t_{i-1}, t_i]$ and the time-interval of interest $(0,T]$ is divided into p sub-intervals such that $0 = t_1 < t_2 \cdots < t_i < \cdots < t_p = T$) and $t_i^* \in (t_{i-1}, t_i]$. $\sum_{i=1}^{p} \sigma_r^{(j)}(t_i^*, X(t_i^*))\left[B_r(t_i) - B_r(t_{i-1})\right]$ approaches $\int_0^t \sigma_r^{(j)}(s, X(s))dB_r(s)$ in $L^2(P)$ as $p \to \infty$. This approximation of the above integral converges to different values for different t_i^*. If the integral is approximated based on $t_i^* = t_{i-1}$, then it leads to Ito integral. On the other hand, if the approximation is based on the mid-point $t_i^* = (t_{i-1} + t_i)/2$, then it is referred to as Stratonovich integral. Ito integral is a martingale because it does not depend on events occurring after time t_{i-1}. It satisfies $E\left[\int_0^t \sigma_r^{(j)}(s, X)dB_r(s)\right] = 0$ and the Ito isometry:

$$E\left[\left(\int_0^t \sigma_r^{(j)}(s, X)dB_r(s)\right)^2\right] = E\left[\int_0^t \left(\sigma_r^{(j)}\right)^2 (s, X)ds\right] \tag{41}$$

If a scalar function $f(t, X)$ is twice continuously differentiable in $\{t, X\} \in [0, \infty) \times \mathbb{R}^n$ (where the vector process $X := \left\{X^{(j)} \mid j \in [1,n]\right\}$ is governed by the SDE in (40)), then the following incremental equation can be written for $Y(t) = f(t, X)$ using the chain rule of differentiation:

$$dY(t) = \frac{\partial[f(t,X)]}{\partial t}dt + \sum_{j=1}^{n} \frac{\partial[f(t,X)]}{\partial X^{(j)}} dX(t) + \sum_{j,k=1}^{n} \frac{1}{2} \frac{\partial^2[f(t,X)]}{\partial X^{(j)} \partial X^{(k)}} \left(dX^{(j)}\right)\left(dX^{(k)}\right) \tag{42}$$

Using the quadratic co-variation of a BM, we get (δ_{rl} is the Kronecker delta):

$$\begin{aligned}\left(dX^{(j)}\right)\left(dX^{(k)}\right) &= \sigma_r^{(j)}(t,X)\sigma_l^{(k)}(t,X)\left(dB_r^{(j)}\right)\left(dB_l^{(k)}\right) \\ &= \sigma_r^{(j)}(t,X)\sigma_l^{(k)}(t,X)\delta_{rl}\ dt\end{aligned} \tag{43}$$

Ito's formula can be derived from Eq. (42), by appropriately replacing its quadratic co-variance terms from Eq. (43). Then $f(t,X)$ is also an Ito process whose SDE may be written as:

$$dY(t) = \sum_{r=1}^{q} \Lambda_r f(t,X) dW_r(s) + Lf(t,X)ds \tag{44}$$

Here the operators Λ_r and L are given by:

$$\Lambda_r f(X,t) = \sum_{r=1}^{q}\sum_{j=1}^{n} \sigma_r^{(j)}(X,t)\frac{\partial f(X,t)}{\partial X^{(j)}} \tag{45}$$

$$\begin{aligned}Lf(X,t) &= \frac{\partial f(X,t)}{\partial t} + \sum_{j=1}^{n} a^{(j)}(X,t)\frac{\partial f(X,t)}{\partial X^{(j)}} \\ &+ \frac{1}{2}\sum_{r=1}^{q}\sum_{j,k=1}^{n}\left\{\sigma_r^{(j)}(X,t)\sigma_r^{(k)}(X,t)\frac{\partial^2 f(X,t)}{\partial X^{(j)}\partial X^{(k)}}\right\}\end{aligned} \tag{46}$$

Ito's formula applied to $\Lambda_r f(X,t)$ and $Lf(X,t)$ in Eq. (44) yields:

$$\begin{aligned}f(X(t_i),t_i) &= f(X(t_{i-1}),t_{i-1}) + \sum_{r=1}^{q}\Lambda_r f(X(t_{i-1}),t_{i-1})(B_r(t_i) - B_r(t_{i-1})) \\ &+ Lf(X_{i-1},t_{i-1})h + \sum_{r,l=1}^{q}\int_{t_{i-1}}^{t_i}\int_{t_{i-1}}^{s}\Lambda_l\Lambda_r f(X,t)dB_l(s_2)dB_r(s_1) \\ &+ \sum_{r=1}^{q}\int_{t_{i-1}}^{t_i}\int_{t_{i-1}}^{s} L\Lambda_r f(X,t)ds_2\, dB_r(s_1) + \int_{t_{i-1}}^{t_i}\int_{t_{i-1}}^{s} L^2 f(X,t)ds_2\, ds_1 \\ &+ \sum_{r=1}^{q}\int_{t_{i-1}}^{t_i}\int_{t_{i-1}}^{s}\Lambda_r Lf(X,t)dB_r(s_2)ds_1\end{aligned} \tag{47}$$

The above expansion involves *multiple stochastic integrals* (MSIs). An MSI has the typical form:

$$I_{j_1,j_2,\cdots,j_k} = \int_{t_i}^{t_{i+1}}\int_{t_i}^{s_{k-1}}\cdots\int_{t_i}^{s_1} dB_{j_k}(s)dB_{j_{k-1}}(s_1)\cdots dB_{j_1}s_{k-1} \tag{48}$$

Here $j_1, j_2, \ldots, j_k$ are integers taking values in the set $\{0,1,\ldots,q\}$ and $I_{j_1,j_2,\cdots,j_k}$ is referred to as the k-th (Ito) MSI. Moreover, $dB_0(s)$ is taken to indicate ds.

3.4 *Kushner-Stratonovich equation*

A nonlinear SDE can be typically written in the incremental form as given below.

$$dY_t = h(X_t)dt + K(t)dB(t) \tag{49}$$

Now, assuming that $K^{-1}(t)$ exists, this can be rewritten as

$$d\overline{Y}_t = K^{-1}(t)h(X_t)dt + dB(t) \tag{50}$$

Let f be a bounded C^2 function such that all its derivatives are also bounded. In stochastic filtering, the optimal filtered estimate $\pi_t(f) = E_P(f(X_t) \mid F_t^Y)$ are computed using the observation filtration F_t^Y (presently assuming that $Y(t)$ denotes the observation process). For computational convenience, $\pi_t(f)$ originally under a measure P is expressed explicitly in a convenient reference measure Q, where Q is chosen in such a way that X_t and F_t^Y are statistically independent under Q. Now $\pi_t(f)$ can be represented under Q as

$$\pi_t(f) = E_P(f(X_t) \mid F_t^Y) = \frac{E_Q(f(X_t)\Lambda_t \mid F_t^Y)}{E_Q(\Lambda_t \mid F_t^Y)} = \frac{\sigma_t}{\sigma_1} \tag{51}$$

By Ito's formula applied to $\sigma_t(f)$, one obtains Zakai equation:

$$\sigma_t(f) = \sigma_0(f) + \int_0^t \sigma_s(L_s^0 f)ds + \int_0^t \sigma_s(K^{-1}(s)h_s f)^* d\overline{Y}_s \tag{52}$$

As a corollary to Zakai equation, Kushner-Stratonovich equation for the filtering distribution can also be derived. The final form of Kushner-Stratonovich equation is given by:

$$\pi_t(f) = \pi_0(f) + \int_0^t \pi_s(L_s^0 f)ds +$$

$$\int_0^t \{\pi_s(K^{-1}(s)h_s f) - \pi_s(f)\pi_s(K^{-1}(s)h_s f)\}^* \{d\overline{Y}_s - \pi_s(f)\pi_s(K^{-1}(s)h_s f)ds\} \tag{53}$$

Terms like $\pi_s(K^{-1}(s)h_s f)$ in Kushner-Stratonovich equation cannot be mapped to $\pi_t(f)$ for a general functional form of f and this leads to the well known 'closure' problem. While by choosing a suitable class of functions $\{f_i\}$ for f, those terms can be mapped to $\pi_t(f_i)$, but unfortunately any finite dimensional representation of such a mapping may not exist in most cases. This 'closure' problem makes the K-S equation infeasible to be applied directly in deriving numerical schemes for stochastic filtering. Nevertheless, in the linear case, K-S equation actually boils down to Kalman- Bucy equation which is widely used.

3.5 *Euler approximation*

Because of the infeasibility in directly dealing with Kushner-Stratonovich equation, nonlinear stochastic integrals are generally approximated by their time discretized counterpart by using Euler approximations. Fortunately the error that comes from Euler approximation can be shown to be bounded in some sense. More specifically, convergence of such discretization is shown in (Gobet et al., 2006) on a generalized nonlinear stochastic integral (Zakai equation). Initially, the global time interval $[0,T]$ is discretized as $t_i = i\Delta t,\ i = 0,....,n$. Here $\Delta t = T/n$ is the uniform time step. For notational convenience, the following definition is introduced: $\phi(t) = \sup\{t_i : t_i \le t\}$. Now any nonlinear stochastic partial differential equation (SPDE) corresponding to a measurable process V can be represented in a weak sense as:

$$\langle V_t, f\rangle = \langle \mu_0, f\rangle + \int_0^t \langle V_s, Lf\rangle ds + \int_0^t \langle V_s, \mathcal{H} f + \gamma^T \nabla f\rangle dW_s \tag{54a}$$

where $f\ (\in C_b^2(\mathbb{R}^d))$ is a class of test functions. L is the second order differential operator. $\langle . \rangle$ denotes the inner product in $L^2(P)$ sense. W is a q dimensional Brownian motion, γ is a $d \times q$ dimensional matrix (providing a coupling of process and observation models) and $\mathcal{H}$ is a q dimensional vector. μ_0 is the initial probability measure. Solution to such generalized nonlinear SPDE of V can be characterized as (Gobet and Munos, 2005):

$$< V_t^{\Delta t}, f >= E_W[f(X_t^{\Delta t})\Lambda_t^{\Delta t}] \tag{54b}$$

$$X_t^{\Delta t} = X_0 + \int_0^t \beta(X_{\phi(s)}^{\Delta t})ds + \int_0^t \sigma(X_{\phi(s)}^{\Delta t})dB_s + \int_0^t \gamma(X_{\phi(s)}^{\Delta t})dW_s \tag{55}$$

$$\Lambda_t^{\Delta t} = \exp\left(\int_0^t \mathcal{H}(X_{\phi(s)}^{\Delta t})dW_s - \frac{1}{2}\int_0^t |\mathcal{H}(X_{\phi(s)}^{\Delta t})|^2 ds\right) \tag{56}$$

Here $X_t^{\Delta t}$ is an approximation to the exact solution X_t (*d*-dimensional signal) with initial condition taken as X_0. $B \in \mathbb{R}^d$ is a Brownian motion independent of $W\ (\gamma \neq 0)$, which essentially represents the observation process. E_W is the conditional expectation under the filtration generated by $W\ (\mathcal{F}_t^W)$. Observation equation is taken as:

$$W_t = \int_0^t \mathcal{H}(X_s)ds + B_m(t) \tag{57}$$

$B_m(t)$ is a Brownian motion independent of B. Now nonlinear filtering problem is basically the estimation of distribution of $X(t)$ given $\mathcal{F}_t^W$. The measure valued process π_t is first transformed from original measure P to a new measure Q under which B is independent of W.

$$\langle \pi_t, f \rangle = E^P[f(X_t) \mid \mathcal{F}_t^W] \tag{58a}$$

By Kallianpur-Striebel formula, Eq. (58a) can be written as:

$$\langle \pi_t, f \rangle = \frac{\langle V_t, f \rangle}{\langle V_t, 1 \rangle} \tag{58b}$$

Hence the difficulty in getting a solution of the nonlinear filtering problem lies in estimating $\langle V_t, f \rangle$. Now, for notational convenience, the following abbreviations are made.

$$X_t^{\Delta t} = \bar{X},\ V_t^{\Delta t} = \bar{V},\ \Delta\bar{B}_{t_i} = B_{t_i} - B_{t_{i-1}},\ \Delta\bar{W}_{t_i} = W_{t_i} - W_{t_{i-1}}$$

Euler approximation at the time instant t_i can be written as:

$$\bar{X}_{i+1} = \bar{X}_i + \beta(\bar{X}_i)\Delta t + \sigma(\bar{X}_i)\Delta\bar{B}_{i+1} + \gamma(\bar{X}_i)\Delta\bar{W}_{i+1} \tag{59a}$$

Now $\langle V_t^{\Delta t}, f \rangle$ is also approximated through Euler discretized format as:

$$\langle V_t^{\Delta t}, f \rangle = E_W[f(\bar{X}_k)\exp(\sum_{j=0}^{i-1} \mathcal{H}(\bar{X}_j)\Delta\bar{W}_{j+1} - \frac{1}{2}|\mathcal{H}(\bar{X}_j)|^2 \Delta t)] \tag{59b}$$

Function f is defined under a unit ball topology ($BL_1(\mathbb{R}^d)$) ($\{f : \mathbb{R}^d \to \mathbb{R}\}$). It is assumed that f satisfies the following requirement:

$$|f(x)| \le 1 \cap |f(x) - f(y)| \le |x - y| \,\forall x, y$$

Now on the function space, $\mathcal{M}(\mathbb{R}^d)$ the following new metric is defined:

$$\rho(V_1, V_2) = \sup\{|\langle V_1, f \rangle - \langle V_2, f \rangle|,\ f \in BL_1(\mathbb{R}^d),\ V_1, V_2 \in \mathcal{M}(\mathbb{R}^d)\} \tag{60a}$$

For simplicity it is assumed that the coefficients are smooth and that they satisfy the ellipticity condition. β, $\mathcal{H}$, σ and γ are C^∞ functions and their derivatives are bounded for some $\sigma\sigma^T > \varepsilon$, for $\varepsilon > 0$. Then it can be shown that (Gobet et al., 2006):

$$\| \rho(V_T, V_T^{\Delta t}) \|_2 \le C\Delta t + C\sqrt{\Delta t} \tag{60b}$$

for some constant $C > 0$. Given these convergence characteristics, such time-discrete approximations can be incorporated within the nonlinear filtering solution strategy.

3.6 *Dynamic SSI using particle filters*

The potential of particle filters can be demonstrated by considering the state estimation of an n-degrees-of-freedom (DOF) non-linear oscillator represented (formally) by the following SDE:

$$[M]\ddot{X} + C\dot{X} + KX + \alpha\left(\bar{X},t\right) = \sum_{r=1}^{q} \sigma_r(t)\dot{W}_r + F(t) \tag{61}$$

Here $\bar{X} := \left\{X^T, \dot{X}^T\right\}^T \in \mathbb{R}^{2n}$ and $X := X_1 = \left\{X_1^{(j)}\right\} \in \mathbb{R}^n$, $\dot{X} := X_2 = \left\{X_2^{(j)}\right\} \in \mathbb{R}^n$ ($j \in [1,n]$) are respectively the displacement and velocity vectors. Equation (61) is subject to the initial condition $\bar{X}(t=0) := \bar{X}_0 = \left\{\left\{X_{10}^{(j)}\right\}^T, \left\{X_{20}^{(j)}\right\}^T\right\}^T$. $[M]$, $[C]$, $[K] \in \mathbb{R}^{n \times n}$ are constant mass, damping and stiffness matrices respectively, $\alpha\left(\bar{X},t\right)$ is a vector function (not necessarily smooth) that ensures a unique solution of Eqn. (61) at least in the weak sense, $B_r(t) \in \mathbb{R}^n$ is the rth element of the set of (additive) diffusion vectors, $\left\{W_r(t) \mid r \in [1,q]\right\}$ is the set of standard (zero-mean) scalar Wiener processes and $F(t) = \left\{F^{(j)}(t)\right\}$ is a deterministic force vector. Eq (61) may be recast as a system of $2n$ first order SDEs in the following incremental form:

$$\begin{aligned} dX_1^{(j)} &= X_2^{(j)} dt \\ dX_2^{(j)} &= a^{(j)}(X,\dot{X},t)dt + \sum_{r=1}^{q} g_r^{(j)}(t) dW_r(t), \quad j = 1,\dots,n \end{aligned} \tag{62}$$

where, $a^{(j)}\left(X,\dot{X},t\right) = \underbrace{-\sum_{k=1}^{n} \hat{C}_{jk}\dot{X}^{(k)} - \sum_{k=1}^{n} \hat{K}_{jk}X^{(k)}}_{a_l^{(j)}} \underbrace{-\hat{\alpha}^{(j)}\left(X,\dot{X},t\right) + \hat{F}^{(j)}(t)}_{a_{nl}^{(j)}}$

$$\begin{aligned} &\left[\hat{C}\right] = \left[M^{-1}\right][C], \left[\hat{K}\right] = \left[M^{-1}\right][K], \left\{\hat{F}\right\} = \left[M^{-1}\right]\{F\}, \\ &\{\hat{\alpha}\} = \left[M^{-1}\right]\{\alpha\}, G_r(t) = \left\{g_r^{(j)}(t)\right\} = \left[M^{-1}\right]\sigma_r(t) \end{aligned} \tag{63}$$

Let the solution $\bar{X}(t)$ be a stochastic process defined with respect to the triplet $(\Omega, \mathcal{F}_t, \mathbb{P})$. The drift vector $A = \{a^{(j)} \mid j = 1,\dots,n\}$ is decomposed into two constituent parts as $a^{(j)} = a_l^{(j)} + a_{nl}^{(j)}$ (*i.e.*, $A = A_l + A_{nl}$), where A_l denotes the linear part and A_{nl} the nonlinear part of the vector field. To ensure boundedness of $\bar{X}(t)$ almost surely and uniqueness in a weak sense, the drift and diffusion vectors $A = \left\{a^{(j)}\right\}$ and $G_r = \left\{g_r^{(j)}\right\}$ are assumed to be measurable and the initial conditions are assumed to be mean square bounded, i.e., $E\left[\left\|\bar{X}(t_0)\right\|^2\right] < \infty$ (Kallianpur, 1980). The time interval $[0,T]$ of interest is ordered (discretized) as $0 = t_0 < t_1 < \cdots < t_i < \cdots < t_L = T$, $h_i = t_i - t_{i-1}$ and $T_i = (t_{i-1}, t_i]$; $i \in \mathbb{Z}^+$. Without a loss of generality, $h_i = h \ \forall i$ can be assumed to be uniform. The measurement equation is taken as:

$$Z = \mathcal{H}(\bar{X}, t) + \eta(t) \tag{64}$$

where $Z \in \mathbb{R}^{nz}$ denotes the vector of measurements and $\eta(t) \in \mathbb{R}^{nz}$ refers to Gaussian (Wiener) measurement noise.

In addition to states, all the parameters, i.e., those defining the damping (C) and stiffness (K) matrices and the nonlinear function χ can also be estimated by defining the damping, stiffness and the 'nonlinearity' parameters as additional states, denoted respectively by $\breve{X}_1 \in \mathbb{R}^{n_1}$, $\breve{X}_2 \in \mathbb{R}^{n_2}$ and $\breve{X}_3 \in \mathbb{R}^{n_3}$ in a combined state-parameter estimation problem. Let their scalar components be given by $\{\breve{X}_i^{(p)} \mid p = 1, 2, \ldots, n_i\}$ for $i = 1, 2, 3, \ldots$. Let $\bar{\bar{X}} \stackrel{\Delta}{=} \left\{\breve{X}_1^T, \breve{X}_2^T, \breve{X}_3^T\right\}^T \in \mathbb{R}^{n_1+n_2+n_3}$ denote the entire parameter vector. Including parameters as additional states and assuming $\breve{n} = n_1 + n_2 + n_3$, the parameter-augmented state vector is defined as $\tilde{X} := \left\{\{\bar{X}\}^T, \{\bar{\bar{X}}\}^T\right\}^T \in \mathbb{R}^{2n+\breve{n}}$. Within this framework, the damping and stiffness matrices, $C\left(\breve{X}_1\right)$ and $K\left(\breve{X}_2\right)$, are functions of states. Following the standard procedure, the parameter states $\breve{X}_1$, $\breve{X}_2$ and $\breve{X}_3$ are assumed to evolve according the following zero-drift (persistent process) SDEs:

$$\left.\begin{aligned} d\breve{X}_1 &= \breve{\sigma}_1 d\breve{W}_1(t) \\ d\breve{X}_2 &= \breve{\sigma}_2 d\breve{W}_2(t) \\ d\breve{X}_3 &= \breve{\sigma}_3 d\breve{W}_3(t) \end{aligned}\right\} \tag{65}$$

$\breve{\sigma}_1, \breve{\sigma}_2$ and $\breve{\sigma}_3$ are (square) diffusion matrices defining artificial evolutions of the parameter states as local martingales. $\breve{W} := \{\breve{W}_i, i = 1, 2, 3\}$ are zero-mean Gaussian white noise vector processes (of appropriate dimensions) uncorrelated with W_r. Temporal (artificial) evolutions of the elements of $\bar{\bar{X}}$ convert the problem of parameter estimation as one of state estimation only. But, such a modification makes even the parameter estimation problem of a linear mechanical system to one of non-linear filtering.

3.7 *Bootstrap filter (BS)*

Bootstrap filter (BS) is one of the basic particle filters. The parameter-augmented state variables $\tilde{X} := \{\overline{X}^T, \overline{\overline{X}}^T\}^T$ are replaced by X for notational convenience. It is recalled that the PF aims at recursively obtaining the filtering density $p(X_i \mid Z_{1:i})$ (or the associated marginal pdf-s), where $Z_{1:i} := \{Z_1, \ldots, Z_i\}^T$ denotes the observation vector up to time t_i. The conditional mean (estimate) $\hat{X}_i$ and conditional variance Σ_i are obtained as:

$$\hat{X}_i = E[X_i \mid Z_{1:i}] = \int X_i p(X_i \mid Z_{1:i}) dX_i \tag{66a}$$

$$\begin{aligned}\Sigma_i &= E\left[(X_i - \hat{X}_i)(X_i - \hat{X}_i)^T\right] \\ &= \int (X_i - \hat{X}_i)(X_i - \hat{X}_i)^T p(X_i \mid Z_{1:i}) dX_i\end{aligned} \tag{66b}$$

For obtaining the estimates recursively, the PDF $p(X_i \mid Z_{1:i})$ conforming to the current state is constructed using the last available filtering PDF $p(X_{i-1} \mid Z_{1:i-1})$, the process SDE (62) and the current measurement data Z_i. In practice, to get $p(X_i \mid Z_{1:i})$ recursively, first the prediction density (i.e., forward projection of particles over $(t_{i-1}, t_i])$ is found using a transition kernel (obtainable through the solution of the process SDE) and then it is updated using current measurement Z_i. Given a set of i.i.d. particles $\{X_{i-1,(u)} : u \in [1, N]\}$ sampled from $p(X_{i-1} \mid Z_{1:i-1})$, the bootstrap filter carries over and updates these particles to obtain a new set of i.i.d. particles $\{X_{i,(u)} : u \in [1, N]\}$, which are distributed according to $p(X_i \mid Z_{1:i})$. The bracketed subscript u denotes the uth particle of the random variable X_i (or X_{i-1}, as applicable). Since a closed-form analytical solution is difficult to find, the locally linearized process SDEs are integrated semi-analytically using $\{X_{i-1,(u)} : u \in [1, N]\}$ as the (ensemble of) initial conditions and thus obtain the (ensemble of) predicted particles $\{X^*_{i,(u)}\}$. Let the corresponding kernel function be denoted by $\mathbb{K}(X_i \mid X_{i-1})$. A set of weights $\omega_{i,(u)}$

calculated from measurement equation, are assigned to the predicted particles so as to account for the currently available observation Z_i. The (normalized) importance weights $\omega_{i,(u)}$ are proportional to the ratios of the true posterior and the prediction densities. These weights are well defined if the two densities are absolutely continuous with respect to each other and they may be interpreted as the Radon-Nikodym derivatives of the associated changes of measures. They are normalized such that $\sum_{u=1}^{N} \omega_{i,(u)} = 1$. Thus the prediction density at t_i may be expressed as (Ristic *et al.*, 2004; Doucet, 1998):

$$q(X_i \mid Z_{1:i-1}) = \int p(X_{i-1} \mid Z_{1:i-1}) \mathbb{K}(X_i \mid X_{i-1}) dX_{i-1} \tag{67}$$

The posterior density $p(X_i \mid Z_{1:i})$ is then given by:

$$p(X_i \mid Z_{1:i}) = \frac{\mathbb{G}(Z_i \mid X_i) q(X_i \mid Z_{1:i-1})}{\int \mathbb{G}(Z_i \mid X_i) q(X_i \mid Z_{1:i-1}) dX_i} \tag{68}$$

where $\mathbb{G}(Z_i \mid X_i)$ is the likelihood of the observation Z_i given the particle X_i. Thus we have the following proportionality of the (unnormalized) weights and $\mathbb{G}$:

$$\tilde{\omega}_{i,(u)} \propto \mathbb{G}(Z_i \mid X^*_{i,(u)}) \tag{69}$$

Within the Monte Carlo setup required by a PF, the filtering distribution at $t = t_{i-1}$ is approximately represented through the following empirical measure:

$$\mathcal{P}(dX_{i-1} \mid Z_{1:i-1}) \approx \mathcal{P}^{(N)}(dX_{i-1} \mid Z_{1:i-1}) = (1/N) \sum_{j=1}^{N} \delta_{X_{i-1}^{(j)}}(dX_{i-1}) \tag{70}$$

where $\delta_X(dX)$ denotes the Dirac delta measure located at X. Accordingly, the prediction and posterior distributions are respectively given by:

$$\mathcal{Q}(dX_i \mid Z_{1:i-1}) \approx \mathcal{Q}^{(N)}(dX_i \mid Z_{1:i-1}) = (1/N) \sum_{u=1}^{N} \delta_{X^*_{i,(u)}}(dX_i) \tag{71a}$$

where $X^*_{i,(u)} = X^*_{i,(u)}(X_{i-1,(u)}, h)$ is the locally linearized solution of the process SDEs and

$$\mathcal{P}(dX_i \mid Z_{1:i}) \approx \mathcal{P}^{(N)}(dX_i \mid Z_{1:i}) = \frac{\mathbb{G}(Y_i \mid X_i) \mathcal{Q}^{(N)}(dX_i \mid Z_{1:i-1})}{\int \mathbb{G}(Y_i \mid X_i) \mathcal{Q}^{(N)}(dX_i \mid Z_{1:i-1})}$$

$$= \sum_{u=1}^{N} \omega_{i,(u)} \delta_{X^*_{i,(u)}}(dX_i) \tag{71b}$$

such that $\sum_{u=1}^{N} \omega_{i,(u)} = 1$. After resampling, the unweighted empirical distribution is obtained as $\mathcal{P}^{(N)}(dX_i \mid Z_{1:i}) = (1/N)\sum_{u=1}^{N} \delta_{X_{i,(u)}}(dX_i)$. Note that, using the form of the observation Eq. (57), one may readily obtain $\mathbb{G}(Z_i \mid X_i)$ as $\mathbb{G}(Z_i \mid X_i) = r_{\eta_i}(Z_i - \mathcal{H}(X_i, t_i))$, where r_{η_i} is the zero-mean Gaussian density of the vector Wiener process $\eta(t)$ at $t = t_i$. Precise computations of $q(X_i \mid Z_{1:i-1})$ (or the kernel $\mathbb{K}(X_i \mid X_{i-1})$) and hence the weights $\tilde{\omega}_{i,u}$ remain intractable as the time-stepping algorithm for the nonlinear SDEs (62) is almost always approximate. Hence approximate computations of $\tilde{\omega}_{i,u}$ must be either via a linearization of Eq. (62) or based on a discrete time stepping scheme through truncated stochastic Taylor expansions.

3.8 *Semi-analytical particle filter* (*SAPF*)

Semi-analytical particle filter (Sajeeb *et al.*, 2010) makes use of local linearizations based on explicit Ito-Taylor expansions to transform the given nonlinear system/observation equations to an ensemble of linearized systems so that their conditionally Gaussian posterior pdf may be obtained analytically (e.g. through a Kalman filter), thus assimilating the most recent observation. Utilizing this information within the framework of a conventional particle filter, samples (particles) from the true posterior pdf can be generated. The advantage comes in the form of a significant reduction in the sample variance of the estimate. As before, we represent the (marginal) filtering density by $p(X_{i+1} \mid Z_{1:i+1})$, which is the pdf of the discretized state vector at t_{i+1} conditioned on the observations available up to the same time instant. The conditional mean $\hat{X}_{i+1}$ and conditional variance Σ_{i+1} may be obtained as:

$$\hat{X}_{i+1} = E[X_{i+1} \mid Z_{1:i+1}] = \int X_{i+1} p(X_{i+1} \mid Z_{1:i+1}) dX_{i+1} \tag{72a}$$

$$\begin{aligned} \Sigma_{i+1} &= E[(X_{i+1} - \hat{X}_{i+1})(X_{i+1} - \hat{X}_{i+1})^T] \\ &= \int (X_{i+1} - \hat{X}_{i+1})(x_{i+1} - \hat{x}_{i+1})^T p(X_{i+1} \mid Z_{1:i+1}) dX_{i+1} \end{aligned} \tag{72b}$$

When the process and measurement equations are linear and the noises are additively Gaussian, the KF provides the exact solution. In the nonlinear case, the PF provides an optimal solution to the problem through MC simulation strategies. Recall that the conditional pdf of the states is represented here through a set of random particles (instantaneous realizations of states) drawn according to a prior pdf. Each particle is assigned a weight, which is a scalar ratio (Radon-Nikodym derivative) of the true posterior to the prior PDFs computed at the particle location. Consider, for instance, that the posterior PDF $p(X_{i+1} | Z_{1:i+1})$ is represented by an ensemble of pairs $\left\{X_{i+1}^{*(j)}, \omega_{i+1}^{(j)}\right\}_{j=1}^{N}$, where $X_{i+1}^{*(j)}$ represents a point in the multi dimensional state space and $\omega_{i+1}^{(j)}$, the normalized weight given by $\omega_{i+1}^{(j)} = \tilde{\omega}_{i+1}^{(j)} \Big/ \sum_{u=1}^{N} \tilde{\omega}_{i+1}^{(u)}$. One readily observes that the unnormalized weight $\tilde{\omega}_{i+1}^{(j)}$ is proportional to $p(X_{i+1}^{*(j)} | Z_{1:i+1})$. The posterior density at t_{i+1} may be expressed as:

$$\begin{aligned} p(X_{i+1} | Z_{1:i+1}) &= \int p(X_{i+1}, X_i | Z_{1:i+1}) dX_i \\ &= \int p(X_{i+1} | X_i, Z_{1:i+1}) p(X_i | Z_{1:i+1}) dX_i \end{aligned} \tag{73}$$

Noting that $p(X_i | Z_{1:i+1}) = \dfrac{p(X_i, Z_{1:i+1})}{p(Z_{1:i+1})} = \dfrac{p(Z_{i+1} | X_i, Z_{1:i}) p(X_i | Z_{1:i})}{p(Z_{i+1} | Z_{1:i})}$ and $p(Z_{i+1} | X_i, Z_{1:i}) = p(Z_{i+1} | X_i)$, Eq. (73) may be rewritten as

$$p(X_{i+1} | Z_{1:i+1}) = \int p(X_{i+1} | X_i, Z_{1:i+1}) \frac{p(Z_{i+1} | X_i) p(X_i | Z_{1:i})}{p(Z_{i+1} | Z_{1:i})} dX_i \tag{74}$$

i.e.

$$p(X_{i+1} | Z_{1:i+1}) \propto \int p(X_{i+1} | X_i, Z_{1:i+1}) p(Z_{i+1} | X_i) p(X_i | Z_{1:i}) dX_i \tag{75}$$

Thus for a particle $X_{i+1}^{*(j)}$ to represent a point in the posterior pdf $p(X_{i+1} | Z_{1:i+1})$, the unnormalized weight is given by

$$\tilde{\omega}_{i+1}^{(j)} = \int p(X_{i+1}^{*(u)} | X_i, Z_{1:i+1}) p(Z_{i+1} | X_i) p(X_i | Z_{1:i}) dX_i \tag{76}$$

It may be observed that, the conditional posterior density function $p(X^{*}_{i+1} | X_i, Z_{1:i+1})$ is a Gaussian pdf which can be obtained analytically using Kalman filter, since conditioned on X_i, the system becomes linear as a result of the local linearization. Hence the first term in the RHS of Eq. (76) can be evaluated analytically. The second term $p(Z_{i+1} | X_i)$ is also analytically deducible if one expresses the measurement equation as $Z_{i+1} = H(X_{i+1}) + v_{i+1}$. The term $p(X_i | Z_{1:i})$ represents the posterior density at the time instant t_i, and, if one has an ensemble $\left\{X_i^{(u)}\right\}_{u=1}^{N}$ from this density, $\tilde{\omega}_{i+1}^{(j)}$ may be calculated as:

$$\tilde{\omega}_{i+1}^{(j)} = \frac{1}{N} \sum_{u=1}^{N} p(X_{i+1}^{*(j)} | X_i^{(u)}, Z_{1:i+1}) p(Z_{i+1} | X_i^{(u)}) \;; j \in [1, N] \tag{77}$$

Now, the ensemble $\left\{X_{i+1}^{*(j)}\right\}_{j=1}^{N}$ along with the normalized weights $\left\{\omega_{i+1}^{(j)}\right\}_{j=1}^{N}$ represent a discrete form of the posterior density $p(X_{i+1} | Z_{1:i+1})$ and one can easily generate samples $\left\{X_{i+1}^{(j)}\right\}_{j=1}^{N}$ from this discrete distribution to estimate the conditional mean and conditional variance as:

$$\hat{X}_{i+1} \cong \frac{1}{N}\sum_{j=1}^{N} X_{i+1}^{(j)} \tag{78a}$$

$$\Sigma_{i+1} \cong \frac{1}{N}\sum_{j=1}^{N} (X_{i+1}^{(u)} - \hat{X}_{i+1})(X_{i+1}^{(j)} - \hat{X}_{i+1})^{T} \tag{78b}$$

The algorithm for implementing the semi analytical particle filter is given below:

1. Set $i=0$. Draw sample $\left\{X_{0}^{(u)}\right\}_{u=1}^{N}$ from the initial pdf $p(X_0)$ and set $i=1$.
2. Using the process equation find $\left\{X_{i+1}^{*(j)}\right\}_{j=1}^{N}$.
3. Consider $X_{i+1}^{*(j)}$. When Z_{i+1} arrives, for each of $X_{i}^{(u)}$, obtain $p(X_{i+1}^{*(j)} | X_{i}^{(u)}, Z_{1:i+1})$ using Kalman filter. Also find $p(Z_{i+1} | X_{i}^{(u)})$. Find the unnormalized weight $\tilde{\omega}_{i+1}^{(j)}$ using Eq. (77) and normalize it to obtain $\omega_{i+1}^{(j)}$. Do these computations for $j=1,2,\ldots,N$.
4. Generate sample $\left\{X_{i+1}^{(j)}\right\}_{j=1}^{N}$ from the discrete pdf defined by $\left\{X_{i+1}^{*(j)}, \omega_{i+1}^{(j)}\right\}_{j=1}^{N}$. Estimate the conditional mean and conditional variance using Eq. (78 a & b)
5. Increment $i+1$ and repeat steps 2 through 5 till the terminal time is reached. It may be emphasized here that in the linearization of the nonlinear vector fields in the process as well as measurement equations, any formal order of accuracy can be achieved, in principle, by considering adequate number of terms in the Ito-Taylor expansions. Also, the local linearization considerably simplifies the procedure to discretize the governing system SDEs.

3.8.1 *Numerical examples*

The performance of SAPF is compared with a conventional SIS filter on a Duffing oscillator (Sajeeb *et al.*, 2010). Both state and parameters are estimated for this SDOF system. The results of state estimation for the Duffing oscillator are shown in fig. 5. It shows the measurement and the estimate (i.e., the mean of the marginal posterior PDF of the state) via the proposed and SIS filters. A fairly good convergence may be observed through the SAPF. Phase plane plots are shown in fig. 6. A sample trajectory of the solution of the system of SDEs, used for generating the measurement, is taken as the 'true' response. A fairly good convergence may be observed via the SAPF. A reduction in the sample variance by SAPF (fig. 7) indicates that sampling fluctuations are less with the present scheme. In Fig. 8 all the three parameters in the Duffing oscillator are estimated. Good convergence is attained as can be observed in fig. 8.

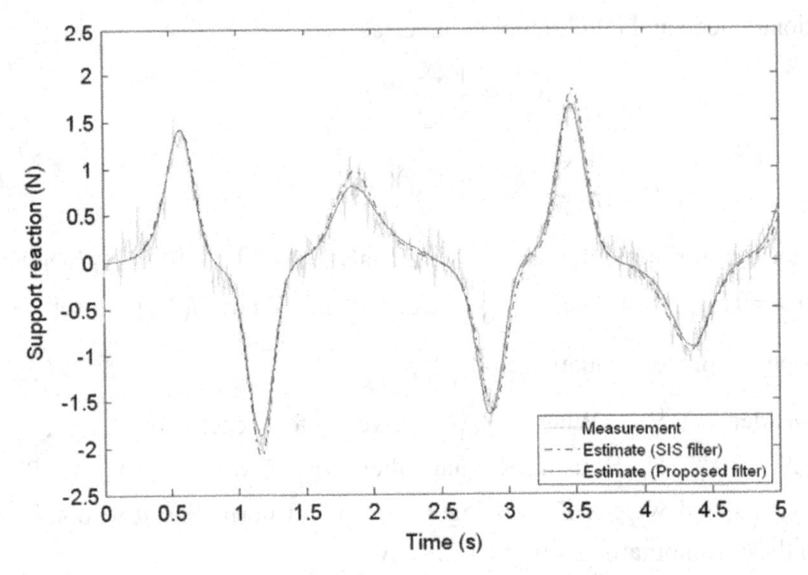

Fig. 5. State estimation of Duffing oscillator for nonlinear measurement.

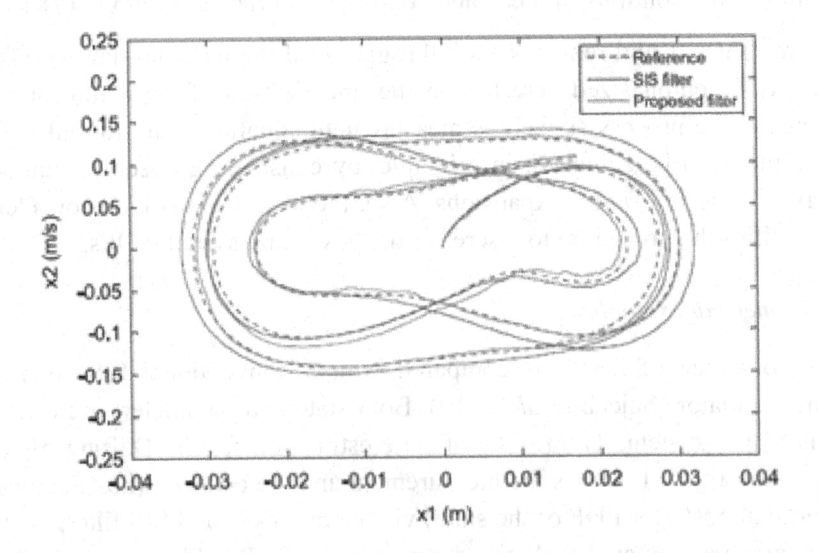

Fig. 6. Phase plane plots of the Duffing oscillator for nonlinear measurements.

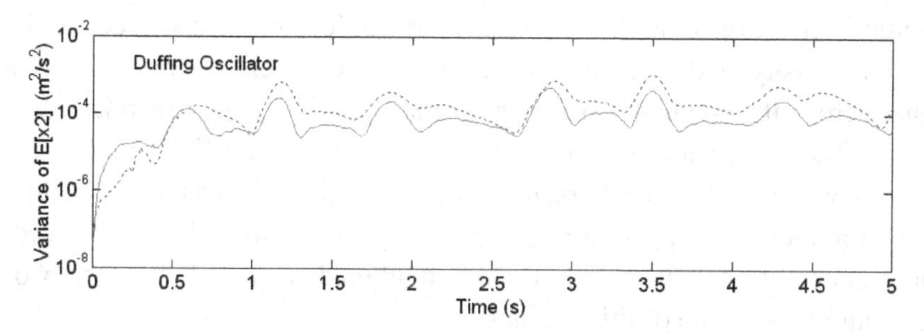

Fig. 7. Time history of the variance of the estimate of X_2 for nonlinear measurement (Red :: SIS and dotted blue :: SAPF)

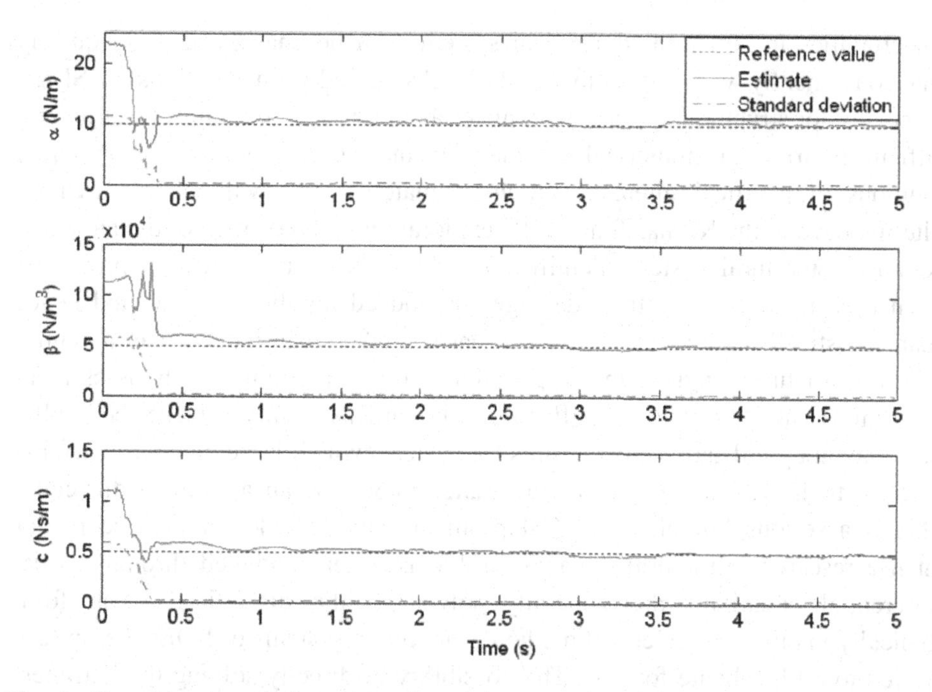

Fig. 8. Time histories of the estimated parameters of the Duffing oscillator, using SAPF (c = damping, β = nonlinearity and α = stiffness).

3.9 *Girsanov corrected particle filter*

By exploiting some of the features of the GCLM (described superficially in introduction) a given particle filtering scheme can be non-trivially improved upon so as to achieve faster convergence and reduction in sampling variance in dynamic structural system identification problems. Specifically, the nonlinear process dynamics is first locally linearized and an appropriate gain-multiplied innovation term added to the linearized dynamics by way of reconciling the prediction density with the latest observation (i.e. to ensure that the computed weights across the predicted particles in the updating stage are relatively more uniform). The linearization errors and those due to the artificial innovations are then weakly corrected via a Girsanov transformation of measures. The updating stage of producing the posterior density using the likelihood ratio derived from the measurement SDE however remains unaltered. Detailed discussion on this issue is out of the scope of this chapter.

4. Conclusions

Uncertainties in structural dynamical systems can be characterized adequately and conveniently with the help of SDEs. As closed-form solutions of SDEs, especially for structural systems with material and/or geometric non-linearity, are difficult to arrive at, numerical schemes become the only options for obtaining solutions of practical interest. But unfortunately most available research on filtering beyond the KF has found little exploration in the context of dynamic and nonlinear structural system identification. Many such identification problems, often used to detect structural damage, are indeed highly nonlinear and hence characteristically non-Gaussian. Hence particle filters can indeed play a central role in structural engineering applications (e.g. structural health assessment problems). But PF-s typically suffer from the dimensionality curse. Specifically, they may not yield satisfactory results for systems with a large number of DOF-s wherein tuning of process noise covariance may pose an additional difficulty. This is a serious limitation for SSI problems and provides enough scope for intense research. Structural dynamic problems, often modelled through partial differential equations, are generally solved in the discretized weak form (typically via the finite element method), yielding a systems of SDEs that in turn are fed to a PF scheme for SSI. The possibility of directly solving the Kushner-Stratonovich equations (which are in the stochastic partial differential form) via such finite element based discretization strategies, possibly coupled with time marching scheme, provides yet another avenue for future research. Finally, as mathematical models for damaged structures often have non-differentiable drift fields, derivative free filters may be more desirable in many such cases and this must also form part of the focus of future research.

References

Arulampalam, S., Maskell, N., Gordon, N. and Clapp, T. (2002). A tutorial on particle filters for online nonlinear/non-Gaussian Bayesian tracking, *IEEE Trans. on Sig. Processing*, 50, pp. 174-188.

Banerjee, B., Roy, D. and Vasu, R. M. (2009). A pseudo-dynamic sub-optimal filter for elastography under static loading and measurements, *Physics in Medicine and Biology*, 54, pp. 285-305.

Banerjee, B., Roy, D. and Vasu, R.M. (2011). A pseudo-dynamical systems approach to a class of inverse problems in engineering, *Proc. of the Royal Soc. A*, 465, pp. 1561-1579.

Beskos, A. and Roberts, G. O. (2005). Exact simulation of diffusions, *The Annals of Applied Probability*, 15(4), pp. 2422-2444.

Brigo, D., Mauri, G., Mercurio, F. and Sartorelli, G. (1998). The CIR++ model: A new extension of the Cox-Ingersoll-Ross model with analytical calibration to bond and option prices, Internal Report, Banca IMI, Milan.

Brown, R.G. and Hwang, P.Y.C. (1992). Introduction to random signals and applied Kalman filtering, Wiley, New York.

Ching, J., Beck, J. L. and Porter, K. A. (2006). Bayesian state and parameter estimation of uncertain dynamical systems, *Probabilistic Engrg. Mech.*, 21, pp. 81-96.

Corigliano, A. and Mariani, S. (2004). Parameter identification in explicit structural dynamics: performance of the extended Kalman filter, *Comp. Meth. Appl. Mech. Engng.*, 193, pp. 3807-3835.

Doucet, A. (1998). On sequential simulation-based methods for Bayesian filtering, Technical Report CUED/F-INFENG/TR.310(1998), Department of Electrical Engineering, University of Cambridge, UK.

Doucet, A., de Freitas, N. and Gordon, N. (2001). Sequential Monte Carlo methods in practice, Springer.

Doucet, A., Godsill, S. and Andrieu, C. (2000). On sequential Monte Carlo sampling methods for Bayesian filtering, Statistics and Computing, 10, pp. 197-208.

Elishakoff, I. and Falsone, G. (1993). Some recent developments in stochastic linearization technique, Computational Stochastic Mechanics, *Elsevier Applied Science*, London, pp. 175-194.

Evensen, G. (1994). Sequential data assimilation with a nonlinear quasi-geostrophic model usingMonte-Carlo methods to forecast error statistics, *J. Geophys.*, 99, pp. 10143-10162.

Evensen, G. (2003). The ensemble Kalman filter: Theoretical formulation and practical implementation, *Ocean Dyn.*, 53, pp. 343-367.

Evensen, G. (2004). Sampling strategies and square root analysis schemes for the EnKF, *Ocean Dyn.*, 54, pp. 539-560.

Ewins, D. J. (2000). Modal testing: theory, procedures and applications, Research studies Press.

Frishwell, M. L., Mottershead, J. E. and Ahmadian, A. (2011). Finite element model updating using experimental test data: parametrization and regularization, *Phil. Trans. of The Royal Soc. A*, 359, pp. 169-186.

Gard, T. C. (1988). Introduction to stochastic differential equations, Marcel Dekker, Inc., New York.

Ghosh, S. J., Manohar, C. S. and Roy, D. (2008). A sequential importance sampling filter with a new proposal distribution for state and parameter estimation of nonlinear dynamical systems, *Proc. R. Soc. A* 464, 25-47.

Gobet, E. and Munos, R. (2005). Sensitivity analysis using Ito-Malliavin calculus and martingales, and application to stochastic optimal control, *SIAM J. Control Optim.*, 43, pp. 1676 -1713.

Gobet, E., Pages, G., Pham, H. and Printems, J. (2006). Discretization and simulation of the Zakai equation, *SIAM J. Numer. Anal.*, 44(6), pp. 2505-2538.

Gordon, N. J., Salmond, D. J. and Smith, A. F. M. (1993). Novel approach to nonlinear/non-Gaussian Bayesian state estimation, *Radar and Sig.Processing, IEE Proceedings F*, 140, pp. 107-113.

Grewal, M. S. and Andrews, A. P. (2001). Kalman filtering: theory and practice using MATLAB, John Wiley and Sons.

Handschin, J. E. and Mayne, D. Q. (1969). Monte Carlo techniques to estimatethe conditional expectation in multi-state nonlinear filtering, *Int. J. contr.*, 9, pp.547- 559.

Hoshiya, M. and Saito, E. (1984). Structural identification by extended Kalman filter, *J. of Engrg. Mech.*, 110(12), paper no. 19331.

Jimenez, J. C. (2002). A simple algebraic expression to evaluate the local linearizationschemes for stochastic differential equations, *Appl. Math. Lett.*, 15, pp. 775-780.

Kallianpur, G. (1980). Stochastic filtering theory, Springer-Verlag, New York.

Kalman, R. E. (1960). A new approach to linear filtering and prediction problems, *Trans. of the ASME- Journal of Basic Engineering*, 82(1), pp. 35-45.

Klebaner, F. C. (2001). Introduction to stochastic calculus with applications, London, UK: Imperial College Press.

Kloeden, P. E. and Platen, E. (1999). Numerical solution of stochastic differential equations, Springer, New York.

Kotecha, J. H. and Djuric, P. M. (2001). Gaussian sum particle filtering for dynamic state space models, *Proceedings of IEEE International Conference on Acoustics, Speech, and Signal Processing*, pp. 3465-3468.

Kotecha, J. H. and Djuric, P. M. (2003). Gaussian sum particle filtering, *IEEE Trans. on Signal Processing*, 51, pp. 2602-2612.

Lieven, N. A. and Ewins, L. D. J. (2001), Experimental modal analysis (theme issue), *Philos. Trans. R. Soc. London A*, 359, pp. 1-219.

Ljung, L. (1997). System identification: theory for the user, Englewood Cliffs, Prentice-Hall, NJ.

Lorentzen, R. J. and Naevdal, G. (2011), An iterative ensemble Kalman filter, *IEEE Trans. on Automatic Control*, 56(8), pp. 1990-1995.

Maruyama, G. (1955). Continuous Markov processes and stochastic equations, *Rend. Circ. Mat. Palermo*, 4, pp. 48-90.

Milstein, G. N. (1995). Numerical integration of stochastic differential integration, Kluwer Academic Publishers.

Oksendal, B. K. (2003). Stochastic differential equations - an introduction with applications, Sixth ed., Springer, New York.

Peeters, B. and Roeck, G. D. (2001). Stochastic system identification for operational modal analysis, *ASME J. Dyn. Syst. Measure. Control*, 123, pp. 1-9.

Pintelon, R. and Schoukens, J. (2001). System identification: a frequency domain approach, New York: IEEE Press.

Raveendran, T., Roy, D. and Vasu, R. M. (2011). A Nearly exact reformulation of the Girsanov linearization for stochastically driven nonlinear oscillators, submitted.

Ristic, B., Arulampalam, N. and Gordon, (2004). Beyond the Kalman filter - particle filters for tracking applications, *Artech House*, Boston.

Robert, C. P. and Casella, G. (2004). Monte Carlo statistical methods, Springer, NewYork.

Roy, D. (2000). Exploration of the phase-space linearization method for deterministic and stochastic nonlinear dynamical systems, *Nonlinear Dyn.*, 23, pp. 225-258.

Roy, D. (2001). A numeric-analytic technique for non-linear deterministic and stochastic dynamical systems. *Proc. Roy. Soc. A*, 457, 539-566.

Roy, D. and Manohar, C. S. (2006). Monte Carlo filters for identification of nonlinear structural dynamical systems, Sadhana, Academy Proceedings in Engineering Sciences, Indian Academy of Sciences.

Rubinstein, R. Y. (1981). Simulation and the Monte Carlo method, Wiley, New York.

Saha N. and Roy, D. (2007). The Girsanov linearization method for stochastically driven nonlinear oscillators, *J. of Applied Mechanics*, 74, pp. 885-897.

Saha, N. and Roy, D. (2009). A Girsanov particle filter in nonlinear engineering dynamics, *Physics Letters A*, 373, pp. 627-635.

Saha, N. and Roy, D. (2009). Extended Kalman Filters Using Explicit and Derivative-Free Local Linearizations, *Applied Mathematical Modelling*, 33, pp. 2545-2563.

Sajeeb, R., Manohar, C. S. and Roy, D. (2010). A semi-analytical particle filter for identification of nonlinear oscillators, *Probabilistic Engrg. Mech.*, 25, pp. 35-48.

Sakov, P. and Oke, P. R. (2008). A deterministic formulation of the ensemble Kalman filter: an alternative to ensemble square root filters, *Tellus A*, 60, pp. 361-371.

Tanizaki, H and Mariano, R. S. (1998). Nonlinear and non-Gaussian state-space modeling with Monte Carlo simulations, *J. of Econometrics*, 83, pp. 263-290.

Tanizaki, H. (1996). Nonlinear filters: estimation and applications, Springer, Berlin, 1996.

Worden, K. and Manson, G. (2001). Nonlinearity in experimental modal analysis, *Phil. Trans. of The Royal Soc. A*, 359, pp. 113-130.

Yun, C. B. and Shinozuka, M. (1980). Identification of nonlinear dynamic systems, *J. Struct. Mech.*, 8(2), pp. 187-203.

Chapter 6

A Novel Health Assessment Method for Large Three-Dimensional Structures

Ajoy Kumar Das[1] and Achintya Haldar[2]

Dept. of Civil Engineering & Engineering Mechanics, University of Arizona
P.O. Box 210072, Tucson, AZ 85721, USA
E-mails: [1]akdas@email.arizona.edu & [2]haldar@u.arizona.edu

1. Introduction

Health assessment of structures (buildings, bridges, and other infrastructures) during their normal operating conditions or just after a natural (strong ground motion, high wind, etc.) or man-made hazard (blasts, explosions, etc.) is very important to maintain the economic activities of a society. Defects of many forms and types (cracks, slippage, corrosions, loss of areas at specific locations, or even broken members) can be expected to develop in them during the natural aging process. Locations and severities of these defects need to be identified and necessary appropriate remedial actions need to be initiated during the normal maintenance process. Ideally, well-maintained infrastructures should be replaced after their design lives are over. However, due to severe world-wide shortage of funds, replacement may not be an attractive alternative; extending their life through proper maintenance actions without compromising the underlying risk has become an attractive choice. For this option, it is essential to establish the current health. It is to be noted that even all the resources are made available; it may be unrealistic to replace them all when needed. When structures are exposed to severe natural or unexpected man-made hazards, their health need to be assessed immediately after the incidence to establish the suitability of their uses causing economic hardship to their owners. In all these scenarios, consequences of improper diagnosis can be very severe, both economically and the loss of trust of the general public.

Widely used subjective assessment of structural health, say by using visual inspection, may not be acceptable considering the economical consequences. The outcome of an inspection will depend on the education and experience of the inspector (Fritzen, 2006). If the location and type of defects are known, the profession has technological sophistication to study their behavior using appropriate instrument and technique. Unfortunately, the location and type of defect will be unknown in most cases, particularly when dealing with large structural systems. Objective, performance-based assessment tools must be used. Problem-specific inspection tools must be developed if they are not available. The general related areas are collectively known as structural health assessment (SHA) or structural health monitoring (SHM). It is a multi-disciplinary area consisting of mathematical theory, sensing technology, signal processing, and assessment norms.

Conceptually, all nondestructive SHA/SHM procedures require measurement of responses and then post-processing them to extract information on location and severity by tracking damage sensitive parameters in the most efficient way. It could be pattern-based or model-based. Rytter (1993) commented that the expected outcomes of an efficient SHA procedure could be the following: (a) determination of existence of damage in structure, (b) determination of the location of damage, (c) quantification of the severity of damage, and (d) prediction of future life of the structure. Considering capabilities, currently available procedures can be classified into four categories: (i) Level 1 – methods that only determine the occurrence of damage, (ii) Level 2 – methods that determine the occurrence as well as location of damage, (iii) Level 3 – methods that determine the occurrence, location, and severity of damage, and (iv) Level 4 – methods that determine the remaining life of the structure considering information on damage extracted at the time of inspection. Issues related to the mathematical aspects of nondestructive Level 3 SHA techniques are emphasized in this chapter.

The mathematical concept most commonly used for Level 3 SHA, is widely known as the system identification (SI). Several methods with various degrees of sophistication are discussed in state-of-the-art documents by Lew et al. (1993), Doebling *et al.* (1996), Salawu (1997), Carden and Fanning (2004), Sohn *et al.* (2004), Humar *et al.* (2006), Kerschen *et al.* (2006), Farrar and Worden (2007), Nasrellah (2009), Fan and Qiao (2011).

2. Concept of System Identification (SI)

The concept of system identification (SI) is essentially the determination of the mathematical model of a system or a process by observing the relationship between the input and output quantities (Hsia, 1977). The general concept was initiated by system and control engineers for space mission programs during mid sixties. At present, it is widely used in the civil, mechanical, and aerospace engineering applications. A basic SI procedure consists of three essential components: (1) excitation information that generated the responses, (2) the system to be identified, represented in algorithmic forms using finite elements (FEs) and governing equations of motion, etc., and (3) measured response information. Using information on excitation and response, the information on the unknown parameters of the system can be established. This is often referred to as parametric SI. The SI technique is a black box procedure when no information on the basic properties of the system is available. Otherwise it is a gray box procedure, when some information on the basic characteristics of the system is available. Structural SI problems can be considered as the gray box type since the information on mathematical models (linear or nonlinear) to be used to satisfy widely accepted classical governing dynamic equations of motion is often available (Kozin and Natke, 1986).

To assess current health of civil engineering structures, both static and dynamic responses have been used. SI-based SHA procedures using static and dynamic information are separately discussed below.

3. SHA Using Static Responses

SHA using static responses received considerable interest because of its mathematical simplicity (Sheena *et al.*, 1982; Hajela and Soeiro, 1989, 1990; Sanayei and Scampoli, 1991; Sanayei and Onipede, 1991; Hjelmstad *et al.*, 1992; Banan *et al.* 1994a, b; Sanayei and Saletnik, 1996; Liang and Hwu, 2001). Static responses are generally measured in terms of displacements, rotations, or strains and the damage detection problems are generally formulated in an optimization framework employing minimization of error between the analytical and measured quantities. Structures are generally represented by FE model. Three classes of error functions are reported in the literature: displacement equation error function, output error function, and strain output error function (Sanayei *et al.*, 1997). Bernal (2002) proposed flexibility-based damage localization method, denoted as the Damage Locating Vector (DLV) method. The basic approach is the determination of a set of vectors (i.e. the DLVs), which when applied as static forces at the sensor locations, no stress will be induced in the damaged elements. The method can be a promising damage detection tool as it allows locating damages using limited number of sensor responses. It was verified for truss elements, where axial forces remain constant throughout its length. However, the verification of the procedure for real structures using noise-contaminated responses has yet to be completed.

There are several advantages to SHA using static responses including that the amount of data needed to be stored is relatively small and simple, and no assumption on the mass or damping characteristics is required. Thus, less errors and uncertainties are introduced into the model. However, there are several disadvantages including that the number of measurement points should be larger than the number of unknown parameters to assure a proper solution. Civil engineering structures are generally large and complex with extremely high overall stiffness. It may require extremely large static load to obtain measurable deflections. Fixed reference locations are required to measure deflections which might be impractical to implement for bridges, offshore platforms, etc. Also, static response-based methods are sensitive to measurement errors (Aditya and Chakraborty, 2008; Anh, 2009).

4. SHA Using Dynamic Responses

Various dynamic response-based SHA procedures were developed in order to eliminate the disadvantages of static response-based procedures. They are computationally intensive, requiring large amount of data acquisition, storage, pre- and post-processing, and so on. However, they have several advantages including that it is possible to excite structures by dynamic loadings of small amplitude relative to static loadings. In some cases the ambient responses caused by natural sources such as wind, earthquake, moving vehicle, etc. can also be used. If acceleration responses are measured, they eliminate the need for fixed physical reference locations. They perform well even in presence of high level of measurement errors.

Dynamic response-based SHA procedures separate into two main categories: frequency-domain and time-domain procedures. Frequency-domain procedures consider spectral estimates and frequency characteristics for the identification of structural damages. Time-domain procedures are generally based on least-squares methods, maximum likelihood techniques, recursive techniques, and related techniques. Modal-based approaches have many desirable features. Instead of using enormous amount of data, the modal information can be expressed in countable form in terms of frequencies and mode shape vectors. Since structural global properties are evaluated, there may be an averaging effect, reducing the effect of noise in the measurements. However, the general consensus is that modal-based approaches fail to evaluate the health of individual structural elements; they indicate overall effect, i.e., whether the structure is defective or not (Ibanez, 1973; McCann, et al., 1998; Ghanem and Ferro, 2006). For complicated structural systems, the higher order calculated modes are unreliable and the minimum numbers of required modes to identify the system parameters is problem dependent, limiting their applicability. It was reported that even when a member breaks, the natural frequency may not change more than 5%. This type of change can be caused by noises in the measured responses. A time-domain approach will be preferable.

5. Time-Domain SI-Based SHA Procedures

Time-domain SI-based SHA procedures are appropriate for calculating element-level structural parameters i.e., mass, stiffness, damping, etc., particularly when multi-degree of freedom (MDOF) structural systems are represented by FEs. The governing differential equation of motion of a MDOF system with *ne* number of elements and *N* number of dynamic degrees of freedom (DDOFs) can be written in matrix notations as:

$$\mathbf{M}\ddot{\mathbf{x}}(t) + \mathbf{C}\dot{\mathbf{x}}(t) + \mathbf{K}\mathbf{x}(t) = \mathbf{f}(t) \tag{1}$$

where **M**, **K**, and **C** are the $N \times N$ global mass, stiffness, and damping matrices, respectively. They are assumed to be time-invariant for linear structures. $\mathbf{x}(t)$, $\dot{\mathbf{x}}(t)$, and $\ddot{\mathbf{x}}(t)$ are the $N \times 1$ vectors containing measured information on displacement, velocity, and acceleration at all the DDOFs at time t. $\mathbf{f}(t)$ is the $N \times 1$ excitation force vector. The mass matrix **M** is generally assumed to be known or available or it can be formed from the as-built construction drawings. Thus, if n samples are measured for each of the responses at Δt time increment for total duration of T seconds, then *ne* number of unknown stiffness and *ne* number of damping parameters can be calculated by solving $n \times N$ (where $n \times N \geq 2ne$) number of equations, provided input excitation $\mathbf{f}(t)$ is known. A brief discussion on these procedures can be found in Wang and Haldar (1994).

After a considerable literature review, the research team at the University of Arizona decided to develop a novel time-domain SI procedure for SHA. By measuring acceleration time histories and successively integrating them to obtain velocity and displacement time histories, Vo and Haldar (2003) proposed a method. They verified it by conducting laboratory experiments. However, they observed that the concept discussed above needs a considerable amount of improvements before it can be used to assess health of real structural systems.

They also noted the comments made by Maybeck (1979) that deterministic mathematical model and control theories do not appropriately represent the behavior of a physical system and thus the SI-based method may not be appropriate for SHA.

Outside the controlled laboratory environment, measuring input excitation force(s) can be very expensive and problematic during health assessment of an existing structure. It will be desirable if a system can be identified using only measured response information, and completely ignoring the excitation information. Responses, even measured by smart sensors, are expected to be noise-contaminated. Depending on the amount of noise, the SI-based approach may be inapplicable. The basic concept also assumes that responses will be available at all DDOFs. For large structural systems, it may be practically impossible or uneconomical to instrument the whole structure; only a part can be instrumented. Thus, the basic challenge is to identify stiffness parameters of a large structural system using limited noise-contaminated response information measured at a small part of the structure. The research team successfully developed such a method in steps, as discussed next.

6. Time-Domain SHA Procedures with Unknown Input (UI)

Available SHA procedures with unknown input are mostly based on least-squares concept (Chen and Li, 2004; Sandesh and Shankar, 2009; Kun *et al.*, 2009; Xu *et al.*, 2011). Other procedures include iterative gradient-based model updating method based on dynamic response sensitivity (Lu and Law, 2007), least-squares method with statistical averaging algorithm (LSM-SAA) (Wang and Cui, 2008), sequential non-linear least-square estimation procedure for damage identification of structures with unknown inputs and unknown outputs (Yang and Huang, 2007), and a genetic algorithm (GA)-based output only procedure (Perry and Koh, 2008). The research group at the University of Arizona proposed several least-squares-based procedures. Wang and Haldar (1994) developed the conceptual framework of Iterative Least-Squares with Unknown Input (ILS-UI) considering viscous-type damping in Eq. 1. The algorithm calculates *ne* number of stiffness parameters, *ne* number of damping parameters, and the time-history of unknown input excitation force(s). The responses could be noise-contaminated. Since information on damping cannot be completely correlated with the damage state of a structure, to improve the efficiency of ILS-UI, Ling and Haldar (2004) proposed modified ILS-UI or MILS-UI considering Rayleigh-type proportional damping, as:

$$\mathbf{C} = \alpha \mathbf{M} + \beta \mathbf{K} \tag{2}$$

where α and β are the mass and stiffness proportional damping coefficients, and are related to the first two natural frequencies f_1 and f_2 of a structure (Clough and Penzien, 1993). Using MILS-UI, *ne* number of stiffness parameters, two damping parameters (α and β), and the time-history of unknown input excitation force(s) can be identified. Katkhuda *et al.* (2005) improved the efficiency further by modifying the iterative steps and denoted the procedure as generalized ILS-UI or GILS-UI. The GILS-UI procedure was extensively verified for one-dimensional beams and two-dimensional frames using both theoretical and experimental responses (Katkhuda *et al.*, 2005; Katkhuda and Haldar, 2006; Martinez-Flores and Haldar, 2007).

The main drawback of GILS-UI is that it requires responses to be measured at all the DDOFs, defeating the basic objective of the team as discussed earlier. When response information is limited and noise-contaminated but excitation information is available, Kalman filter (KF)-based algorithm can be used. For unknown excitation and limited response information (linear or mildly nonlinear), Wang and Haldar (1997) proposed a conceptual framework for Iterative Least-Squares Extended Kalman Filter with Unknown Input (ILS-EKF-UI) and verified it for shear-type buildings. It went through several stages of improvement including modified ILS-EKF-UI (MILS-EKF-UI) (Ling and Haldar, 2004), generalized ILS-EKF-UI (GILS-EKF-UI) (Katkhuda and Haldar, 2008), and three dimensional GILS-EKF-UI (3-D GILS-EKF-UI) (Das and Haldar, 2012). The basic concept of Kalman filter used in the SHA of civil infrastructures is discussed very briefly as below.

7. The Kalman Filter Concepts and its Application for SHA

The Kalman filter concept is a powerful mathematical tool developed during the space mission programs by Kalman (1960). Later the concept was applied in many applications in electrical, mechanical, aerospace, civil, and chemical engineering. It was also used in navigation, robotics, economics, biology, ecology, and many other fields. Its application extends to wide range of research areas including missile tracking, economical modeling and forecasting, computer vision problems including feature tracking, cluster tracking, fusing data from radar, laser scanner, stereo camera for depth and velocity measurement, etc.

As mentioned earlier, SI concepts refer to the development of system mathematical model to relate the system variables, inputs to the system, and outputs from the system. The analyses are generally performed using deterministic procedures and control theories, which may not be completely sufficient. Mayback (1979) pointed out three basic reasons behind this. First, no mathematical model is perfect to represent the behavior of a system, since there could be various parameters that are left unaccounted because of the lack of our knowledge, or just to obtain simple, approximate mathematical representation. Second, dynamic systems are driven not only by control inputs, but also by disturbances that can neither be controlled nor modeled using deterministic formulations. Third, sensors used for data measurements cannot be perfect; they are expected to be corrupted with various sources of uncertainties. Thus, an appropriate mathematical model (may not be exact) should include capabilities to (a) account for the uncertainties from different sources, (b) optimally estimate data from noise-contaminated and incomplete sensor measurements, and (c) optimally control the system to perform in a desired manner.

The Kalman filter is an optimal linear estimator that performs recursively i.e. it does not require all the past data to be kept in storage and processed every time a new data is obtained. However, it is able to process all the information provided to it regardless of their precision. It can calculate the current values of the variables considering (i) the knowledge of the system and sensor dynamics, (ii) statistical description of system and measurement noises, and uncertainty in the dynamical model, and (iii) any information on the initial condition of the variables (Maybeck, 1979).

In a mathematical sense, the Kalman filter provides efficient recursive computational means to estimate the state of a discrete-time process, governed by linear stochastic difference equation of the following form:

$$\mathbf{x}(k+1) = \mathbf{F}(k)\mathbf{x}(k) + \mathbf{G}(k)\mathbf{u}(k) + \mathbf{w}(k) \tag{3}$$

The measurement model considers the following form:

$$\mathbf{z}(k+1) = \mathbf{H}(k+1)\mathbf{x}(k+1) + \mathbf{v}(k+1) \tag{4}$$

where $\mathbf{x}(k+1)$ and $\mathbf{x}(k)$ are the state vectors at time k+1 and k, respectively. Random vectors $\mathbf{w}(k)$ and $\mathbf{v}(k)$ represent the process and measurement noises, respectively. They are considered to be independent, zero-mean, and white with normal probability distributions. The matrix $\mathbf{F}(k)$ relates the two state vectors in absence of either a driving function or process noise. The matrix $\mathbf{G}(k)$ relates the optimal control input $\mathbf{u}(k)$ to the state, and matrix $\mathbf{H}(k+1)$ in the measurement model relates the state vector to the measurement vector $\mathbf{z}(k+1)$. Based on the linear process and measurement models, the Kalman filter aims to set up mathematical equations to minimize the mean of squared error (Welch and Bishop, 2006). The algorithmic process includes two steps; (1) prediction i.e.

projecting the current states and error covariances to obtain the priori estimate in the next time step, and (2) updating i.e. incorporating the measurements in the priori estimates to obtain the posteriori estimates. The prediction and updating at each time is commonly known as *local iteration*. The real problems are not always linear; they may contain some amount of nonlinearity caused by large deformation due to strong excitation or severe damage. To consider nonlinear behavior, the extended Kalman filter (EKF) concept (Kopp and Orford, 1963; Cox, 1964) was introduced. It is an extension of KF to consider piecewise linear approximation of the nonlinear responses. Using EKF, Eq. 3 is written as a nonlinear stochastic difference equation of the following form:

$$\mathbf{x}(k+1) = f\left[\mathbf{x}(k), \mathbf{u}(k), \mathbf{w}(k)\right] \tag{5}$$

The measurement Eq. 4 is written in the following form:

$$\mathbf{z}(k+1) = h\left[\mathbf{x}(k+1), \mathbf{v}(k+1)\right] \tag{6}$$

where nonlinear function f relates the state at time k to the current state at time k+1, and it includes parameters, any driving force $\mathbf{u}(k)$ and process noise $\mathbf{w}(k)$. The nonlinear function h relates the state $\mathbf{x}(k+1)$ to the measurement $\mathbf{z}(k+1)$. Again, $\mathbf{w}(k)$ and $\mathbf{v}(k)$ are the independent, zero-mean, and white with normally distributed random vectors representing the process and measurement noises, respectively. It is important to mention here that the EKF estimates the state by linearizing (using Taylor series expansion) the process and measurement equations (Eq. 5 and 6) about the current state and covariances (Welch and Bishop, 2006). For systems with well-developed mathematical model, such as for civil engineering structures, the process noise can be ignored. To improve the efficiency, Hoshiya and Saito (1984) suggested using a weighted *global iteration* (WGI) technique after the completion of local iterations and the whole procedure is denoted as EKF-WGI. The WGI technique is generally used to ensure (but not necessarily) the stability and convergence in the estimation of system parameters. The EKF-WGI procedure will be discussed later in detail.

Civil engineering structures are often represented into smaller finite elements (FEs). Since the mass is generally assumed to be known in Eq. 1, the unknown stiffness and damping parameters for all the elements can be estimated using EKF-WGI procedure, and the structural health can be assessed by comparing the current element-level stiffness parameters with those of the reference defect-free state or previous state if periodic inspections are performed or significant deviation from other similar elements. The location of defect spot can be indentified more accurately by discretizing the defective element into smaller elements.

To conclusively verify the novel concept developed by the team at the University of Arizona, a one-bay three-story steel frame built to one-third scale to fit the laboratory facility was tested. Initially, the health of the defect-free frame was successfully assessed using sinusoidal and impulsive excitations. Subsequently, defects in various forms and severity levels including removal of a broken member, single or multiple cracks in a member, loss of area at a specific location in a member, etc. were introduced for verification. A pictorial view of a structural damage due to loss of area over a finite length is shown Figure 1. Flange and web thicknesses were removed by grinding over a finite length. The measured acceleration time histories were expected to be noise-contaminated. The GILS-EKF-UI procedure successfully indentified all damage scenarios by locating the defect spot and the severity of the defect (Martinez-Flores and Haldar, 2007; Martinez-Flores *et al.*, 2008).

8. Extension of GILS-EKF-UI for 3-D Structures

Real civil structures are expected to be three dimensional (3-D) and relatively large. They need to be represented by a large number of finite elements. It will be impractical to install accelerometers at all node points; only a small part of the structure, generally known as substructure, will be instrumented. The defects may not be always visible; they may be behind obstructions. Also, to limit economic burden, the structures may need to be inspected during their normal operation.

For economic reason, the size substructure should be kept to a minimum. Detail procedure for the selection of substructure is discussed by Katkhuda and Haldar (2008). For large structures, there could be multiple substructures so that at least one of them will be close to defect location to increase the detection potential. Past maintenance history and experience of the inspector can be used to select substructure(s) during real inspections. For earthquake excitation, any part can be considered as substructure as the inertia force will act at all the structural nodes. The objective is to assess the health of the whole structure using only responses measured at the substructure(s). As briefly discussed in Section 6, 3-D GILS-EKF-UI can be used to assess the health, as discussed next.

To implement the 3-D GILS-EKF-UI concept, the information on the unknown excitation time-history and the initial state vector needs to be generated. The research team proposed the following two-stage approach to generate the required information.

- Stage 1: Based on the available responses, the substructure(s) are selected and the 3-D GILS-UI procedure is used to generate information on the input excitations, stiffness of all the elements in the substructure, and the two Rayleigh damping coefficients.

- Stage 2: Information extracted from Stage 1 will satisfy all the requirements to implement EKF-WGI and the health of the whole structure can be assessed, providing the 3-D GILS-EKF-UI procedure.

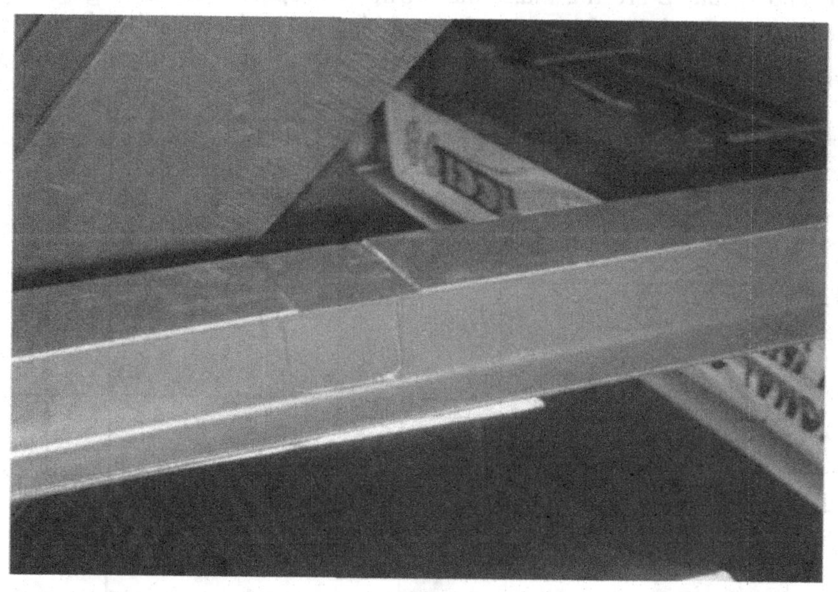

Fig. 1 Reduction in cross-sectional area over a finite length in a defective member.

The whole process can be explained with the help of the following example. Suppose the health of a 3-D frame shown in Figure 2 needs to be assessed. During the inspection, the structure is excited dynamically at node 1 and the responses are measured at nodes 1, 2, 3, and 5. Based on the measured response information, the selected substructure is shown in double lines, as shown in Figure 2. The substructure consists of two beams and one column and will have 24 DDOFs. The whole frame consists of 24 elements; 12 beams and 12 columns, and a total of 72 DDOFs. The frame is assumed to be rigidly connected to the base at nodes 13, 14, 15, and 16. The task is to assess the heath of the frame using 24 responses measured at nodes 1, 2, 3, and 5, without knowing the excitation information.

8.1. *Stage 1 – concept of 3-D GILS-UI*

Without losing the generality, the governing differential equation of motion for the substructure with Rayleigh damping can be expressed as:

$$\mathbf{M}_{sub}\ddot{\mathbf{x}}_{sub}(t) + (\alpha\mathbf{M}_{sub} + \beta\mathbf{K}_{sub})\dot{\mathbf{x}}_{sub}(t) + \mathbf{K}_{sub}\mathbf{x}_{sub}(t) = \mathbf{f}_{sub}(t) \tag{7}$$

where $\mathbf{M}_{sub}$ is the global mass matrix; $\mathbf{K}_{sub}$ is the global stiffness matrix; $\ddot{\mathbf{x}}_{sub}(t)$, $\dot{\mathbf{x}}_{sub}(t)$, and $\mathbf{x}_{sub}(t)$ are the vectors containing the acceleration, velocity, and displacement, respectively, at time t; $\mathbf{f}_{sub}(t)$ is the input excitation force vector at time t; and α and β are the mass and stiffness proportional Rayleigh damping coefficients, respectively. The subscript '*sub*' is used to denote substructure.

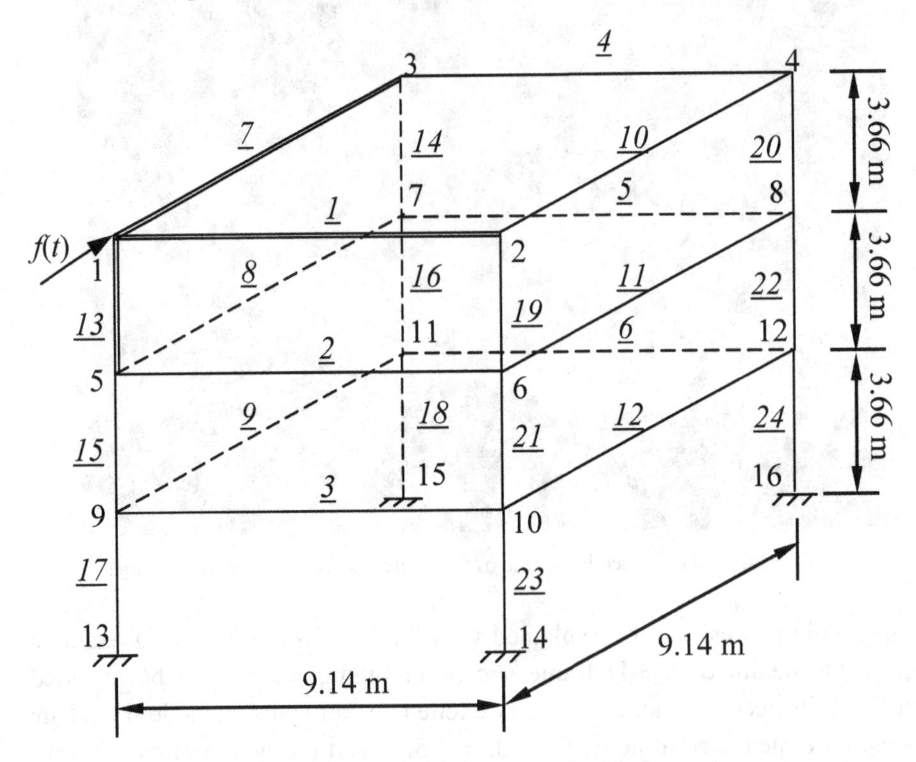

Fig. 2 Health assessment of a 3-D frame.

For a 3-D frame with rigid connections, the mass matrix $\mathbf{M}_i$ and stiffness matrix $\mathbf{K}_i$ for i^{th} element can be expressed in the local coordinate system as (Paz, 1985):

$$\mathbf{M}_i = \frac{\bar{m}_i L_i}{420}\left[f(L_i, \bar{m}_i, I_{\bar{m}_i}) \right]_{12\times 12} \tag{8}$$

$$\mathbf{K}_i = \frac{E_i I_{z_i}}{L_i}\left[f(A_i, L_i, I_{y_i}, I_{z_i}, J_i, \upsilon) \right]_{12\times 12} \tag{9}$$

where L_i, A_i, E_i, and $\bar{m}_i$ are the length, cross-sectional area, material Young's modulus of elasticity, and mass per unit length, respectively, of the ith frame element; I_{z_i} and I_{y_i} are the cross-sectional moment of inertias with respect to the major and minor principal axis, respectively; J_i is the torsional moment of inertia; υ is the Poisson's ratio of the material, and $I_{\bar{m}_i}$ is the polar mass moment of inertia per unit length of the member. Denoting the coefficient matrix in the square bracket by $\mathbf{S}_i$, Eq. 9 can be written as $\mathbf{K}_i = k_i\mathbf{S}_i$, where $k_i = E_i I_{z_i}/L_i$ is defined as the stiffness parameter for the ith element. Matrices $\mathbf{M}_i$ and $\mathbf{K}_i$ in the element local coordinate system need to be transformed to the global coordinate system using the appropriate transformation matrix $\mathbf{T}_i$. Thus, the global mass and stiffness matrices, $\bar{\mathbf{M}}_i$ and $\bar{\mathbf{K}}_i$, for the ith element are obtained as (Paz, 1985):

$$\bar{\mathbf{M}}_i = \mathbf{T}_i^T \mathbf{M}_i \mathbf{T}_i \qquad \bar{\mathbf{K}}_i = k_i \mathbf{T}_i^T \mathbf{S}_i \mathbf{T}_i = k_i \bar{\mathbf{S}}_i \tag{10}$$

For *nesub* number of elements in the substructure, the global mass and stiffness matrices for the substructure can be assembled as:

$$\mathbf{M}_{sub} = \sum_{i=1}^{nesub} \bar{\mathbf{M}}_i \qquad \mathbf{K}_{sub} = \sum_{i=1}^{nesub} \bar{\mathbf{K}}_i = \sum_{i=1}^{nesub} k_i \bar{\mathbf{S}}_i \tag{11}$$

Introducing Eq. 11 into Eq. 7, one can equivalently write it as:

$$\begin{aligned} \mathbf{f}_{sub}(t) - \mathbf{M}_{sub}\ddot{\mathbf{x}}_{sub}(t) = [&\bar{\mathbf{S}}_1\mathbf{x}_{sub}(t)\ \bar{\mathbf{S}}_2\mathbf{x}_{sub}(t)\ \cdots\ \bar{\mathbf{S}}_{nesub}\mathbf{x}_{sub}(t) \\ &\bar{\mathbf{S}}_1\dot{\mathbf{x}}_{sub}(t)\ \bar{\mathbf{S}}_2\dot{\mathbf{x}}_{sub}(t)\ \cdots\ \bar{\mathbf{S}}_{nesub}\dot{\mathbf{x}}_{sub}(t)\ \mathbf{M}_{sub}\dot{\mathbf{x}}_{sub}(t)]\mathbf{P} \end{aligned} \tag{12}$$

where $\mathbf{P} = [k_1\ k_2\ \cdots\ k_{nesub}\ \beta k_1\ \beta k_2\ \cdots\ \beta k_{nesub}\ \alpha]^T$ contains total *Lsub* number of unknown stiffness and damping parameters in the substructure. Here *Lsub* = 2*nesub*+1. Eq. 12 consists of N_{dkey}×n number of equations at the key node(s) and can be reorganized as (Katkhuda and Haldar, 2008):

$$\mathbf{F}_{Ndkey.n\times 1} = \mathbf{A}_{Ndkey.n\times Lsub}\mathbf{P}_{Lsub\times 1} \tag{13}$$

where *Ndkey* is the total number of DDOFs at the key node(s) in the substructure. Vector **F** on the left hand side of Eq. 13 contains the time-history of unknown input excitation and the inertia force for all *n* time points. Rectangular matrix **A** on the right hand side contains the response information on displacement and velocity at all *n* time points. The least-squares technique can be used to estimate the unknown parameters **P** using the following expression:

$$\mathbf{P}_{Lsub\times 1} = (\mathbf{A}^T_{Lsub\times Ndkey.n}\mathbf{A}_{Ndkey.n\times Lsub})^{-1}\mathbf{A}^T_{Lsub\times Ndkey.n}\mathbf{F}_{Ndkey.n\times 1} \tag{14}$$

The solution for the unknown parameters **P** using Eq. 14 can be relatively simple provided the force vector **F** is known. But for unknown excitation, the vector **F** is partially unknown and so an iterative procedure will be necessary. Wang and Haldar (1994) initially proposed an iterative scheme assuming the excitation information to be zero for first few time points. Later Katkhuda et al. (2005) improved the iterative scheme by assuming the excitation information to be zero for all the time points. The iteration process is continued until the identified time-history of input excitation in two successive iterations satisfy a predetermined convergence tolerance level δ (assumed to be 10^{-6} in this study). At the completion of Stage 1, the time history of the excitation, stiffness parameter of the column and the two beams and the two Rayleigh coefficients will be generated. The damping in the substructure can be assumed to be applicable to the whole structure. Structural members in a structure are expected to have similar cross sectional properties. The identified stiffness parameters of the beam in the substructure can be assigned to all the beams in the structure. Similarly, the identified stiffness parameters of the column can be assigned to all the columns. This will give information on the initial state vector. With the information on initial state vector and excitation information, the EKF-WGI procedure can be initiated to identify the whole structure in Stage 2.

8.2. *Stage 2 – concept of EKF-WGI*

In the EKF concept, the mathematical model of dynamic system is represented by a set of first-order nonlinear differential equations in the state-space form. The mathematical models for linear or mildly nonlinear structural dynamics are well-developed. Without losing any generality, the process noise term in Eq. 5 can be dropped, and it can be written in continuous state-space form as (Saridis, 1995):

$$\dot{\mathbf{Z}}(t) = f\left[\mathbf{Z}(t), t\right] \tag{15}$$

where $\mathbf{Z}(t)=[\mathbf{x}(t)\ \ \dot{\mathbf{x}}(t)\ \ \tilde{\mathbf{K}}\ ]^T$ is the augmented state vector containing the vectors of displacement $\mathbf{x}(t)$ and velocity $\dot{\mathbf{x}}(t)$ at time t, and the vector of unknown stiffness parameters $\tilde{\mathbf{K}}$, which is considered to be time-invariant during the whole identification process. For a structure consisting of *ne* numbers of elements, $\tilde{\mathbf{K}}$ is defined as $\tilde{\mathbf{K}} = [k_1\ k_{2 \ldots}\ k_{ne}]^T$, where k_i is the stiffness parameter of the ith element. f is the nonlinear function of the state vector. Using the governing differential equation of motion in Eq. 1 with Rayleigh damping expressed in Eq. 2, Eq. 15 can be conveniently expressed as:

$$\dot{\mathbf{Z}}(t)=\begin{bmatrix}\dot{\mathbf{x}}(t)\\ \ddot{\mathbf{x}}(t)\\ 0\end{bmatrix}=\begin{bmatrix}\dot{\mathbf{x}}(t)\\ \mathbf{M}^{-1}\left[\mathbf{f}(t)-\alpha\mathbf{M}\dot{\mathbf{x}}(t)-\beta\mathbf{K}\dot{\mathbf{x}}(t)-\mathbf{K}\mathbf{x}(t)\right]\\ 0\end{bmatrix} \tag{16}$$

The above continuous differential equation needs to be solved using numerical procedure. The measurement equation is linear in our case and noise contaminated, and can be expressed at any discrete time k as:

$$\mathbf{Y}(k)=\mathbf{H}\cdot\mathbf{Z}(k)+\mathbf{V}(k) \tag{17}$$

where **H** is the measurement matrix; for measured responses it becomes a unit diagonal matrix. **V**(k) is a zero-mean, white random process having normal probability distributions indicating noise in the measurements. It is generally presented as:

$$\mathbf{V}(k)\sim N\left[0,\mathbf{R}(k)\right] \tag{18}$$

where **R**(k) is the discrete measurement noise matrix consisting of covariance values considering noise source in each measurement at any time k. The noise covariance matrix for the basic EKF is assumed to be diagonal and the values remain constant with time. In this study, it is assumed to be of the same magnitude of 10^{-4} for all responses.

Before starting the EKF algorithm, it is necessary to assign the initial values to the uncertain state vector. They are generally assumed to be Gaussian with mean $\hat{\mathbf{Z}}(0\,|\,0)$ and an initial error covariance matrix $\mathbf{P}(0\,|\,0)$. The information on the initial values for the stiffness parameters for the whole structure was generated after the completion of Stage 1, as discussed earlier. The initial error covariance matrix represents the uncertainty in the assumption of the initial state vector and is generally defined to be diagonal as (Hoshiya and Saito, 1984; Koh *et al.*, 1991; Wang and Haldar, 1997):

$$\mathbf{P}(0\,|\,0)=\begin{bmatrix}\mathbf{I}_1 & 0\\ 0 & \lambda\mathbf{I}_2\end{bmatrix} \tag{19}$$

where $\mathbf{I}_1$ and $\mathbf{I}_2$ are unit matrices and λ is a large positive number and its value will depend on the magnitude of the element stiffness parameters and the unit system used. In this study, it is assumed to be of the order of 10^6 and 10^8, respectively, for the frame and truss elements. Thus, $\mathbf{I}_1$ and $\lambda\mathbf{I}_2$ represent the covariance values for the responses and the stiffness parameters, respectively. Jazwinski (1970) commented that the large positive numbers for the covariance values for the system parameters accelerate the convergence of the local iteration.

After assigning appropriate values to the uncertain initial conditions, the filtering process is performed using the following steps:

(i) Prediction of new state mean and its new error covariance - The state mean $\hat{\mathbf{Z}}(k|k)$ and error covariance $\mathbf{P}(k|k)$ at time k are propagated one step forward in time to predict the new state mean $\hat{\mathbf{Z}}(k+1|k)$ and new error covariance $\mathbf{P}(k+1|k)$ at time k+1 by numerically solving the differential equations as:

$$\hat{\mathbf{Z}}(k+1|k) = \hat{\mathbf{Z}}(k|k) + \int_{k\Delta t}^{(k+1)\Delta t} f(\mathbf{Z}(t|k),t)dt \tag{20}$$

$$\mathbf{P}(k+1|k) = \mathbf{\Phi}(k+1|k)\mathbf{P}(k|k)\mathbf{\Phi}^T(k+1|k) \tag{21}$$

where $\mathbf{\Phi}(k+1|k)$ is the state transition matrix of the system. For small time increment Δt, it can be approximately defined as:

$$\mathbf{\Phi}(k+1|k) = \mathbf{I} + \Delta t \left[\frac{\partial f\left[(\mathbf{Z}(t),t\right]}{\partial \mathbf{Z}(t)} \right]_{\mathbf{Z}(t)=\hat{\mathbf{Z}}(k|k)} \tag{22}$$

where $\mathbf{I}$ is a unit matrix. All the other terms were defined earlier. Numerical solution of Eq. 20 will introduce the piece-wise linear approximation in the formulation.

(ii) Updating the predicted state mean and its error covariance – Using the available measurements at time k+1, the predicted state mean and error covariance are updated using the Kalman gain matrix $\mathbf{K}(k+1)$ as:

$$\hat{\mathbf{Z}}(k+1|k+1) = \hat{\mathbf{Z}}(k+1|k) + \mathbf{K}(k+1)[\mathbf{Y}(k+1) - \mathbf{H}\cdot\mathbf{Z}(k+1|k)] \tag{23}$$

$$\begin{aligned}\mathbf{P}(k+1|k+1) = {} & [\mathbf{I} - \mathbf{K}(k+1)\mathbf{H}]\mathbf{P}(k+1|k)[\mathbf{I} - \mathbf{K}(k+1)\mathbf{H}]^T \\ & + \mathbf{K}(k+1)\mathbf{R}(k+1)\mathbf{K}^T(k+1)\end{aligned} \tag{24}$$

Where

$$\mathbf{K}(k+1) = \mathbf{P}(k+1|k)\mathbf{H}^T[\mathbf{H}\mathbf{P}(k+1|k)\mathbf{H}^T + \mathbf{R}(k+1)]^{-1} \tag{25}$$

Prediction and updating operations described in Eqs. 20 through 25 are carried out locally for each of the time points, i.e., k = 1, 2, … , n, and is termed as the local iterations. When the local iterations are performed for all the time points in the entire time-history, it is denoted as the completion of first global iteration. Jazwinski (1970) suggested that the global iteration procedure should be continued until two successive states become essentially identical. The popularly used weighted global iteration (WGI) scheme, suggested in Hoshiya and Saito (1984), is carried out by scaling up the covariance values of the system parameters (obtained at the end of the previous global iteration) by a sufficiently large weight factor (w). The weight factor in this study is considered to be 10 for all the examples. The weighted global iterations are continued until the error (ε) in identified system parameters converge within a predetermined convergence tolerance level. In this study, the error is defined as the absolute percentage change in the identified stiffness parameters in two successive global iterations, and the tolerance level is considered to be 1%. It is generally expected that the WGI procedure will provide overall stability and convergence in the identified system parameters towards the true solutions. However, this is not guaranteed. The stability and convergence of the estimated system parameters can also be judged by the nature of an objective function ($\overline{\theta}$) (not presented here) suggested in Hoshiya and Saito (1984). The same objective function is implemented in this study. In cases when the parameters tend to diverge, the best estimates are judged by observing the value of minimum objective function.

It is important to mention here that the SHA potential of the 3-D GILS-EKF-UI procedure depends on many factors including duration of responses, sampling time, number and location of measured responses, initial information on the state vector and corresponding error covariance matrix, noises in the measured responses, location of defect with respect to measurement locations, defect orientation, weight factor (w). Some of them will be discussed in the following examples.

9. Application Examples

Application of the 3-D GILS-EKF-UI procedure for the SHA of 3-D structures will be verified for two types of structural configurations; frame and truss-frame. Defect-free state and several defective states will be assessed.

9.1. *Example 1 - health assessment of a 3-D frame*

A single-bay three-story 3-D steel frame shown in Figure 2 will be considered first to verify the procedure.

9.1.1. *Description of the frame*

The frame shown in Figure 2 has square base of sides 9.14 m and floor to floor height of 3.66 m. The frame is made of Grade 50 steel members. The beams and the columns are made of W21×68 sections and W14×61 sections, respectively. In the FE representation, the frame is modeled by 16 nodes and 24 elements; 12 beams and 12 columns. Each node has six DDOFs; three translational and three rotational. The frame is fixed at the base at nodes 13, 14, 15, and 16. The actual theoretical values of stiffness parameters k_i (in terms of E_iI_{zi}/L_i) are calculated to be 13476 kN-m for the beams and 14553 kN-m for the columns. The first two natural frequencies of the frame are estimated to be f_1 = 2.7229 Hz and f_2 = 3.5717 Hz, respectively. Assuming damping to be the same for the first two natural modes and following procedure suggested in Clough and Penzien (1993), the Rayleigh damping coefficients α and β are calculated to be 0.97077 and 0.0025284, respectively, for an equivalent modal damping of 5% (commonly used in the model codes in the U.S.) of the critical.

To demonstrate the SHA procedure using 3-D GILS-EKF-UI, the responses are generated analytically by applying a harmonic excitation $f(t) = 45$ Sin $(20t)$ at node 1 as shown in Figure 2. A commercial software ANSYS (Ver. 11) is used for this purpose. The responses are recorded at 24 DDOFs (at nodes 1, 2, 3, and 5) only, instead of 72 DDOFs, at the interval 0.0001 s for a total duration of 0.8 s. After the responses are obtained, the information on input excitation is completely ignored. Responses between 0.31 s and 0.63 s providing 3201 samples are used in the subsequent health assessment process.

9.1.2. *Scaling of additional responses*

The aim of the SHA is to identify all the element stiffness parameters using the responses available at nodes 1, 2, 3, and 5. In the FE configuration, the geometric properties at the reference state are assumed to be known based on the values used to generate the necessary responses. In real inspections they will be known based on as-built drawings or previous data if periodic inspections are made. The information on the location of excitation (the time-history is unknown) and the recorded responses will be sufficient for accurate identification of the substructure using 3-D GILS-UI in Stage 1. The generated information can be used to identify the whole structure. One issue yet to be resolved is the absolute minimum size of the substructure and its efficiency in identifying the whole structure. Suppose at the completion of Stage 2, the error in identification of the whole structure appears to be relatively large and our experience suggests that it is due to the small size of the substructure. At the completion of Stage 2, it will be relatively expensive to initiate another test with larger substructure. It is a very challenging issue for large structural systems and is being actively studied by the research team. After extensive numerical study on different frames, the team observed that if the responses are assumed to be linear, as expected during inspections, some responses at a node can be scaled from other nodes suggested by Vo and Haldar (2004). This is also found to be valid for the frame under consideration. In general, this will depend on the type of the structure, direction of excitation, magnitude and orientation of defects, and so on. Measured and analytical responses can be used to extract the scaling information. It is explained in more detail with the help of the example.

9.1.3. *Health assessment of a defect-free frame*

The substructure is identified in Stage 1 using the 24 available responses. The results are presented in column 3 in Table 1(a). The maximum error in identification is 0.009% as shown in column 4 indicating the members are identified accurately. About 10% error in identification was reported to be acceptable even when excitation information was used in the identification process (Hoshiya and Sato, 1984; Toki *et al.*, 1989; Koh *et al.*, 1991). The time history of unknown input excitation and damping parameters (not shown here) are also identified accurately.

The results obtained in Stage 1 are then used in Stage 2 and the whole frame is identified using the 24 responses. The results presented in Column 3 and 4 of Table 1(b) show wide variation having maximum error of 14.4%. For the defect-free frame, the wide variation may not be sufficient to assess the health. It is to be noted that the EKF algorithm estimates the stiffness parameters for the whole frame by predicting the responses at all the DDOFs and then correcting them using the available measurements. When the number of responses is insufficient for identification, the results deviate from the expected or true solution. Thus, additional responses will be necessary to improve the accuracy. For the frame considered in this study, three additional translational responses are scaled at node 4. The in-plane translation in the horizontal direction is scaled from that at node 3. The in-plane translation in the vertical direction and out-of-plane translation are scaled from those at node 2. These observations are observed to be valid for both defect-free and defective states.

Table 1. Stiffness parameter k_i (= $E_i I_{zi}/L_i$) identification of defect-free frame using 3-D GILS-EKF-UI

(a) Stage 1 – Identification of the substructure

Element (no.)	Theoretical k_i values (kN-m)	Identified k_i values (kN-m)	Error (%)
(1)	(2)	(3)	(4)
1-2 (*1*)	13476	13476	-0.002
1-3 (*7*)	13476	13475	-0.009
1-5 (*13*)	14553	14552	-0.007

(b) Stage 2 – Identification of the whole frame

Element (no.)	Theoretical k_i values (kN-m)	Identified k_i values using responses at 24 DDOFs		Identified k_i values using responses at 27 DDOFs	
		Identified (kN-m)	Error (%)	Identified (kN-m)	Error (%)
(1)	(2)	(3)	(4)	(5)	(6)
1-2 (*1*)	13476	13497	0.2	13485	0.1
5-6 (*2*)	13476	12803	-5.0	13330	-1.1
9-10 (*3*)	13476	15060	11.8	14034	4.1
3-4 (*4*)	13476	12972	-3.7	13449	-0.2
7-8 (*5*)	13476	14088	4.5	13362	-0.8
11-12 (*6*)	13476	15414	14.4	14462	7.3
1-3 (*7*)	13476	13390	-0.6	13446	-0.2
5-7 (*8*)	13476	13850	2.8	13577	0.7
9-11 (*9*)	13476	13040	-3.2	13453	-0.2
2-4 (*10*)	13476	13083	-2.9	13280	-1.5
6-8 (*11*)	13476	13887	3.0	13785	2.3
10-12 (*12*)	13476	14287	6.0	13474	0.0
1-5 (*13*)	14553	14426	-0.9	14507	-0.3
3-7 (*14*)	14553	14164	-2.7	14431	-0.8
5-9 (*15*)	14553	14177	-2.6	14349	-1.4
7-11 (*16*)	14553	12815	-11.9	13934	-4.3
9-13 (*17*)	14553	14700	1.0	14650	0.7
11-15 (*18*)	14553	16438	13.0	14900	2.4
2-6 (*19*)	14553	14745	1.3	14702	1.0
4-8 (*20*)	14553	14029	-3.6	15036	3.3
6-10 (*21*)	14553	15274	5.0	14802	1.7
8-12 (*22*)	14553	14692	1.0	14224	-2.3
10-14 (*23*)	14553	13770	-5.4	14240	-2.2
12-16(*24*)	14553	12921	-11.2	14353	-1.4

The frame is then identified using 27 responses; and the results improved considerably, with a maximum error of identification 7.3% as can be seen in column 6. Identified stiffness parameters did not vary significantly from member to member, indicating that the 3-D GILS-EKF-UI procedure accurately identified the whole frame. This exercise indicates the benefit of additional response information; in fact, if all the responses are made available, the order of error will be similar to that of the substructure. Considering the accuracy in identification, 27 responses are considered to be the absolute minimum responses required for this frame. The defective frames will also be identified using the same number of responses.

9.1.4. *Health assessment of defective frames*

After successfully assessing the defect-free state, the following three defective states are considered: (a) defect 1 - thickness of the cross-section of member 2 is reduced by 30%, (b) defect 2 - thickness of the cross-section of member <u>*12*</u> is reduced by 30%, and (c) defect 3 - thicknesses of the cross-sections of members <u>*2*</u>, <u>*7*</u>, and <u>*12*</u> are reduced by 30%, 20%, and 50%, respectively. It is important to note that in the first two cases, the defective member is not in the substructure. In the third case one defective member is in the substructure. These types of defects are very common in real structures. Martinez-Flores and Haldar (2007) and Martinez-Flores *et al.* (2008) experimentally verified a two-dimensional frame in presence of this type of defect. Inspired by the experience gained during experimental verification, the thicknesses of the flanges and web are reduced over the entire length by the different magnitudes. The reduction of 30%, 20%, and 50% in the web and flange thicknesses caused the reductions in the principal moment of inertias about major axis (I_z) by 32.26%, 22.15%, 52.14%, respectively; principal moment of inertias about the minor axis (I_y) by 30.07%, 20.05%, 50.07%, respectively; and cross-sectional areas (A) by 30.65%, 20.74%, 50.47%, respectively. The FE representation of the frame remains the same as for defect-free state. For the generation of responses, the modified cross-sectional properties of the defective member(s) are considered. Using the similar procedure as defect-free state, the responses at 24 DDOFs are simulated using ANSYS. Three additional responses are scaled at node 4 during identification of the whole frame. The identified results are presented in Table 2(a) and (b).

The results clearly indicate the locations of the defective members and the severity of defects. This example establishes that the 3-D GILS-EKF-UI method can identify single or multiple defective members and the levels of degradations in them. The defective member need not be in the substructure. However, the defect predictability and accuracy increase if the defect is in the substructure or close to it and the number of substructures should be more than one since the defect(s) will be close to one of the substructures. The area is under extensive investigation by the research team. The results also indicate that the response scaling approach works well for the defective cases also.

Table 2. Stiffness parameter k_i (= $E_i I_{zi}/L_i$) identification of defective frames using 3-D GILS-EKF-UI

(a) Stage 1 – Identification of the substructure

Element (no.)	Theoretical k_i values (kN-m)	Identified k_i values					
		Defect-1		Defect-2		Defect-3	
		Identified (kN-m)	Error (%)	Identified (kN-m)	Error (%)	Identified (kN-m)	Error (%)
(1)	(2)	(3)	(4)	(5)	(6)	(7)	(8)
1-2 (<u>*1*</u>)	13476	13476	-0.002	13476	-0.002	13572	0.712
1-3 (<u>*7*</u>)	13476	13475	-0.009	13475	-0.008	10598	**-21.36**
1-5 (<u>*13*</u>)	14553	14552	-0.007	14552	-0.006	14697	0.987

(b) Stage 2 – Identification of the whole frame

Element (no.)	Theoretical k_i values (kN-m)	Identified k_i values using responses at 27 DDOFs					
		Defect-1		Defect-2		Defect-3	
		Identified (kN-m)	Error (%)	Identified (kN-m)	Error (%)	Identified (kN-m)	Error (%)
(1)	(2)	(3)	(4)	(5)	(6)	(7)	(8)
1-2 (1)	13476	13532	0.4	13530	0.4	13687	1.6
5-6 (2)	13476	9209	**-31.7**	13339	-1.0	9227	**-31.5**
9-10 (3)	13476	13647	1.3	13682	1.5	12761	-5.3
3-4 (4)	13476	13458	-0.1	13605	1.0	13130	-2.6
7-8 (5)	13476	13317	-1.2	13093	-2.8	15190	12.7
11-12 (6)	13476	14680	8.9	14250	5.7	12457	-7.6
1-3 (7)	13476	13425	-0.4	13426	-0.4	10467	**-22.3**
5-7 (8)	13476	13633	1.2	13627	1.1	13859	2.8
9-11 (9)	13476	13479	0.0	13191	-2.1	13082	-2.9
2-4 (10)	13476	13244	-1.7	13394	-0.6	13470	0.0
6-8 (11)	13476	13964	3.6	13361	-0.9	13201	-2.0
10-12 (12)	13476	13285	-1.4	9437	**-30.0**	6699	**-50.3**
1-5 (13)	14553	14491	-0.4	14499	-0.4	14481	-0.5
3-7 (14)	14553	14363	-1.3	14393	-1.1	14197	-2.4
5-9 (15)	14553	14557	0.0	14101	-3.1	16691	14.7
7-11 (16)	14553	13863	-4.7	13623	-6.4	15481	6.4
9-13 (17)	14553	14459	-0.6	15025	3.2	13431	-7.7
11-15 (18)	14553	14881	2.3	15558	6.9	14353	-1.4
2-6 (19)	14553	14604	0.4	14685	0.9	14955	2.8
4-8 (20)	14553	14932	2.6	15099	3.8	16421	12.8
6-10 (21)	14553	14715	1.1	15009	3.1	16254	11.7
8-12 (22)	14553	14162	-2.7	14295	-1.8	15716	8.0
10-14 (23)	14553	14507	-0.3	14054	-3.4	13858	-4.8
12-16 (24)	14553	14372	-1.2	14008	-3.7	13375	-8.1

9.2. *Example 2 - health assessment of a 3-D truss-frame*

Real 3-D structures often consist of some truss members to provide lateral stability, for example braced buildings, bridges, towers, etc. They are sometimes denoted as truss-frame structures. The following sections will consider the SHA of this type of structures using 3-D GILS-EKF-UI procedure.

9.2.1. *Description of the truss-frame*

The configuration of the frame used in the previous example is modified by introducing three bracings on the front face as shown Figure 3. Thus the new truss-frame configuration will consist of 27 members; 24 frame members and 3 truss members. The cross-sectional properties of the frame members remain the same. The cross-sectional areas for the truss members are considered to be

100 cm^2. Accordingly their actual theoretical stiffness parameters k_i defined in terms of E_iA_i/L_i are calculated to be 203084 kN/m. The first two natural

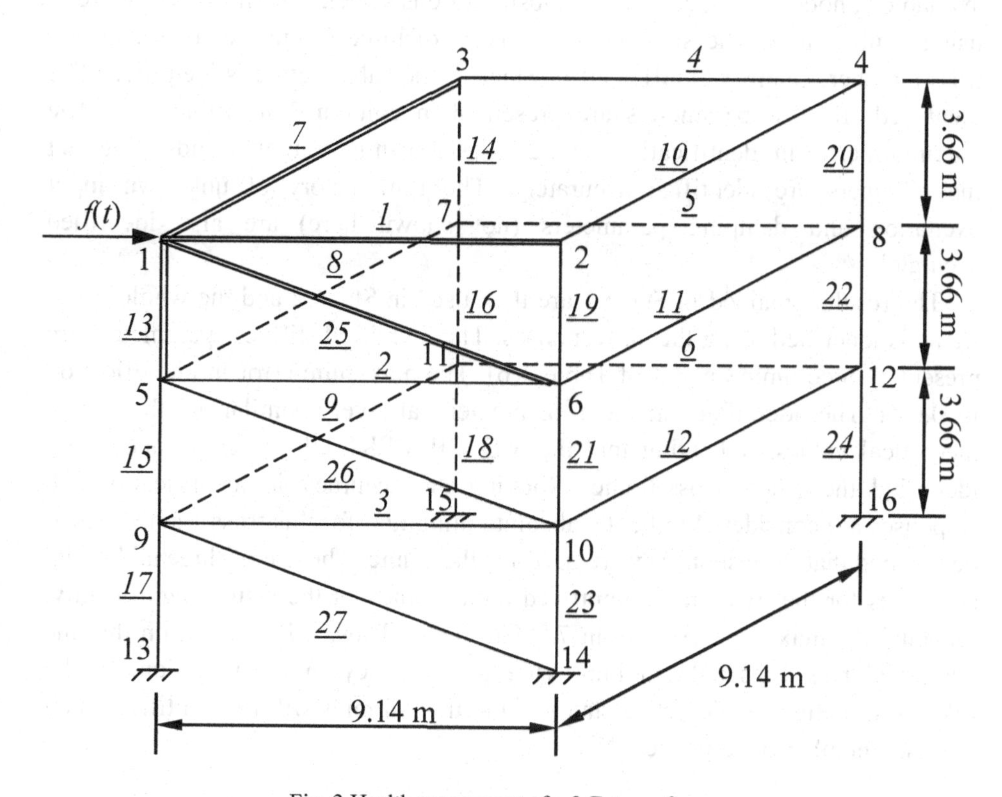

Fig. 3 Health assessment of a 3-D truss-frame.

frequencies of the truss-frame are estimated to be f_1 = 3.1048 Hz and f_2 = 4.5286 Hz, respectively. Following the same procedure mentioned before, the Rayleigh damping coefficients α and β are calculated to be 1.1573 and 0.002085, respectively, for an equivalent modal damping of 5%.

Theoretical responses for the truss-frame are generated by applying a sinusoidal load $f(t)$ = 45 Sin(40t) kN at node 1 in the horizontal direction as shown in Figure 3. Following the similar procedure as for the frame, the responses are obtained at nodes 1, 2, 3, 5, and 6 at every 0.0001 s time interval for duration of 0.8 seconds. The information on the input excitation is completely ignored after obtaining the responses. Responses between 0.16 s and 0.46 s providing 3001 samples are used in the subsequent health assessment processes.

9.2.2. *Health assessment of a defect-free truss-frame*

Defect-free state of the truss-frame is considered first. Based on 30 responses available at node 1, 2, 3, 5, and 6, a substructure is selected as shown in Figure 3, using double lines. The substructure consists of three frame members and one truss member. Using 3-D GILS-UI in Stage 1, the substructure is identified. The identified stiffness parameters are presented in column 3 in Table 3(a). The maximum error in identification is 0.021% as shown in column 4, indicating that the members are identified accurately. The time-history of unknown input excitation and damping parameters (not shown here) are also identified accurately.

The results obtained in Stage 1 are then used in Stage 2 and the whole truss-frame is identified using the 30 responses. The identified stiffness parameters are presented in column 3 and 4 of Table 3(b). The maximum error in identification is 0.6%. The identified stiffness parameters are very similar to the actual theoretical values, indicating that the 3-D GILS-EKF-UI procedure accurately identified the whole truss-frame. Noticing the accuracy in identification, 30 responses are considered to be the absolute minimum for this truss-frame. It is to be recalled that 27 responses were used for the frame. The use of three additional responses for the truss-frame improved the accuracy of the results significantly, reducing the maximum error from 7.3% to 0.6%. Thus, it is clear again that the additional responses help in improving the accuracy in identification. In the following section the defective states of the truss-frames will be identified using the same number of responses.

9.2.3. *Health assessment of defective truss-frames*

After successfully assessing the defect-free state, the following two defective states are considered: (i) Defect 1 - thickness of the cross-section of member *1* is reduced by 90%, and (ii) Defect 2 - thickness of the cross-section of member *26* is reduced by 10%. It is important to note that Defect 1 in the frame member is in the substructure. Defect 2 in the truss member is very close to the substructure. The reduction of 90% of the web and flange thicknesses of the frame member caused the reductions in the principal moment of inertia about major axis (I_z) by 90.64%; principal moment of inertia about the minor axis (I_y) by 90.02%; and cross-sectional area (A) by 90.09%. And for 10% reduction of the thickness of truss member, the cross-sectional area of the member is reduced by 9.71%. The FE representation of the truss-frame remains the same as for defect-free state. Using the similar procedure, the responses are obtained at 30 DDOFs and the defective truss-frames are identified. The identified results are presented in Table 3(a) and (b).

The results clearly indicate the locations of the defective members and the severity of defects. This exercised demonstrated that the 3-D GILS-EKF-UI method can identify defects in truss-frame structures.

10. Conclusions

A novel structural health assessment technique is under development by the research team at the University of Arizona by combining several desirable features of an extended Kalman filter (EKF)-based system identification (SI) procedure and an iterative least-squares-based procedure with unknown input developed by the team earlier. The structure is represented by finite elements and the health of each element can be assessed by comparing stiffness of each element by comparing with its previous value if periodic inspections were conducted, or expected value obtained from the design drawings, or relative value compared to the other elements. The most desirable feature of the method is that it can assess structural health using few noise-contaminated dynamic response information measured only a small part(s) of the structure, denoted as substructure(s), and without using any information on input excitation. Different stages of development of the procedure are presented to consider various implementation issues, emphasizing most recent advances. Recently, the procedure was extended for SHA of large real three-dimensional (3-D) structures. It is denoted as 3-D Generalized Iterative Least-Squares Extended Kalman Filter with Unknown Input (3-D GILS-EKF-UI). The application potential of the procedure is demonstrated with examples of 3-D structures consisting of frame and truss systems. Dynamic responses at a small part of the structures, were

Table 3. Stiffness parameter k_i (= E_iI_{zi}/L_i or E_iA_i/L_i) identification of truss- frame using the 3-D GILS-EKF-UI

(a) Stage 1 — Identification of the substructure

Element (no.)	Theoretical k_i values (kN, m)	Identified k_i values					
		Defect-free		Defect-1		Defect-2	
		Identified (kN, m)	Error (%)	Identified (kN, m)	Error (%)	Identified (kN, m)	Error (%)
(1)	(2)	(3)	(4)	(5)	(6)	(7)	(8)
1-2 (*1*)	13476	13473	-0.021	1267	**-90.60**	13473	-0.021
1-3 (*7*)	13476	13475	-0.012	13507	0.229	13475	-0.012
1-5 (*13*)	14553	14552	-0.007	14595	0.291	14552	-0.007
1-6 (*25*)	203084	203067	-0.008	203678	0.293	203067	**-0.008**

(b) Stage 2 — Identification of the whole truss-frame

Element (no.)	Theoretical k_i values (kN, m)	Identified k_i values using responses at 30 DDOFs					
		Defect-free		Defect-1		Defect-2	
		Identified (kN, m)	Error (%)	Identified (kN, m)	Error (%)	Identified (kN, m)	Error (%)
(1)	(2)	(3)	(4)	(5)	(6)	(7)	(8)
1-2 (*1*)	13476	13482	0.0	1341	**-90.1**	13549	0.5
5-6 (*2*)	13476	13472	0.0	13521	0.3	13485	0.1
9-10 (*3*)	13476	13481	0.0	13521	0.3	13263	-1.6
3-4 (*4*)	13476	13465	-0.1	13464	-0.1	13357	-0.9
7-8 (*5*)	13476	13529	0.4	13560	0.6	13787	2.3
11-12 (*6*)	13476	13488	0.1	13517	0.3	12797	-5.0
1-3 (*7*)	13476	13476	0.0	13404	-0.5	13454	-0.2
5-7 (*8*)	13476	13459	-0.1	13529	0.4	13446	-0.2
9-11 (*9*)	13476	13475	0.0	13412	-0.5	13506	0.2
2-4 (*10*)	13476	13475	0.0	13415	-0.5	13453	-0.2
6-8 (*11*)	13476	13465	-0.1	13505	0.2	13457	-0.1
10-12 (*12*)	13476	13468	-0.1	13418	-0.4	13482	0.0
1-5 (*13*)	14553	14545	-0.1	14676	0.8	14501	-0.4
3-7 (*14*)	14553	14549	0.0	14550	0.0	14686	0.9
5-9 (*15*)	14553	14514	-0.3	14547	0.0	14154	-2.7
7-11 (*16*)	14553	14515	-0.3	14528	-0.2	14154	-2.7
9-13 (*17*)	14553	14584	0.2	14655	0.7	14829	1.9
11-15 (*18*)	14553	14588	0.2	14521	-0.2	14897	2.4
2-6 (*19*)	14553	14548	0.0	14677	0.9	14555	0.0
4-8 (*20*)	14553	14565	0.1	14564	0.1	14214	-2.3
6-10 (*21*)	14553	14534	-0.1	14552	0.0	14168	-2.6
8-12 (*22*)	14553	14583	0.2	14571	0.1	14543	-0.1
10-14 (*23*)	14553	14567	0.1	14647	0.6	14753	1.4
12-16 (*24*)	14553	14570	0.1	14580	0.2	14783	1.6
1-6 (*25*)	203084	202988	0.0	203603	0.3	202998	0.0
5-10 (*26*)	203084	201792	-0.6	202159	-0.5	175018	**-13.8**
9-14 (*27*)	203084	204201	0.6	205871	1.4	214894	5.8

numerically simulated under harmonic excitation for defect-free and several defective states. The procedure accurately identified the defect-free and defective structures, indicated the defective element(s), and identified various stages of degradations. It was observed that single or multiple members can be defective at the same time, and they may not be in the substructure. However, the defect predictability increases when the defects are in substructure or close to it. The study advanced several areas of extended Kalman Filter and iterative least-squares procedures.

Acknowledgements

This chapter is based on work partly supported by University of Arizona Foundation, and the Department of Civil Engineering and Engineering

Mechanics of the University of Arizona. Some of the graduate students also received financial supports from many sponsors. Some of the collaborators of the work presented here are Drs. Duan Wang, Peter H. Vo, Xiaolin Ling, Hasan Katkhuda, Flores Martinez-Flores, Ajoy Kumar Das. Any opinions, findings, conclusions, or recommendations expressed in this chapter are those of the authors and do not necessarily reflect the views of the sponsors.

References

Aditya, G. and Chakraborty, S. (2008). Sensitivity based health monitoring of structures with static responses, *Scientia Iranica*, 15(3), pp. 267-274.

Anh, T.V. (2009). Enhancements to the damage locating vector method for structural health monitoring, PhD Dissertation, National University of Singapore, Singapore.

ANSYS version 11.0, The Engineering Solutions Company, 2007.

Banan, M.R., Banan, M.R. and Hjelmstad, K.D. (1994a). Parameter estimation of structures from static response. I. Computational aspects, *J. Struct. Engrg.*, 120 (11), pp. 3243—3258.

Banan, M.R., Banan, M.R. and Hjelmstad, K.D. (1994b). Parameter estimation of structures from static response. I. Numerical simulation studies, *J. Struct. Engrg.*, 120(11), pp. 3259—3283.

Bernal. D. (2002). Load vectors for damage localization, *ASCE J. Engrg. Mech.*, 128(1), pp. 7-14.

Carden E.P. and Fanning P. (2004). Vibration based condition monitoring: a review, *Struct. Health Monitoring*, 3(4), pp. 355-377.

Chang P.C., Flatau A. and Liu S.C. (2003). Review paper: health monitoring of civil infrastructure, *Struct. Health Monitoring*, 2(3), pp. 257—267.

Chen, J. and Li, J. (2004). Simultaneous identification of structural parameters and input time history from output-only measurements, *Comput. Mech.*, 33, pp. 365—374.

Clough, R.W. and Penzien, J. (2003). Dynamics of structures, Third edition. Computers & Structures, Inc. CA.

Cox, H. (1964). On the estimation of state variables and parameters for noisy dynamic systems, *IEEE Trans. Automatic Cont.*, 9(1), 5—12.

Das, A.K. and Haldar, A. (2012). Health assessment of three dimensional large structural systems – a novel approach, *Life Cycle Reliability and Safety Engrg.*, 1, pp. 1-14.

Doebling S.W., Farrar C.R., Prime M.B. and Shevitz D.W. (1996). Damage identification and health monitoring of structural and mechanical systems from changes in their vibration characteristics: a literature review, Los Alamos National Laboratory, Report LA-13070-MS.

Fan, W. and Qiao, P. (2011). Vibration-based damage identification methods: a review and comparative study, *Struct. Health Monitoring*, 10(1), pp. 83—111.

Farrar, C.R. and Worden, K. (2007). An introduction to structural health monitoring, *Phil. Trans. R. Soc. A*, 365, pp. 303—315.

Fritzen C-P. (2006). Vibration-based techniques for structural health monitoring, *Structural Health Monitoring*, Edited by Balageas, D., Fritzen, C-P., Güemes, A., ISTE Ltd, CA.

Ghanem, R. and Ferro, G. (2006). Health monitoring for strongly non-linear systems using the ensemble Kalman filter, *Struct. Control and Health Monitoring*, 13, pp. 245-259.

Hajela, P. and Soeiro, F.J. (1989). Structural damage detection based on static and modal analysis, *AIAA J.*, 28(6), pp. 1110—1115.

Hajela, P. and Soerio, F.J. (1990). Recent developments in damage detection based on system identification methods, *Struct. Optim.*, 2, pp. 1—10.

Hjelmstad. K.D., Wood, S.L. and Clark, S.J. (1992). Mutual residual energy method for parameter estimation in structures, *J. Struct. Engrg.*, 118(1), pp. 223—242.

Hoshiya, M. and Saito, E. (1984). Structural Identification by extended Kalman filter, *ASCE J. Engrg. Mech.*, 110(12), pp. 1757—1770.

Hsia, T.C. (1977). System identification: Least-Squares Methods, D. C. Health and Company, Lexington, Massachusetts, Toronto.

Humar, J., Bagchi, A. and Xu, H. (2006). Performance of vibration-based techniques for the identification of structural damage, *Struct. Health Monitoring*, 5(3), pp. 215—241.

Ibanez, P. (1973). Identification of dynamic parameters of linear and non-linear structural models from experimental data, *Nucl. Engrg. and Design*, 25, pp. 30-41.

Jazwinski, A.H. (1970). Stochastic process and filtering theory, Academic Press, Inc. New York.

Kalman, R.E. (1960). A new approach to linear filtering and prediction problems, *Trans. of the ASME—J. Basic Engrg.*, pp. 35—45.

Katkhuda, H. and Haldar, A. (2006). Defect identification under uncertain blast loading, *Optim. Engrg.*, 7, pp. 277–296.

Katkhuda, H. and Haldar, A. (2008). A novel health assessment technique with minimum information, *Struct. Cont. & Health Monitoring*, 15(6), pp. 821—838.

Katkhuda, H., Martinez-Flores, R. and Haldar, A. (2005). Health assessment at local level with unknown input excitation, *ASCE J. Struct. Engrg.*, 131(6), pp. 956—965.

Kerschen, G., Worden, K., Vakakis, A.F., Golinval, J.C. (2006). Past, present and future of nonlinear system identification in structural dynamics, *Mech. Systems and Sig. Processing*, 20(3), pp. 505—592.

Koh, C.G., See, L.M. and Balendra, T. (1991). Estimation of structural parameters in time domain: a substructural approach, *Earthquake Engrg. and Struct. Dyn.*, 20, pp. 787—801.

Kopp. R.E. and Orford, R.J. (1963). Linear regression applied to system identification for adaptive control systems, *AIAA Journal*, 1(10), pp. 2300—2306.

Kozin, F., Natke, H.G. (1986). System identification techniques, *Struct. Safety*, 3, pp. 269—316.

Kun, Z. and Law, S.S. and Zhongdong, D. (2009). Condition assessment of structures under unknown support excitation, *Earthq. Eng. & Eng. Vib.*, 8, pp.103—114.

Lew J.S., Juang J.N. and Longman R.W. (1993). Comparison of several system identification methods for flexible structures, *J. Sound and Vib.*, 167(3), pp. 461—480.

Liang, Y.C. and Hwu, C. (2001). On-line identification of holes/cracks in composite structures, *Smart Mat. and Struct.*, 10(4), pp. 599—609.

Ling, X. and Haldar, A. (2004). Element level system identification with unknown input with Rayleigh damping, *ASCE J. Engrg. Mech.*, 130(8), pp. 877—885.

Lu, Z.R. and Law, S.S. (2007). Identification of system parameters and input force from output only, *Mech. Sys. and Sig. Process.*, 21, pp. 2099—2111.

Martinez-Flores, R. and Haldar, A. (2007). Experimental verification of a structural health assessment method without excitation information, *J. Struct. Engrg.*, 34(1), pp. 33—39.

Martinez-Flores, R., Katkhuda, H. and Haldar, A. (2008). A novel health assessment technique with minimum information: verification, *Int. J. Performability Engrg.*, 4(2), pp. 121—140.

Maybeck, P.S. (1979). *Stochastic models, estimation, and control theory*. Academic Press, Inc. New York.

McCaan, D., Jones, N.P. and Ellis, J.H. (1998). Toward consideration of the value of information in structural performance assessment, *Structural Engineering World Wide*, Paper No. T216-6, CD-ROM.

Nasrellah, H.A. (2009). Dynamic state estimation techniques for identification of parameters of finite element structural models, PhD Dissertation, IISc, Bangalore, India.

Paz, M. (1985). Structural dynamics: theory and computation, Second edition. Van Nostrand Reinhold Company, New York.

Perry, M.J. and Koh, C.G. (2008). Output-only structural identification in time domain: numerical and experimental studies, *Earthq. Engrg. and Struct. Dyn.*, 37, pp. 517—533.

Rytter. A. (1993). Vibration based inspection of civil engineering structures, PhD Dissertation, Department of Building Technology and Structural Engineering, Denmark.

Salawu O.S. (1997). Detection of structural damage through changes in frequency: a review, *Engrg. Struct.*, 19(9), pp. 718—723.

Sanayei, M. and Onipede, O. (1991). Damage assessment of structures using static test data, *AIAA J.*, 29(7), pp. 1174—1179.

Sanayei, M. and Saletnik, M.J. (1996). Parameter estimation of structures from static strain measurements. I: Formulation, *J. Struct. Engrg.*, 122(5), pp. 555—562.

Sanayei, M. and Scampoli, S.F. (1991). Structural element stiffness identification from static test data, *J. Engrg. Mech.*, 117(5), pp. 1021—1036.

Sanayei, M., Imbaro, G.R., McClain, J.A.S. and Brown, L.C. (1997). Structural model updating using experimental static measurements, *ASCE J. Struct. Engrg.*, 123(6), pp. 792-798.

Sandesh, S. and Shankar, K. (2009). Time domain identification of structural parameters and input time history using a substructural approach, *Int. J. Struct. Stability and Dyn.*, 9(2), pp. 243-265.

Saridis, G.N. (1995). Stochastic processes, estimation, and control: the entropy approach, John Wiley & Sons, Inc., New York.

Sheena, Z., Unger, A. and Zalmanovich, A. (1982). Theoretical stiffness matrix correction by using static test results, *Israel J. Tech.*, 20, pp. 245—253.

Sohn, H., Farrar, C.R., Hemez, F.M., Shunk, D.D., Stinemates, D.W., Nadler, B.R., Czarnecki, J.J. (2004). A review of structural health monitoring literature: 1996-2001, Los Alamos National Laboratory, LA-13976-MS.

Toki, K., Sato, T. and Kiyono, J. (1989). Identification of structural parameters and input ground motion from response time histories, *Struct. Engrg./Earthq. Engrg.*, 6(2), pp. 413-421.

Vo, P.H. and Haldar, A. (2003). Post processing of linear accelerometer data in system identification, *J. Struct. Engrg.*, 30(2), pp. 123—130.

Vo, P.H. and Haldar, A. (2004). Health assessment of beams - theoretical and experimental investigation, Special issue on Advances in Health Monitoring/Assessment of Structures including Heritage and Monument Structures, *J. Struct. Engrg.*, 31(1), pp. 23-30.

Wang, D. and Haldar, A. (1994). An element level SI with unknown input information, *ASCE J.* Engrg. *Mech.*, 120(1), pp. 159—176.

Wang, D. and Haldar, A. (1997). System identification with limited observations and without input, *ASCE J. Engrg. Mech.*, 123(5), pp. 504—511.

Wang, X.J. and Cui, J. (2008). A two-step method for structural parameter identification with unknown ground motion, *The 14th world conference on earthq. Engrg.*, Beijing, China.

Welch. G. and Bishop, G. (2006). An introduction to the Kalman filter, Department of Computer Science, University of North Carolina , Chapel Hill, Tech. Rep. TR95-041 2006.

Xu, B., He, J., Rovekamp, R. and Dyke, S.J. (2011). Structural parameters and dynamic loading identification from incomplete measurements: approach and validation, *Mech. Sys. and Sig. Process.*, doi:10.1016/j.ymssp.2011.07.008.

Yang, J.N. and Huang, H. (2007). Sequential non-linear least-square estimation for damage identification of structures with unknown inputs and unknown outputs, *Int. J. Non-linear Mech.*, 42, pp. 789—801.

Chapter 7

Wavelet-Based Techniques for Structural Health Monitoring

Z. Hou[1], A. Hera[2], and M. Noori[3]

[1,2]*Mechanical Engineering Department*
Worcester Polytechnic Institute, 100 Institute Road, Worcester MA 01609, USA.

[3]*Mechanical Engineering Department*
California Polytechnic State University, San Luis Obispo, CA 93407, USA,
Email: [1]hou@wpi.edu, [3]mnoori@calpoly.edu

1. Introduction

Damage is often observed in infrastructural systems during their service life. The damage may be caused by sudden breakage of a structural member due to an excessive response during a severe natural event such as strong earthquake, or by progressive stiffness degradation due to material fatigue under cyclic loading or chemical corrosion in a hazardous environment. Assessment of structural condition is critically important to ensure the structural functionality and safety.

Over the past two decades structural health monitoring (SHM) has emerged as a reliable and economical approach to monitor the system performance, detect damages if incurred, asses/diagnose the structural health condition, and make corresponding maintenance decisions. A SHM system consists of two major components: a network of sensors to collect the response data and data-mining algorithms to extract information on the structural health condition. If damage is detected, or the structural performance becomes unsatisfactory, appropriate control/maintenance actions can be taken. Recent advances in SHM can be found in Doebling, *et al.* (1996), Sohn *et al.* (2003), Carden and Fanning (2004), Chang (2009 and 2011), and Fan and Qiao (2011).

Wavelet analysis has become a promising tool for SHM based on its ability of multi-scale analysis of transient data. Viewed as extended traditional Fourier transform with adjustable window location and size, local data can be examined by a "zoom lens with an adjustable focus". The multi-levels of details and approximations of the original data from discrete wavelet transform (DWT) or the wavelet spectra from continuous wavelet transform (CWT) may reveal rich structural health related information. Applications of wavelet analysis in the SHM area are exemplified by Masuda, *et al.* (1995), Al-Khalidi, *et al.* (1997), Staszewski (1998), Hou *et al.* (2000), Masuda, *et al.* (2002), Hera and Hou 92004), Alonso, *et al.* (2004), Shinde and Hou (2005), Hou *et al.* (2006), and Reda Taha *et al.* (2006), Khatam, *et al.* (2007).

This chapter presents a brief background of wavelet analysis and its SHM applications. The wavelet-based techniques are illustrated for simple structural models with sudden and progressive damage. The wavelet approach is also successfully applied for the ASCE SHM benchmark study structure (Johnson *et al.*, 2004). An innovative wavelet packet based sifting process and its SHM applications are validated by both simulation data and experimental data. Some practical issues are discussed.

2. Brief Background of Wavelet-Based Methodologies for Damage Detection

This section presents a brief overview on wavelet analysis related to wavelet-based techniques for damage detection. General background of wavelet analysis may be found in (Chui, 1992; Daubechies, 1992; Mallat, 1998).

A wavelet is a square-integrable waveform with an effectively finite support in time domain and zero time average. Using a given wavelet function Ψ(t) as the mother wavelet an associated wavelet family can be constructed by two operations: dilation/scaling and translation/shifting, as expressed as

$$\Psi(t; a, b) = \Psi\left(\frac{t-b}{a}\right) \tag{1}$$

where a is the dilation parameter and b is the translation parameter.

Using a selected analyzing or mother wavelet function Ψ(t), the continuous wavelet transform (CWT) of a signal *f(t)* is defined as

$$(Wf)(a,b) = \frac{1}{\sqrt{a}} \int_{-\infty}^{+\infty} f(t)\overline{\Psi}\left(\frac{t-b}{a}\right) dt \tag{2}$$

where a and b are the dilation and translation parameters, respectively. Both are real numbers and a must be positive. The bar over $\Psi(t)$ indicates its complex conjugate. The translation parameter, b, indicates the location of the moving wavelet window. Shifting the wavelet window along the time axis implies examining the signal in the neighborhood of the current window location. Therefore, information in the time domain will still remain, as contrast to the Fourier transform where the time domain information becomes almost invisible after the integration over the entire time domain. The dilation parameter, a, indicates the width of the wavelet window. A smaller value of a implies a higher-resolution filter, i.e. the signal is examined through a narrower wavelet window in a smaller scale.

The signal $f(t)$ may be reconstructed by an inverse wavelet transform of $(Wf)(a,b)$ as defined by

$$f(t) = \frac{1}{C_\psi} \int_{-\infty}^{+\infty} \int_{-\infty}^{+\infty} (Wf)(a,b) \Psi\left(\frac{t-b}{a}\right) \frac{1}{a^2} \, da\, db \qquad (3)$$

where C_Ψ is defined by

$$C_\psi = \int_{-\infty}^{+\infty} \frac{|F_\psi(\omega)|^2}{|\omega|} d\omega \qquad (4)$$

The mother wavelet needs to satisfy the admissibility condition, $C_\psi < \infty$, to ensure existence of the inverse wavelet transform.

In practical signal processing a discrete version of wavelet transform (DWT) is often employed by discretizing the dilation parameter a and the translation parameter b. In general, the procedure becomes much more efficient if dyadic values of a and b are used, i.e.

$$a = 2^j \quad b = 2^j k \quad j,k \in Z \qquad (5)$$

where Z is the set of positive integers. For a special choice of $\psi(t)$, the corresponding discretized wavelets $\{\psi_{j,k}\}$ defined as

$$\Psi_{j,k}(t) = 2^{-j/2} \Psi(2^{-j} t - k) \quad j,k \in Z \qquad (6)$$

constitute an orthogonal basis for $L^2(R)$. Using the orthogonal basis, the wavelet expansion of a function $f(t)$ and the coefficients of the wavelet expansion are defined as follows:

$$f(t) = \sum_j \sum_k \alpha_{j,k} \Psi_{j,k}(t) \tag{7}$$

$$\alpha_{j,k} = \int_{-\infty}^{+\infty} f(t) \overline{\Psi}_{j,k}(t) dt \tag{8}$$

In the discrete wavelet analysis a signal can be represented by its approximations and details. The detail at level j is defined as

$$D_j = \sum_{k \in Z} \alpha_{j,k} \Psi_{j,k}(t) \tag{9}$$

and the approximation at level J is defined as

$$A_J = \sum_{j > J} D_j \tag{10}$$

It becomes obvious that

$$A_{J-1} = A_J + D_J \tag{11}$$

$$f(t) = A_J + \sum_{j \leq J} D_j \tag{12}$$

Equations (11) and (12) provide a tree structure of a signal and also a reconstruction procedure of the original signal. By selecting different dyadic scales, a signal can be broken down into many low-resolution components, referred to as the wavelet decomposition tree. The wavelet tree structure with details and approximations at various levels may reveal valuable information of the signal characteristics that may not be clearly seen in the original data or the results from other approaches. A down-sampling technique (Strang and Nguyen, 1996) can be used to efficiently reduce the data size in the tree and the Mallat's fast wavelet transform algorithm (Mallat, 1988) can be used to greatly reduce the computational efforts involved.

The analysis may be extended to the wavelet packet analysis where at each level of wavelet decomposition, the approximation can also be further decomposed to its own details and approximation. As a result, a wavelet packet tree structure for the original data can be constructed. Details were referred from Chui (1992).

3. Damage Detection Using Simulation Data for a Simple Structural Model

To demonstrate an application of the wavelet analysis for structural damage detection, a set of simulation data from a simple structural model are utilized in this section. The structure is modeled as a single-degree-of-freedom (SDOF) mass-damper-spring system with multiple paralleled springs, as shown in Fig. 1. Each spring in the system is pre-assigned a threshold value and a spring will break if the response exceeds its corresponding threshold value. In the case of accumulated damages, breakage of a spring is governed by the allowable number of cycles of the response based on fatigue testing. The springs are used to model stiffness of structural members such as columns, beams, and joints. Therefore, a broken spring may imply occurrence of structural damage.

The governing equation of motion of the system in Fig. 1 is given by

$$m\frac{d^2}{dt^2}x(t)+c\frac{d}{dt}x(t)+k(t)x(t)=f(t) \tag{13}$$

where *m, c, k* are the system mass, viscous damping, and stiffness coefficients, respectively. x(t) is the displacement response and f(t) is the external excitation.

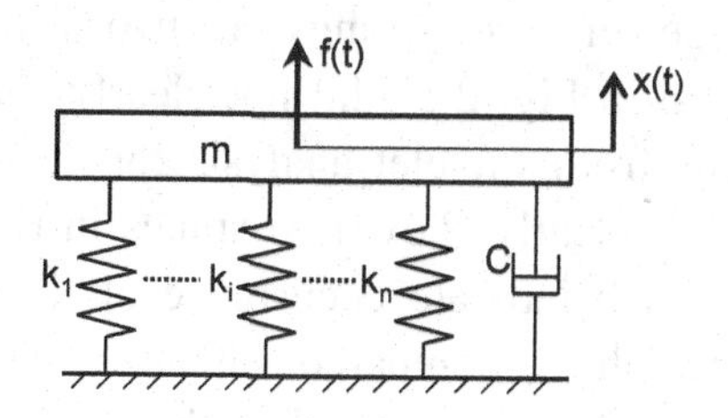

Fig. 1. A simple structural model with multiple parallel breakable springs.

For the sake of simplicity, f(t) is assumed to be a harmonic in this study. The system stiffness k(t) can be expressed by

$$k(t)=\sum_{i=1}^{n}k_i(t) \tag{14}$$

where k_i(t) represents stiffness of the *i*th spring in the system at time t. These springs are breakable due to different types of failure mechanism. For examples, if breakage of a spring is due to an excessive response, k_i(t) can be defined as

$$k_i(t)=\begin{cases}k_{i0}, \text{ if abs}(x(t'))<x_i^* \ \forall t'\le t;\\ 0, \text{ Otherwise.}\end{cases} \tag{15}$$

in which k_{i0} and x_i* denote the initial stiffness and the threshold value for breakage of the *i*th spring, respectively, and both are positive constants. If a spring is broken because of fatigue, k_i(t) may be determined by

$$k_i(t) = \begin{cases} k_{i0}, \text{if } N(t') < N_i^* \ \forall t' \leq t; \\ 0, \ \text{Otherwise.} \end{cases} \quad (16)$$

where N(t) is the total number of cycles of the response in the time interval [0,t] and Ni^* is the allowable number of cycles for the ith spring based on its fatigue testing. The system is nonlinear in general due to possible breakage of springs.

Figures 2 plots the results for the above structural model to a harmonic excitation for the case where the damage is caused by an excessive response. For sake of simplicity, only three springs were used for illustration. An analytical solution does not exist for the nonlinear system expressed by equation (13) with zero initial conditions due to the nonlinearity caused by possible breakage of springs. Therefore, the structural response was numerically calculated by the fourth-order Runge-Kutta integration scheme. The transient stiffness history of the system was traced in the simulation to see whether and when the springs break to validate the wavelet result. While the stiffness history in Fig. 2(c) clearly indicated breakage of two springs at two different moments, the acceleration response curve in Fig. 2(a) did not clearly show sign of structural damage. However, results from wavelet analysis for the acceleration response data, as shown in Fig. 2(b), clearly show two spikes in the level-1 detail of its wavelet decomposition, which indicate occurrence of the structural damage at two different moments, which, as compared with the stiffness history curve in Fig. 2(c), accurately indicate the moments when the structural damage occurred. Note that two spikes at the beginning and the end in Fig. 2(b) are caused by truncation of the response data and, therefore, should be ignored.

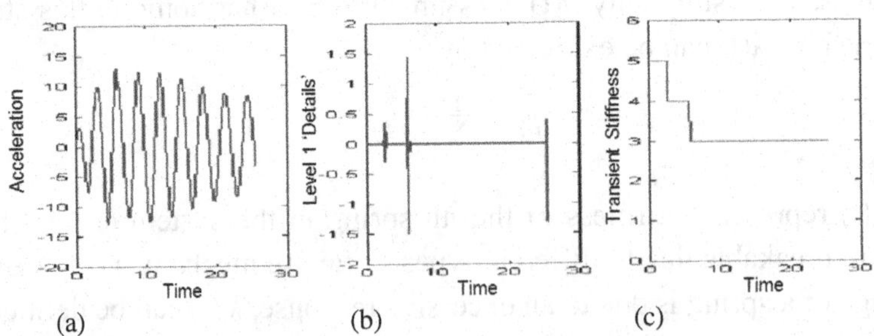

Fig. 2. Response of the simple structural model to a harmonic input and its wavelet analysis. Case 1: abrupt damages due to an excessive response. (a) the acceleration response; (b) its level-1 detail; and (c) the transient stiffness history. [Hou et al., 2000]

Similar results can be obtained for the case of accumulative damage caused by fatigue (Hou *et al.*, 2000). To apply the wavelet approach to on-line health monitoring many practical issues need to be addressed. A concept of detectability map was introduced to investigate the effect of measurement noise and damage severity on the wavelet approach (Hou *et al.*, 2000). It was concluded that damage is more detectable for weaker noise and greater stiffness loss.

Structural damage by mechanical fatigue and chemical corrosion more often is a progressive process and may not necessarily associate with an abrupt stiffness loss and, in turn, a spike in the lower-level details of DWT of the response data may not be seen. CWT may be used in this case to monitor the structural health condition. The ability of wavelet analysis for multi-level analysis in both time and frequency domain may reveal the time-dependent frequency content of the response and therefore, trace the change in structural stiffness.

Figure 3 demonstrates an application of CWT for structural health monitoring of the same SDOF system. The response is simulated for the stiffness history, as reflected in the instantaneous frequency plot in Fig. 3(a) where the middle part models a case of progressive damage. Fig. 3(b) presents the wavelet spectrum for a free-vibration signal where change of the system natural frequency is clearly observed by the bright pattern. Note that a higher scale-value corresponds to a lower frequency. The dominant frequency content in the signal, corresponding to the natural frequency, remains constant first (the undamaged stage), gradually decreases in the middle (the stage of progressive damage), and then remains constant. Similar results can be seen in Fig. 3(c) for a force vibration signal of the same system under a harmonic excitation. Two bright patterns are observed: one for the excitation frequency that remains constant and the other for the system natural frequency that suggests progressive damage in the middle.

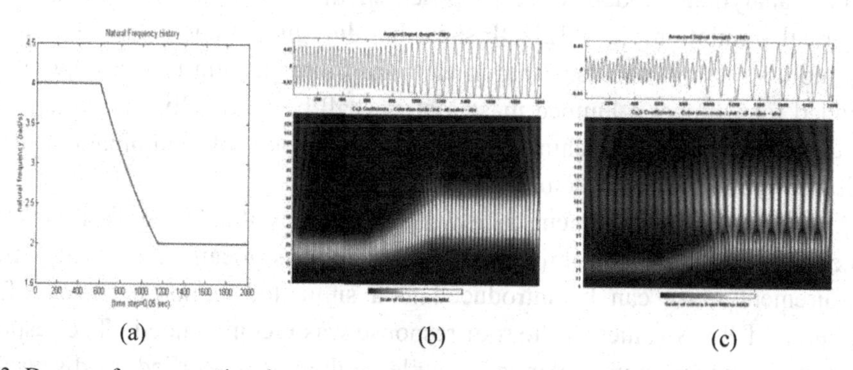

Fig. 3. Damage for progressive damage: (a) the system instantaneous natural frequency calculated from the specified stiffness history in simulation; (b) the wavelet spectrum of a free vibration signal; and (c) the wavelet spectrum of a harmonic response.

4. Wavelet Approach for ASCE SHM Benchmark Study Data

In order to test and compare various damage identification techniques, a benchmark problem was proposed by the ASCE Task Group on Health Monitoring. This section provides an application of the wavelet-based approach for the benchmark study problem. For more detailed information, the reader is referred to Johnson *et al.* (2004).

The structure used in the benchmark study is a four-story, two-bay by two-bay steel-frame scale model structure with a total height of 3.6m and a plan of 2.5m x 2.5 m. Each story is comprised of 9 columns of 0.9 m each in height, 12 beams of 1.25m each in length, 8 lateral braces and 4 floor weights (one per bay). The weights of the first floor are 800 kg and those on the second and the third floor are 600 kg. On the fourth floor there are either four identical weights of 400 kg for a symmetric model or three weights of 400 kg and the fourth one of 550 kg in order to create some asymmetry in the structure. All members are hot rolled steel with nominal yield stress of 300 MPa.

In this study damage is simulated by removing braces in the structural model. Various damage patterns are proposed by the ASCE SHM Task Group. The excitation is either low-level ambient wind loading at each floor in the horizontal y-direction perpendicular to the floors or a shaker force applied on the roof at the center column position. To take the uncertainty of environmental loads into account, the wind loading is modeled as filtered Gaussian white noise processes passed through a 6th order low-pass Butterworth filter with a 100 Hz cut-off frequency.

Two analytical models were proposed in the ASCE benchmark study for numerical simulation, i.e. a 12DOF shear building model and a 120 DOF model with more structural details. Both models are finite element based. Options are provided for whether a lumped mass matrix shall be used. Six simulation cases are classified in the benchmark study as the result of combination of the excitation types and the structural models.

A MATLAB finite element code was provided by the ASCE Task Group to numerically simulate dynamic response at various locations on each floor. Measurement noise can be introduced to a simulated response. 10% of the maximum of RMS values of the roof response was recommended. The response data in the study are non-contaminated unless otherwise specified for discussions on the noise effect.

DWT was used to analyze the ASCE benchmark study data. Different types of mother wavelets were tested and good results were obtained in general. The results presented herein were obtained using the db4 wavelet (Daubechies, 1992).

The level-1 detail of the floor acceleration response data is presented to evidence occurrence of sudden damage. The sampling time used in the following results is 0.001 second.

Damage Detection Fig. 4 presents wavelet results for a case of minor damage where only one brace was broken at time t = 5 seconds. The change in the first natural frequency was about 5%. The acceleration response at a node 15 that is originally connected to the removed brace is used for the analysis. Fig. 4(a) shows the Fourier spectra of the response data before and after the damage at t=5 seconds and Fig. 4(b) plots the level-1 wavelet details for the whole data set. While the damage and the moment when it occurred can be clearly recognized in the wavelet results, the change in the natural frequencies and, in turn, the corresponding damage became somewhat difficult to be identified from the Fourier spectra in Fig. 4(a) in this case of minor damage. For the purpose of comparison, Fig. 4(b) also includes the wavelet results for the acceleration response data at nodes 13 and 14. It is also concluded that the damage can be detected from the acceleration response at node 14 that is close to the damage location but not from the data at node 13 that is relatively further away from the damage location.

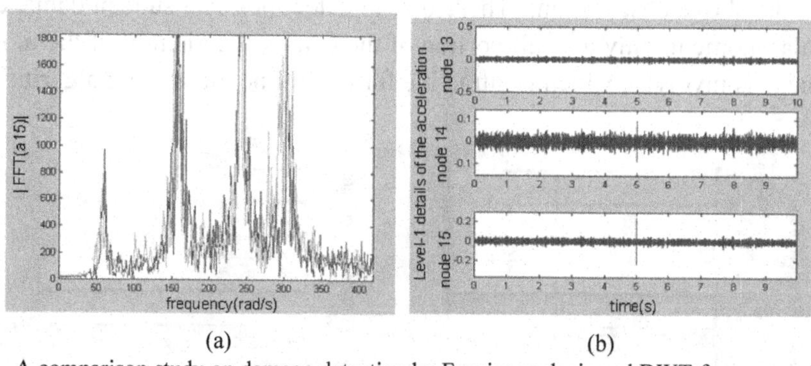

(a) (b)

Fig. 4. A comparison study on damage detection by Fourier analysis and DWT for a case of minor damage: (a) Fourier spectra of the acceleration response at node 15 before and after the damage; (b) the level-1 details of the acceleration responses at node 15 at the bottom and neighboring nodes 13 at top and 14 in the middle (Hera and Hou, 2004).

Locating Damage Region Consider the case of 12-DOF structural model and load applied at all stories in the y-direction. The numerical simulation was implemented for a damage pattern with braces breaking in the following order: (1) all braces of the fourth story were first broken at t = 5 seconds; (2) at t = 10 seconds, all braces of the second stories were then broken; (3) at t = 15 seconds, more braces of the third story were also broken; and finally (4) at t = 20 seconds all braces of the first story were broken. Results from this simulation are plotted in Fig. 5. Note that response in the x-direction should be zero due to the structural symmetry and the load direction.

The spikes in Fig. 5 clearly record occurrence of the damage and the moments when the damage occurred. The pattern of the spikes may also reveal the location of damage. For example, damage at t=5 seconds is detected from the spikes in the wavelet level-1 details of the responses at nodes 31 and 33 on the 3rd floor and nodes 40 and 42 on the 4th floor, which is an indication that some braces connecting the 3rd and 4th floors were removed. It can be concluded from the spike pattern that braces connecting the 3rd and the 4th floors were removed at t=5 seconds, the braces connecting the 1st and the 2nd floors were removed at t=10 seconds, the braces connecting the 2nd and the 3rd floors were removed at t=15 seconds, and the braces connected the 1st floor and the ground were removed at t=20 seconds. The pattern of the spikes agrees exactly with the damage sequence specified in the simulation.

The results in Fig. 5 also demonstrate the potentials of the wavelet approach for on-line monitoring. Note that the wavelet approach is a multi-resolution analysis for a piece of data windowed by shifted and scaled wavelets generated from a mother wavelet. Furthermore, only the higher-resolution details were used in the above observation. Therefore, to detect possible sudden damage at a particular moment, only a small portion of the data in the vicinity of the moment of damage is involved, which is attractive for an on-line monitoring algorithm.

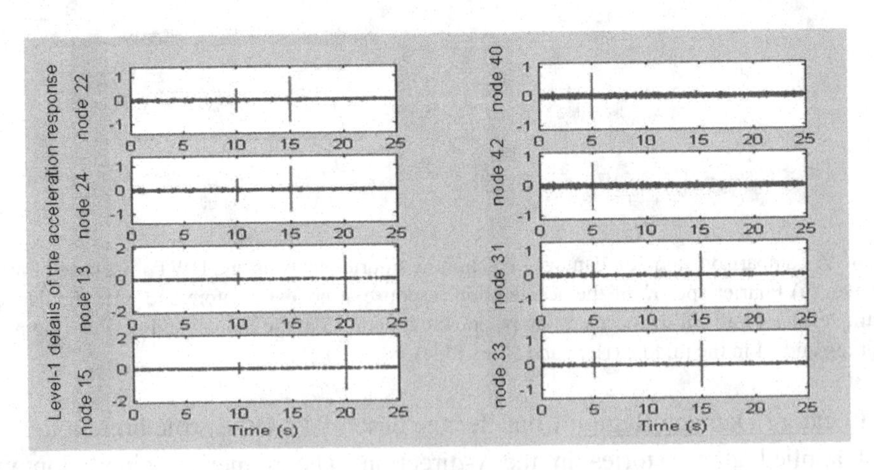

Fig. 5. The level-1 details for acceleration responses at different nodes for the case where braces connecting floors were removed at different times (Hera and Hou, 2004).

The noise effect and the dependence on modeling error of the wavelet approach were also investigated (Hera and Hou, 2004). In general, the wavelet approach for SHM becomes less effective if a vibration signal is contaminated significantly in the case of minor damage. It was also found that the wavelet approach is insensitive to the modeling. Similar results can be obtained for both the relatively complicated 120DOF finite element model and the simpler 12DOF shear beam model. Sudden damage can be detected for both symmetric and asymmetric models. There is only insignificant difference whether a lumped mass matrix or a consistent mass matrix is used in the finite element modeling.

5. SHM by the Wavelet-packet Based Sifting Process

This section presents an innovative sifting process based on the wavelet packet decomposition of a signal. The original signal can be decomposed into its dominant components with nearly distinct frequency contents by performing a wavelet packet analysis with an appropriate mother wavelet based on their percentage energy contribution to the original signal (Shinde and Hou, 2005). The wavelet-packet based sifting process is applied for structural health monitoring, as illustrated in a few study cases using both simulation data and experimental data.

5.1. *Wavelet Packet (WP) Decomposition*

Discrete wavelet transform provides a tree structure of a signal and therefore the signal can be decomposed into different levels of approximations and details with different frequency contents. In certain applications where the information in higher frequency components is important, the frequency resolution of the conventional discrete wavelet decomposition may not be fine enough to extract pertinent frequency information about the signal. The necessary frequency resolution may be achieved by using wavelet packet decomposition, an extension of the discrete wavelet decomposition where the wavelet detail at each level is, in addition to decomposition of only the wavelet approximation in the regular wavelet analysis, further decomposed to its own approximation and detail components. By this process, some lower frequency contents leaked in the wavelet details at the previous level can be further sifted out at the current level and also the frequency resolution for signal analysis increases. As a result, the wavelet packet analysis may provide better accuracy in both higher and lower frequency components of the signal.

An innovative sifting process based on the wavelet packet analysis was proposed in Shinde and Hou (2005). The process starts with interpolation of data with a cubic spline interpolation scheme. Note that as in the case of DWT, the time resolution decreases as the number of decomposition levels increases in the case of wavelet packet decomposition. The interpolated data increases the time resolution of the signal which will in turn increase the regularity of the decomposed components. The cubic spline interpolation assures the conservation of signal data between sampled points.

The interpolated data are then decomposed into different frequency components by using wavelet packet decomposition. A symmetrical wavelet is preferred in the process to guarantee symmetrical and regular shaped decomposed components. Daubechies wavelets of higher order and discretized Meyer wavelet show good symmetry and lead to symmetrical and regularly-shaped components.

To evaluate importance of the wavelet packet components to a signal, the concept of entropy is often applied in signal processing. There are various definitions of entropy in the literature. Among them, two representative ones, i.e. the energy entropy and the Shannon entropy, are used. The wavelet packet node energy entropy at a particular node n in the wavelet packet tree of a signal is a special case of P=2 of the P-norm entropy, defined as

$$e_n = \sum_k \left|\omega_{n,k}\right|^P \qquad (P \geq 1) \tag{17}$$

where $\omega_{n,k}$ denotes the wavelet packet coefficients corresponding to node n at time k. It was demonstrated that the wavelet packet node energy has more potential for use in signal classification as compared to the wavelet packet node coefficients alone (Yen and Lin, 2000). The wavelet packet node energy represents energy stored in a particular frequency band and is mainly used to extract the dominant frequency components of the signal. The Shannon entropy is defined as

$$e_n = -\sum_k \omega_{n,k}^2 \log(\omega_{n,k}^2) \tag{18}$$

Note that one can define his/her own entropy function if necessary.

In case of the binary wavelet packet tree, decomposition at level n results in 2^n components. This number may become very large at a higher decomposition level. To reduce the number, an optimum wavelet packet tree of the signal is obtained by imposing a best entropy criterion in the process. A particular node (N) is split into two nodes N_1 and N_2 if and only if the sum of the entropy of those decomposed nodes N_1 and N_2 is lower than the entropy of node N, thus the entropy of WP decomposition is kept as least as possible (Coifman and Wickerhauser, 1992).

To sift out the significant components of the signal, the percentage contribution of energy entropy of an individual component to the total signal is calculated and evaluated. The components with a higher contribution are sifted out as the dominant mono-frequency components. Other criteria such as the minimum number of zero crossings and the minimum peak value of components can also be applied as stopping criteria of the sifting process. Note that the decomposition is unique if the mother wavelet in the wavelet packet analysis is given and the optimal entropy and the sifting criteria are specified.

5.2. *Instantaneous modal parameters*

By incorporating Hilbert transform with the sifting process, the instantaneous frequency and instantaneous amplitude variation of the sifted components can be found. As a general definition, the instantaneous frequency of a signal at time t can be expressed as the rate of change of phase angle function of the analytic function obtained by Hilbert Transform of the signal (Boashash, 1992). The analytic function $z(t)$ of a signal $f(t)$ is a complex signal having the original signal $f(t)$ as its real part and Hilbert transform of the original signal as its imaginary part. By representing the analytic function in the polar coordinate form one obtains

$$z(t) = f(t) + iH[f(t)] = a(t)e^{i\Phi(t)} \tag{19}$$

where $a(t)$ is the instantaneous amplitude and $Ø(t)$ is the instantaneous phase function. Thus, the instantaneous frequency can be calculated by

$$\omega(t) = \frac{d\Phi}{dt} \tag{20}$$

In general, the concept of instantaneous frequency provides an insightful description as how the frequency content of the signal and varies with the time. Note that when vibration signal is decomposed into its mono-frequency components, these components often represent modal responses. Therefore, an instantaneous frequency history from Hilbert transform of these mono-components may indicate change in the natural frequency and, in turn, the health condition.

If measurement data at all degree-of-freedoms (DOF) of an M-DOF system are available, the modal response of a particular mode can be obtained by first sifting all the response data and then selecting the mono-components corresponding the same instantaneous frequency of this particular mode. Using the instantaneous amplitudes of the ith DOF of the system, $a_i(t)$ (i=1,2,...M) obtained from Hilbert Transform of these mono-components, one for each DOF the amplitude ratios of $a_i(t)/a_k(t), i=1,2,\cdots M$ represent the instantaneous mode shape of this particular mode normalized with respect to the *k*-th DOF. Change in the instantaneous modal shapes may also indicate structural damage.

5.3. *Numerical validation*

The wavelet packet based sifting process is validated by analyzing a vibration signal from a linear three-degree-of-freedom (3DOF) spring-mass-dashpot system, as shown in Fig. 6. The values of structural parameters of the system are M1= M2 = M3 = 300 Kg and K1 = K2 = K3 = 100000 N/m. The system natural frequencies are 1.29, 3.62 and 5.23 Hz, respectively. To demonstrate the basic concepts, zero damping is assumed here without loss of generality. An impact force of intensity 1000 N was applied to the first mass element (M1) at time 0.1 sec for one time-step of *dt* where *dt* is the time-domain resolution of the signal. Dynamic response of the system was numerically simulated with a sampling rate of 100Hz.

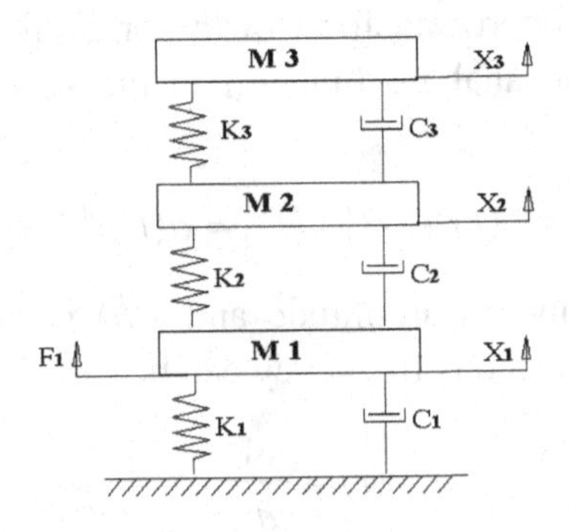

Fig. 6. Sketch of a 3DOF mass-spring-damper system used in the simulation study.

Without loss of generality, acceleration response at the second mass element is selected to illustrate the concept and accuracy of the WP decomposition. By applying the WP sifting process with db36 as mother wavelet, the original signal is decomposed into three dominant components as shown in Fig. 7(a). The time resolution of the signal is increased by using a spline interpolation. For this time invariant system, the Fourier spectra of these components are plotted in Fig. 7(b). The peak frequencies are almost identical to the natural frequencies, as expected.

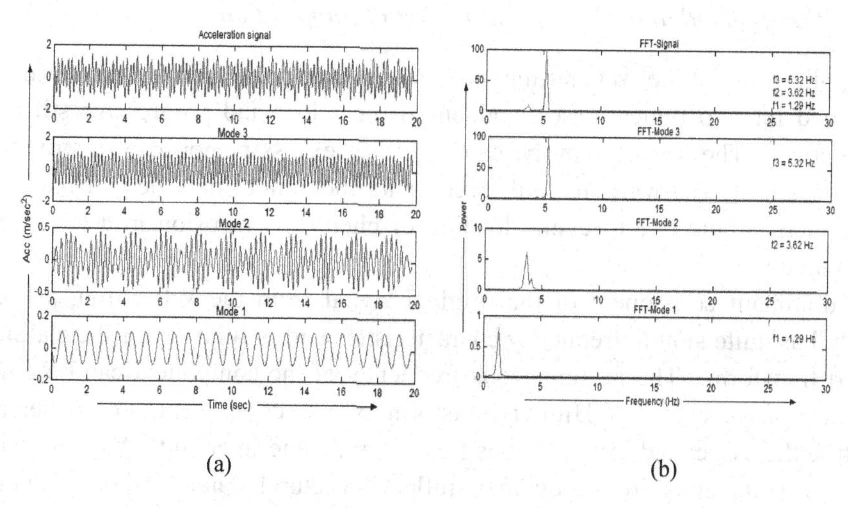

(a) (b)

Fig. 7. WP decomposition of an acceleration response signal of a linear 3DOF system. (a) the WP components and (b) their Fourier spectra. The top is for the original signal and the bottom three are for its dominant components sifted. (Shinde and Hou, 2005).

Table 1 shows the percentage energy contribution of the sifted dominant frequency components to the original acceleration signal at M2. By evaluating the percentage energy contribution of the decomposed components, dominant components of the signal thus can be sifted out.

Table 1. Percentage contribution of energy entropy of dominant frequency components of a vibration signal

Mode No.	Wavelet Packet Node No.	Percentage Contribution of Energy in Wavelet Packet Tree (%)
1	(9,0)	4.1847
2	(9,1)	10.0003
3	(8,1)	83.8203

The wavelet packet based sifting process was compared with the well-recognized sifting process in the EMD method (Huang et al., 1998) for different types of signal (Shinde and Hou, 2005). While the two approaches provide similar results for most of signals, better results were observed for certain type of signals by the present WP based technique.

5.4. *SHM application of the wavelet-packet decomposition*

An application of the WP sifting process for structural health monitoring is illustrated for two typical cases: sudden stiffness loss and progressive stiffness degradation. The former may be caused by an excess response of a structural member during a severe natural event and the latter may be attributed to mechanical fatigue due to cyclic loading or chemical corrosion in a hazardous environment.

A dominant component of the original signal from the WP sifting process usually has quite simple frequency characteristics and is suitable for the classical Hilbert transform. The instantaneous frequency of the component can be found from the phase curve of Hilbert transform of the component. For a healthy structure the associated instantaneous frequency is time-invariant. Any reduction in the instantaneous frequency may reflect structural damage. For a sudden stiffness loss, the change occurs in a very small time interval and for progressive stiffness degradation a gradual change in the instantaneous frequency can be observed.

In this study the same 3DOF structural model in Fig. 6 is employed. A proportional damping is employed here and a damping matrix is proportional to the stiffness matrix with a proportionality factor of 0.0002. Structural damage is introduced by linearly reducing the stiffness of spring K2 down to certain extent. By selecting the rate of change in stiffness reduction, both cases of sudden damage and progressive damage can be simulated. Ideally, sudden damage happens at a particular instant and stiffness of the structure decreases instantly at that moment. However, for the sake of convenience in the numerical simulation a stiffness loss in a sufficiently small time period, five time step in this simulation study, is considered as a sudden loss. In all studied cases the WP sifting process is first applied to the simulated response data to sift out the dominant components and the Hilbert transform is then applied to investigate their transient frequency characteristics for the purpose of structural health monitoring.

Case study 1: Detection of sudden damage

A sudden stiffness loss is introduced at t=15sec by linearly reducing stiffness of the middle spring, i.e. K2 by 10% in a short time interval from t = 15sec to t = 15.05sec. In a practical application, measurement data are collected with certain sampling rate and so a sudden stiffness loss may be treated as linear reduction between two sampling points. Without loss of generality only the dominant component of acceleration response data of M2, which is obtained by the WP sifting process and corresponds to the highest mode of the healthy system, is selected for the analysis.

Figure 8 presents some results of the WP decomposition application for detecting sudden damage and also provides a comparison of the instantaneous frequency and the instantaneous modal shapes of the system for three different levels of damage, i.e. 5%, 10%, and 15% sudden local stiffness losses of K2, respectively, at t = 15 seconds. Figure 8(a) presents change in the damped natural frequency of the highest mode, obtained by sifting the acceleration response of M2. In all three cases, the sudden damage can be clearly identified by the sudden changes of these instantaneous frequencies. Note that the 5%, 10%, and 15% stiffness loss of K2 cause only 1.4%, 2.8%, and 4.1% reduction in frequency, respectively.

It is expected that the abrupt change in the instantaneous frequency becomes less and less recognized for sufficiently small local stiffness loss and severe noise contamination.

Figure 8(b) shows the instantaneous normalized modal shape of the highest mode. Note that the absolute values of the amplitude ratio of M2 and M3 with respect to M1 are used for convenience. Sudden damage at t = 15 seconds was successfully detected. The 5%, 10%, and 15% stiffness losses of K2 result in 7.3%, 14.5%, and 21.6% change in the modal shape respectively. In general, the instantaneous modal shape is more sensitive to a small local stiffness loss. It can be shown that a local stiffness loss of order of ε may result in change in the natural frequency of order ε^2 and change in the modal shape of order ε. Therefore, the modal shape is a more sensitive index for structural damage. However, measurement data at multiple locations must be available if the instantaneous modal shape is to be traced.

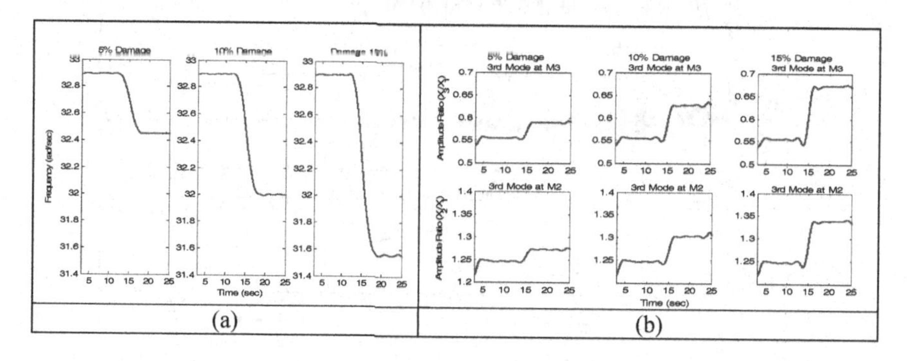

Fig. 8. Comparison of detection of sudden damage of 5%, 10%, and 15% stiffness loss of K2. (a) the instantaneous damped natural frequency and (b) the instantaneous normalized Instantaneous modal shapes, both for the highest mode (Shinde and Hou, 2005).

Case study 2: Monitoring stiffness degradation

To model a progressive stiffness degradation, the K2-value is reduced linearly by 10% from t = 15sec to t = 45sec. The acceleration response from M2 is selected for analysis. Its highest-mode component of the signal obtained by WP decomposition and the associated instantaneous frequency are shown respectively in the middle and the bottom parts of Fig. 9.

A gradual change in the instantaneous frequency is clearly observed in the same time interval as specified in the simulation. The trend and the amount of change in the instantaneous frequency provide valuable information as how stiffness degradation is developed.

Case study 3: Damage assessment by experimental data

To examine the feasibility of the WP sifting process in real-life applications, the approach was applied to experimental data obtained by a shaking table test of a two-story full size wooden frame performed at the Disaster Prevention Research Institute (DPRI), Kyoto University. The NS component of 1940 El Centro earthquake was used as the ground excitation. Several test runs were conducted and in each test run the structure was excited by the original records scaled at a nominal level targeted at certain intensity. Various types of damage were observed during the testing. The acceleration responses of the first floor at load levels of 1, 3, and 6 m/s^2 were analyzed. For detailed information about the test, the reader is referred to (Shimizu *et al.*, 2001; Hou, 2001).

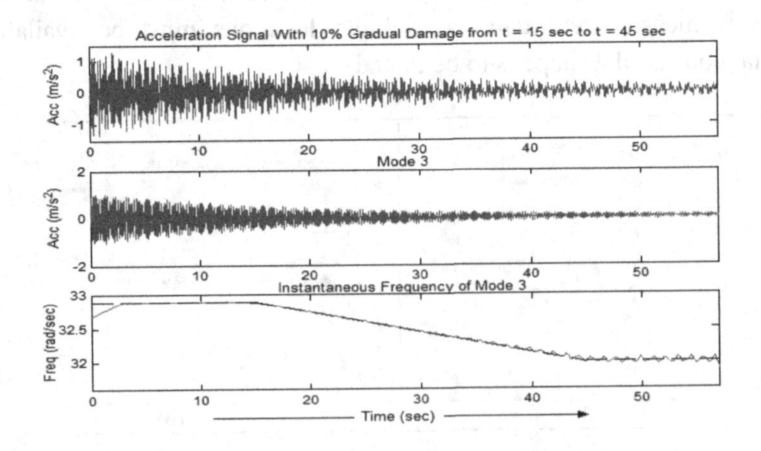

Fig. 9. Results from a case study for monitoring progressive damage (the top: the vibration signal; the middle: one of the sifted component; and the bottom: the instantaneous frequency history) (Shinde and Hou, 2005).

Figure 10 compares the instantaneous frequency histories corresponding to the lower mode of acceleration measurements at the first floor at load levels of 1, 3, and 6 m/s^2, respectively. The plots are adjusted to the same scale such that a trend can be clearly seen. It is observed that average levels of the instantaneous frequency plots are reduced in subsequent test runs, which reflects accumulated damage during these test runs. Note that the shaking table testing was conducted in several runs in a load increment of 0.5 m/s^2 starting from 1 m/s^2. Often a subjective judgment or a best guess is needed to draw conclusions from real vibration measurements for minor damage in a short time period, mainly due to fluctuations in the instantaneous frequency history. However, a trend of stiffness degradation can be observed clearly over a relatively long time period, as suggested by the results in Fig. 10.

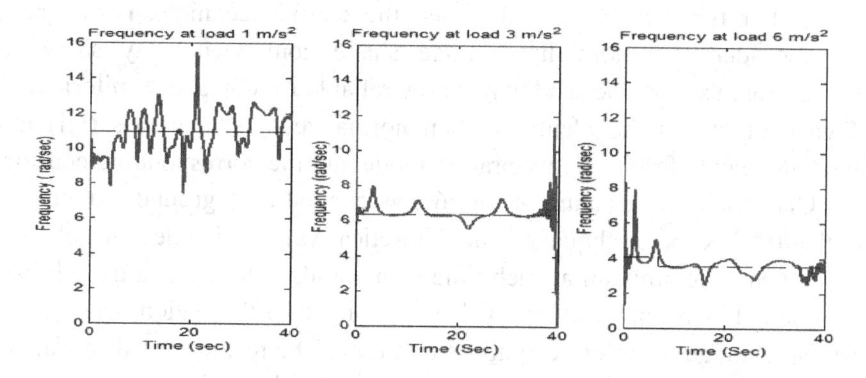

Fig. 10. Comparison of the extracted instantaneous frequency histories of the lower frequency components at load levels of 1, 3, and 6 m/s^2, respectively (Hou and Shinde, 2005).

5.5. *Confidence index for measurement data*

While the results in Figs. 10 illustrate great promises of the approach for real-life applications some practical issues need to be addressed. For example, when a structure is subjected to an excitation of complex nature such as an earthquake ground motion, the decomposition of modal components from a measured seismic response is not as evident as from a free vibration response, especially in the existence of noise for minor damage. The instantaneous modal parameters directly identified from the seismic response may fluctuate and the trend may be difficult to find even with trained eyes or if found the trend may not be fully trusted. A quantity referred as the confidence index was introduced in (Hou *et al.*, 2006) to validate the results.

The confidence index (CI) is defined as

$$CI_{k,p}^{(i)} = \left(\omega_{(k)}^{(i)}(t) \Big/ \omega_{(p)}^{(i)}(t) \right) \tag{21}$$

for the i-th component sifted from the responses measured at the k-th and p-th DOF where the p-th DOF response is used for the normalization. It represents the ratio of the identified instantaneous frequencies of the corresponding sifted components of the vibration responses measured at two different locations.

For a practical implementation, a confidence index which validates the corresponding instantaneous mode shape component needs to be in a range close to 1, generally 0.98 to 1.02, with an ideal value equal to one. The range [0.98:1.02] was chosen based on a sensitivity study, using different types of excitations. The deviation of the CI may also be caused by measurement noise, insignificant participation of the mode of interest, and nonlinear behavior of the structure. For those time intervals when the confidence index is beyond this range, the identified normalized mode shape component may show some deviation from exact values and may not be reliable for damage monitoring.

Figure 11 shows the identified then normalized instantaneous (NI) modal shape components for the first vibration mode and the corresponding confidence index (CI), when the structure in Fig. 6 was excited by a ground motion record from the 1995 Kobe earthquake. The theoretical values obtained by solving the system eigenvalue problem at each time step are also plotted as a dotted line for comparison. The NI modal shape value is changed in the region near t = 12sec indicating damage in structure at about that time. The results for NI modal shape components are in quite good agreement with the theoretical results along the whole time interval except in certain regions where the confidence index is not in the CI range of good confidence between 0.98 and 1.02.

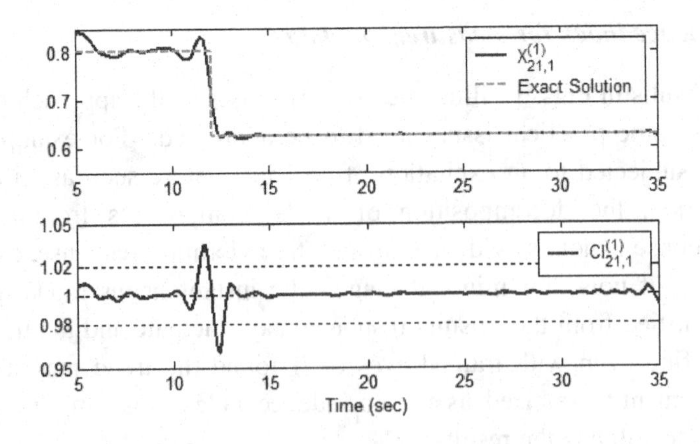

Fig.11. The NI modal shape (top) and the CI (bottom) for the sifted first modal component from the simulated response excited by the 1995 Kobe earthquake (Hou, et al, 2006).

6. Concluding Remarks

This chapter presents some applications of wavelet analysis for SHM, conducted in prior research by the authors. It is concluded that wavelet-based techniques including the conventional wavelet transform and the innovative wavelet-packet sifting can be effectively used to detect abrupt structural damage and monitor progressive stiffness degradation.

Occurrence of a spike in the high-resolution wavelet details of the acceleration response data may associate breakage or sudden damage of structural members. The moment when a spike is observed corresponds to the time instance when sudden damage occurs. The spatial distribution pattern of the spikes can be used to locate the damage region.

The instantaneous modal parameters identified from the dominant WP components of the response data can be used to monitor structural health condition. Structural stiffness degradation may be accompanied by a decreasing trend of the instantaneous frequencies.

The wavelet approach has merits of efficiency in computer implementation, less dependence on structural modeling, and sensitivity to local damage. It has shown great promises for online implementation with today's intelligent structural systems.

The effectiveness of the wavelet approach for SHM depends on the measurement noise level and the damage severity. In general, damage is easier to detect using less contaminated measurements from a severely damaged structure. A detectability map may be created to evaluate the effectiveness of the wavelet approach. A confidence index for dominant components of the response data can be calculated in the wavelet packet sifting process to assess reliability of the wavelet results.

Acknowledgement

The authors acknowledge Professor Suzuki at the Disaster Prevention Research Institute, Kyoto University, Japan for providing the shaking table experimental data and the ASCE Task Group on Health Monitoring for providing the MATLAB codes, both used for the results herein.

References

Alonso, R.J., Noori, M., Saadat, S., Zhikun Hou, (2004). Effects of Excitation Frequency on Detection Accuracy of Orthogonal Wavelet Decomposition for Structural Health Monitoring, *J. of Earthquake Engineering and Vibration, June 2004, Vol. 3, No. 1, pp 101-106.*

Al-Khalidi, Noori, M., Hou, Z., Yamamoto, S., Masuda, A., and Sone, A., (1997). Health Monitoring Systems of Linear Structures Using Wavelet Analysis. *Structural Health Monitoring, Status and Perspectives*, Ed: Fu-Kuo Chang, pp 164-175.

Boashash, B. (1992). Estimating and Interpreting the Instantaneous Frequency of a Signal Part 1: Fundamentals, *Proceedings of the IEEE*, 80-4, 520-537.

Carden, E.P. and Fanning, P. (2004). Vibration based condition monitoring: A review, *Structural Health Monitoring,* 3, pp. 355–377.

Chang, F.K. ed. (2011). Structural Health Monitoring, *Proceedings of the 8th International Workshop on Structural Health Monitoring*, Stanford University, Stanford, CA.

Chang, F.K. ed. (2009). Structural Health Monitoring, *Proceedings of the 7th International Workshop on Structural Health Monitoring*, Stanford University, Stanford, CA.

Coifman,R.R and Wickerhauser, M. V. (1992). Entropy based algorithms for best basis selection, *IEEE Trans. Information Theory,* 38(1), pp.713-718.

Chui C.K. (1992) *Introduction to Wavelets,* Academic Press, Inc., San Diego, CA.

Daubechies, I. (1992). Ten Lectures on Wavelets, Proc., *CBS-NSF Regional Conference in Applied Mathematics*.

Doebling, S. W., Farrar, C. R., Prime, M. B., and Shevitz, D. W. (1996). Damage Identification and Health Monitoring of Structural and Mechanical Systems From Changes in Their Vibration Characteristics: A Literature Review, *Rep. No. LA-13070-MS* , Los Alamos National Laboratory, Los Alamos, N.M.

Fan, W. and Qiao, P. (2011). Vibration-based Damage Identification Methods: A Review and Comparative Study, *Structural Health Monitoring,* 10(1), pp.83-11.

Hera, A. and Hou, Z. (2004). Application of Wavelet Approach for ASCE Structural Health Monitoring Benchmark Studies, *ASCE Journal of Engineering Mechanics*, 130(1), pp. 96-104.

Hou, Z., Noori, M., and St. Amand, R. (2000). A Wavelet-Based Approach for Structural Damage Detection, *ASCE Journal of Engineering Mechanics*, 126(7), 677 - 683.

Hou, Z. (2001) *Wavelet-Based Damage Detection Techniques and Its Validation using Shaking Table Test Data of a Wooden Building Structure*, Disaster Prevention Research Institute, IMDR Research Booklet No. 3A .

Hou. Z., Hera, A., Shinde, A. (2006). Wavelet-based Health Monitoring of Structures under Earthquake Excitation, *International Journal of Computer Aided Civil and Infrastructure Engineering*, 21(4), pp.268-279.

Johnson, E.A., Lam, H.F., Katafygiotis, L.S., and Beck, J.L. (2004). Phase I IASC-ASCE Structural Health Monitoring Benchmark Problem Using Simulated Data, *ASCE Journal of Engineering Mechanics*, 130(1), pp. 3-15.

Huang et al. (1998). The empirical mode decomposition method and the Hilbert spectrum for non-linear and non-stationary time series analysis, *Proc. R. Soc. Lond,* 454, pp. 903-995.

Khatam, H., Beheshti, S., Noori, M., (2007). Harmonic Class Loading for Damage Identification in Beams Using Wavelet Analysis," *Structural Health Monitoring; vol. 6:1, pp. 67 – 80.*

Masuda, A., Noori, M., Sone, A., Hashimoto, Y., (2002). Wavelet-Based Health Monitoring of Randomly Excited Structures, *CD ROM Proceedings of: 15th ASCE Engineering mechanics Conference, Columbia University, N.Y., (Andrew W. Smyth ed.), 6 pages.*

Mallat, S.G. (1988). A theory for Multiresolution Signal Decomposition: the Wavelet Representation, *IEEE Pattern Anal. and Machine Intell.*, 11(7). pp. 674-659.

Mallat, S.G. (1998) *A Wavelet Tour of Signal Processing,* Academic Press.

Masuda A, Nakaoka A, Sone A, Yamamoto S.(1995). Health monitoring system of structures based on orthonormal wavelet. *Seismic Engineering, Transactions* (ASME) 312:161–167.

Reda Taha, M.M., Noureldin, A., Lucero, J.L., and Baca, T.J. (2006), Wavelet Transform for Structural Health Monitoring: A Compendium of Uses and Features, Structural *Health Monitoring*, 5(3), pp. 267-295.

Shimizu et al (2001). Full Scale Vibration Tests of Two Storied Wood Houses by Post and Beam Structure, Experimental Results of Post-and-Beam Frames with Braces, *Summaries of Technical Papers of 2001 Annual Meeting, Architectural Institute of Japan* (in Japanese).

Shinde, A. and Hou, Z. (2005). A Wavelet Packet Based Sifting Process and Its Application for Structural Health Monitoring, *Structural Health Monitoring*, 2(4), pp. 153-170.

Sohn, H., Farrar, C., Hunter, N. and Worden, K. (2003). A Review of Structural Health Monitoring Literature: 1996–2001, *Rep. No. LA-13976-MS*, Los Alamos National Laboratory, Los Alamos, N.M.

Strang G. and Nguyen T. (1996) *Wavelet and Filter Banks*, Wellesley-Cambridge Press.

Staszewski, W.J., (1998). Structural and Mechanical Damage Detection Using Wavelets, *The shock and Vibration Digest*, 30, 457-472.

Yen, G.G. and Lin, K.C. (2000). Wavelet Packet Feature Extraction for Vibration Monitoring, *IEEE Trans. Industrial electronics*, 47(3), pp. 650-667.

Chapter 8

HHT-Based Structural Health Monitoring

Norden E. Huang[1], Liming W. Salvino[2], Ya-Yu Nieh[3], Gang Wang[4] and Xianyao Chen[4]

[1]Research Center for Adaptive Data Analysis
Center of Dynamical Biomarkers and Translational Medicine
National Central University, Zhongli, Taiwan 32001
E-mail: norden@ncu.edu.tw

[2]Office of Naval Research Global
Embassy of the United States of America, 27 Napier Road
Singapore 258508, E-mail: liming.salvino@org.navy.mi

[3]Materials & Electro-Optics Research Division
Chung-Shan Institute of Science & Technology
Longtan Township, Taoyuan County 325, Taiwan, R.O.C.
E-mail: Sophia.nieh@gmail.com

[4]The Key Laboratory of Data Analysis and Applications
First Institute of Oceanography, State Ocean Administration,Qingdao, 266061 China
E-mail: wangg@fio.org.cn, chenxy@fio.org.cn

1. Introduction

Just like living beings, structures have life cycles. The continuous wear and tear and occasional unexpected events would all adversely affect the integrity of structures; they inevitably will deteriorate and decay. Also like living beings, the structure could also suffer infirmities throughout its service life, which could reduce the capacity of its performance, and even lead to catastrophic consequences if the defect is serious. But unlike living beings, the structures can't utilize self-healing processes. Structural health monitoring, therefore, is a necessary step to assess the health condition of the structure continuously. Hopefully, the methods used can identify the health condition of structures as a basis for implementing remedial measures, if necessary, to prevent and avert disastrous failures. As a result, the structure health monitoring has been traditionally viewed as damage detection. To be effective, the methods must be able to provide quantitative information on the following conditions of the structures:

a. The existence of damage
b. The location of the damage
c. The nature of the damage
d. The severity of the damage
e. The prognosis of the damage (i.e., what actions need to be taken?)

as suggested by Rytter (1993) and Doebling *et al.* (1996). With traditional thinking centered on damage, SHM methods emphasize damage detection. As a result, visual inspection was adopted as the operational SHM method, assuming damages should be visible. Other more sophisticated methods such as acoustic and ultra-sonic scanning, magnetic and electric field variations, microwave radar and thermographs have been developed and tried with varying degrees of success. The critical flaw for these methods is the test conditions and the parameters identified from the test are unrelated to the function and safety of the structure. Therefore, all these static tests suffer the implicit difficulty in equating damage to safety. After all, the safety should be the most important consideration in the SHM efforts.

Safety consideration dedicates dynamic SHM tests when the structures are subjected to certain loading conditions. Currently, the most popular dynamic test is the modal analysis, which studies the dynamic properties and deformation characteristics when excited by dynamic forces. A very comprehensive review could be found in a report by Doebling et al (1996). The principles of the modal analysis are to induce a vibration in the structure and measure the changes in vibration frequency, vibration mode shapes and its curvature, and even the whole dynamical flexibility matrix and relate them to structural health condition. The critical step is to relate the changes in mass, damping and stiffness of the structure to the changes in the frequencies, mode shapes and damping rates.

For the modal analysis to be successful, there are many demanding conditions: first, to define the structural mode shape accurately, there must be a large number of sensors deployed throughout the structure. Sometimes, the tests have to be conducted with the same sensors redeployed repeatedly. Then the influence of the environmental condition would be a serious factor. The thermal condition of a bridge in two different time of a day could be so large that all the change due to the damage would be masked. Secondly, we have to induce vibration to the structure. Typically, low frequency vibration yields global information; high frequency yields local information. As most damages are local, we need high frequency signal, but high frequency vibration dissipates too fast and is hard to maintain. Thirdly, we need the undamaged state as a reference point, but in most cases the historical data are not readily available. Finally, the most serious limitation of the modal analysis is that the whole method is based on the same assumptions as in Fourier analysis: linear and stationary. The complicated geometry of the structures would require many Fourier components to represent its vibration modes. The frequency and the damping characteristics changes are hard to analyze for complicated structures. Furthermore, the changes due to local damage would only show up on higher harmonic modes, where noises from all sources would easily overwhelm them.

Additionally, for the modal analysis to work, we should have a structure reacting as a linear system under stationary vibration. We should have a perfect model and a complete history of the structure. During the test, we should have unlimited number of sensors that is deployable at any place of the structure under a friendly environmental condition with no noise. And we should also have unique damage vs. response modal shape relationships. Finally, we should have structures sensitive to damage and a clear cut threshold for damage.

In reality, we would have none of those idealized conditions: The structures are usually nonlinear especially when they are damaged, the test condition are also non-stationary. We would have limited accessibility to the whole structure, with limited knowledge and historical references. The structure model could never be perfect, even construction processes could induce enough non-homogeneity to the final product. The number of sensors is always limited; noises are unavoidable. The most critical limitation is the structures are usually insensitive to small damages, especially when we have to infer the damage from the frequency changes, as the stiffness change is proportional to the squared of the frequency. For example, a 50% reduction of the stiffness would only produce 30% reduction in frequency. As a result, the application of modal analysis would be difficult and limited to simple structures.

Here we will introduce a different approach for structure health monitoring. We will utilize the detection of the nonlinear behavior of the structure, long before it suffers permanent damage, as our first line of defense. It turns out that nonlinearity is far easier to detect than frequency shift. To have this new approach to work, we have to introduce a new data analysis method that would break through the assumption of linear and stationary assumptions. This new method is known as the Hilbert-Huang Transform (HHT) (Huang *et al.*, 1998, 1999) and its subsequent improvements (Wu and Huang, 2009), which is designed to analyze data from nonlinear and non-stationary processes. With the new data analysis method, we can present the results in time-frequency space, quantify nonlinearity, and portray damping and stability spectral analysis under non-stationary loading condition for nonlinear structural systems. In fact, HHT based structure health monitoring had been advocated by many scholars including Huang (1998) first and followed by Xu and Chen (2003), Yang *et al.* (2003, 2004), Huang *et al.* (2005), Chen *et al.* (2004), Chen (2009) and Yu and Ren (2005). Here we will introduce some new approaches and give concrete examples to illustrate the power of HHT in structural health monitoring.

In the following sections, we will introduce the new Time-Frequency data analysis method - the HHT and how to detect nonlinearity using heavy load vs. light load. Then, a numerical model will be presented to illustrate how the nonlinearity in a bilinear system can be evaluated. Finally examples on structural health monitoring of bridges (emphasizing heavy and light transient loads), ship structures (using damping spectral analysis), and airplanes (using stability spectral analysis) will be given.

2. Time-Frequency analysis

Before starting the time-frequency analysis, we have to ask a critical question: how to define frequency ω. The standard definition of frequency is simply,

$$\omega = 1/T \tag{1}$$

where T is the period of a wave. This definition, though physically sound, is too crude to be used in a quantitative time-frequency analysis. Furthermore, this definition is mathematically correct only for regular sinusoidal waves only. For nonstationary and nonlinear waves this definition is too crude and useless, since it gives only a mean value over a whole period.

Mathematically, the commonly accepted definition of frequency in the classical wave theory is based on the existence of a phase function (see, for example, Whitham, 1974; or Infeld and Rowlands, 1990). Here, starting with the assumption that the wave surface is represented by a 'slowly' varying function consisted of an amplitude function $a(x,t)$, and a phase function $\theta\,(x,t)$, such that the wave profile is the real part of the complex valued function,

$$\varsigma(x,t) = \mathrm{Re}\left(a(x,t)\,\mathrm{e}^{\mathrm{i}\theta(x,t)}\right) \tag{2}$$

Then, the frequency, ω, and the wave number, k, are defined as

$$\omega = -\frac{\partial\theta}{\partial t} \quad k = \frac{\partial\theta}{\partial \mathrm{x}} \tag{3}$$

Cross-differentiating the frequency and wave number, one immediately obtains the wave conservation equation as:

$$\frac{\partial k}{\partial t} + \frac{\partial \omega}{\partial x} = 0 \tag{4}$$

This is one of the fundamental laws governing all wave motions, be it acoustic or electromagnetic. The assumption of the classic wave theory is very general: that there exists a 'slowly' varying function such that we can write the complex representation of the wave motion given in Equation (2). If frequency and wave number can be defined as in Equation (3), then they have to be differentiable functions of the temporal and the spatial variables for Equation (4) to hold. Thus, for any wave motion, other than the trivial kind of constant frequency sinusoidal motion, the frequency representation should have instantaneous values. Therefore, there should not be any doubt of the mathematical meaning or the necessity of the existence of instantaneous frequency. Classical wave theory is based on this assumption, and the classical wave theory has proven to be quite rigorous with many of the theoretical results confirmed by observations (Infeld and Rowlands, 1990). This concept can be generalized to all kind of wave phenomena such as in acoustics and optics. The pressing questions are how to define the phase function and the instantaneous frequency for a given data set.

Physically, there is also a real need for instantaneous frequency (IF) in faithfully representing the underlying mechanism for data from nonstationary and nonlinear processes. Obviously, the non-stationarity is one of the key features here, but, as explained by Huang et al. (1998), the concept of IF is even more critical for a physically meaningful interpretation of nonlinear processes. For a non-stationary process, the frequency should be ever changing. Consequently, we need a time-frequency representation for the data, or that the frequency value has to be a function of time. For nonlinear processes, the frequency variation as a function of time is even more drastic. To illustrate the need for IF in the nonlinear cases, let us examine a typical nonlinear system as given by the Duffing equation:

$$\frac{\mathrm{d}^2 x}{\mathrm{d}t^2} + x + \varepsilon x^3 = \gamma \cos \omega t \tag{5}$$

in which ε is a parameter not necessarily small, and the right hand term is the forcing function of magnitude γ and frequency ω. This cubic nonlinear equation can be re-written as:

$$\frac{d^2 x}{dt^2} + x(1 + \varepsilon x^2) = \gamma \cos \omega t, \tag{6}$$

where the term in the parenthesis can be regarded as a single quantity representing the spring constant of the nonlinear oscillator or the pendulum length of a nonlinearly constructed pendulum. As this quantity is a function of position, the frequency of this oscillator is ever changing, even within one oscillation. This intra-wave frequency modulation is the singular most unique characteristics of a nonlinear oscillator as proposed by Huang *et al.* (1998, 1999). The geometric consequence of this intra-wave frequency modulation is the waveform distortion. Traditionally, such nonlinear phenomena are represented by harmonics. As the waveform distortion can be fitted by harmonics of the fundamental wave in Fourier analysis, it is viewed as harmonic distortions. This traditional view, however, is the consequence of imposing a linear structure on a nonlinear system: the superposition of simple harmonic functions with each as a solution for a linear oscillator. One can only assume that the sum and total of the linear superposition would give an accurate representation of the full nonlinear system. Such an assumption has worked well for systems with infinitesimal nonlinearity, but the assumption is certainly false when the nonlinearity is finite and the motion becomes chaotic (see, for example, Kantz and Schreiber, 1997). A natural and logical representation should be the one that can capture the physical meaning: the physical essence of this nonlinear system is an oscillator with variable frequency assuming different values at different time even within one single period. To describe such a motion, we should use instantaneous frequency (IF) to capture this essential physical characteristic of nonlinear oscillators.

In real world experimental and theoretical studies, the conditions of ever changing frequency are common, if not prevailing. Chirp signal is one class of the signals used by human in speech, bats in hunting and navigation as well as in radar. The frequency content in speech, though not exact chirp, is also ever changing. And many of the consonants are produced through highly nonlinear mechanisms such as explosion or friction. Furthermore, for any nonlinear system, the frequency is definitely modulating not only among different oscillation periods, but also within one period as discussed above. To understand the underlying mechanism of these processes, we can no longer rely on the traditional perturbation method or Fourier analysis; we should examine the true physical processes through instantaneous frequency.

There are copious publications in the past on the instantaneous frequency (Boashash, 1992 a, b, c; Kootsookos *et al.*, 1992; Lovell *et al.*, 1993; Cohen,

1995; Flandrin, 1995; Loughlin and Tracer, 1996; and Picinbono, 1997). Most of these publications, however, were concentrated on Wigner-Ville Distribution (WVD) and it variations, where the instantaneous frequency is defined through the mean moment of different components at a given time. Other than the WVD, instantaneous frequency through Analytic Signal (AS) produced by Hilbert transform has also received a lot attention. Boashash (1992b and c), in particular, gave a summary history of the evolution of the definition of instantaneous frequency. Most of the discussions given by Boashash (1992a, b, c) were on mono-component signal. For more complicate signal, he again suggested to utilize the moments of the WVD. But there is no a priori reason to assume that the multi-component signal should have a single valued instantaneous frequency at any given time and still retain some physical significance. Even for mono-component signal, the moment approach still could be problematic as discussed by Boashash. He also suggested cross WVD of the signal with a reference signal. This method will be seriously compromised when the signal to noise ratio is high. We will return to these points and discuss them in more detail later.

One of the most basic confusing points concerning IF stems from the erroneous idea that one must find for each instantaneous frequency value a corresponding frequency in the Fourier Spectrum of the signal. In fact, instantaneous frequency of a signal when properly defined should have very different meanings when compared with the frequency in the Fourier spectrum as discussed in Huang *et al.* (1998). But the divergent and confused viewpoints on IF indicate that the erroneous view is deeply rooted: there are misconceptions and fundamental difficulties in computing IF. Some of the traditional objections on IF stem from the mistaken assumption that a single valued IF exists for any function at any given instant.

Instantaneous frequency witnessed two major advances recently through the introduction of Empirical Mode Decomposition (EMD) method and the Intrinsic Mode Function (IMF) introduced by Huang *et al.* (1998) for data from nonlinear and nonstationary processes. And the alternative by Wavelet based decomposition introduced by Olhede and Walden (2004) for data from linear stationary processes. Huang *et al.* (1999) have also introduced the Hilbert view on nonlinearly distorted waveform, which provided explanations to the many of the paradoxical issues raised by Cohen (1995) on the validity of IF, as discussed in details by Huang *et al.* (2009). Indeed, the introduction of EMD or the wavelet decomposition resolved one key obstacle for computing a meaningful IF from a multi-component signal by reducing it to a collection of mono-component functions. Once we have the mono-component functions, there are still limitations on applying AS for physically meaningful instantaneous frequency as postulated by the well-known Bedrosian (1963) and Nuttall (1966) theorems. Some of the mathematical problems associated with the Hilbert transform of IMFs have also been addressed by Vatchev (2004).

We propose a normalization scheme (Huang, 2003) that will not only remove most of the difficulties associated with AS, but also enable us to compute the quadrature directly: The normalization makes AS to satisfy the limitation imposed by the Bedrosian theorem (Bedrosian, 1963). At the same time, it also provides a sharper and easily computable error index than the one proposed by the Nuttall theorem (Nuttall, 1966), which governs the case when the Hilbert transform of a function is different from its quadrature. More importantly, the normalized IMF also enables us to define the quadrature of the carrier function directly, and then compute IF through direct quadrature function. Additionally, we will also introduce alternative methods based on a generalized zero-crossing and an energy operator to define frequency locally. The details of this result could be found in Huang *et al.* (2009).

With the instantaneous frequency and energy defined, the time-frequency presentation could be easily achieved. We will use three examples to illustrate the usefulness of the adaptive data analysis approach in time series data. The first example demonstrates the prowess of the method to represent chirp signal and the second example demonstrate the application of the method in speech analysis. Finally, we also use a 2-dimensional image to illustrate the Multi-dimensional EMD.

2.1. *The chirp data*

Let us consider the following model equation:

$$x(t) = \sin\left[\left(\frac{2\pi t}{256}\right)\frac{t}{1024}\right], \quad \text{for } t = 0{:}1024. \tag{7}$$

The wave shape and the analyzed results based on Fourier, Morlet wavelet, and Hilbert spectral analysis are given in Figure 1.

The model wave clearly has frequency variation along the time axis. Yet the Fourier spectrum totally missed the temporal variation; in the analysis both amplitude and frequency are constant over the whole time domain. Wavelet certainly could capture the temporal variation to some degree. Yet the leakage and fixed basis limit the power of frequency resolution. Only Hilbert spectral analysis could give a clean and crisp frequency variation as a function of time.

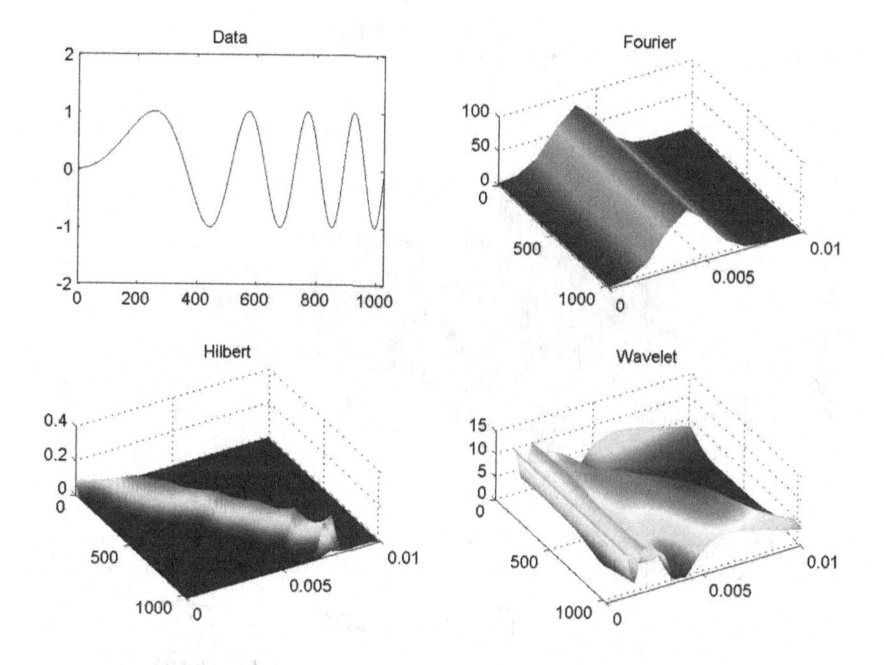

Fig. 1. Comparisons among Fourier, wavelet and Hilbert Spectral analysis for a chirp wave.

2.2. *Speech signal analysis*

Any utterance of speech could be transcribed as time variation of the pressure field. Figure 2 gives the recorded signal when the word, 'Hello,' is uttered. The recording was taken at a sampling rate of 22050 Hz.

Again, various methods of analyses were used on this set of data. The results are given in Figure 3. Here top row are the results from the Fourier Spectrogram for narrow and wide band representations. The narrow band result, on the left, is computed with a window size of 1024 data points; while the wide band result, on the right, is computed with a window size of 64 data points. The narrow band representation can certainly resolve the frequency better, but the time location is smeared. The wide band representation can resolve the time location better, but the frequency is smeared. The inability to represent both time and frequency to an arbitrary degree of precision is known as the uncertainty principle. This uncertainty principle is very different from the counterpart in physics, known as the Heisenberg Uncertainty Principle, which has full physical meaning. The uncertainty principle is purely the result of using integral transformation for frequency determination. It is artificially inflected on the result because improper analysis method is used.

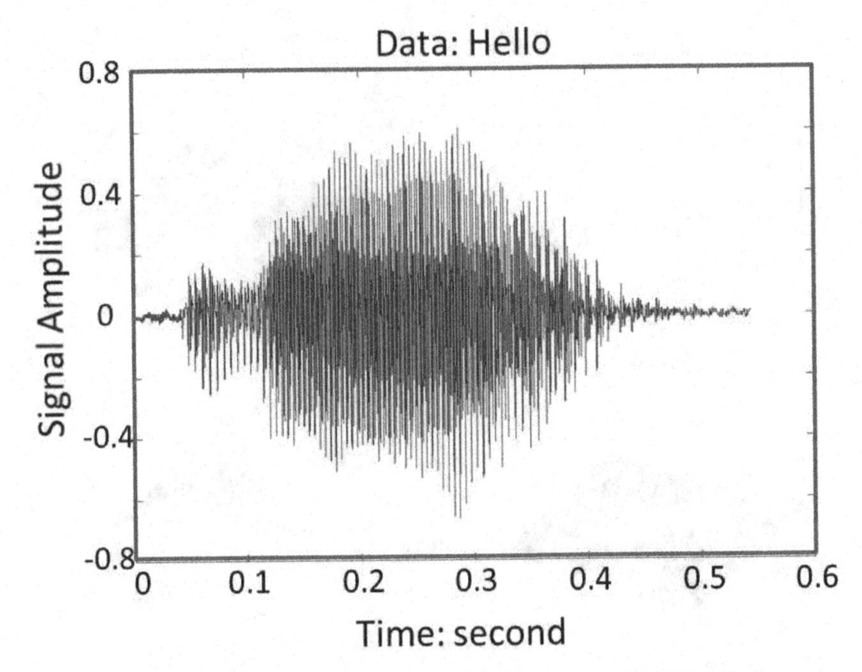

Fig. 2. The signal of the word, 'Hello,' recorded at 22050 Hz.

The Wavelet seems to be an improvement over the Fourier in this representation. Yet the resolution is still poor. Furthermore, the continuous Wavelet analysis used here is over redundant and not energy conserved. The discrete Wavelet analysis, unfortunately, could not be used to recognize or extract physical features of the data. The solution for true time-frequency analysis is undoubtedly the Hilbert Spectral analysis. Here the frequency is determined by differentiation; therefore, its value and location is precise. The adaptive basis also enables the frequency to be a function of time and totally eliminated the need of the spurious harmonics. The intra-wave frequency modulation fully represents the nonlinear characteristics of the speech sound production processes.

2.3. *Comparisons amongst HHT, Wigner-Ville and Wavelet analysis*

In a recent Royal Society publication on Structural Health Monitoring, Staszewski and Robertson (2006) have mentioned time-frequency and time-scale analysis for structural health monitoring. The methods they employed were Wigner-Ville and Wavelet analysis. Let us use the final example to illustrate the difference between those methods and HHT. For details, one should refer to the Huang *et al.* (1998).

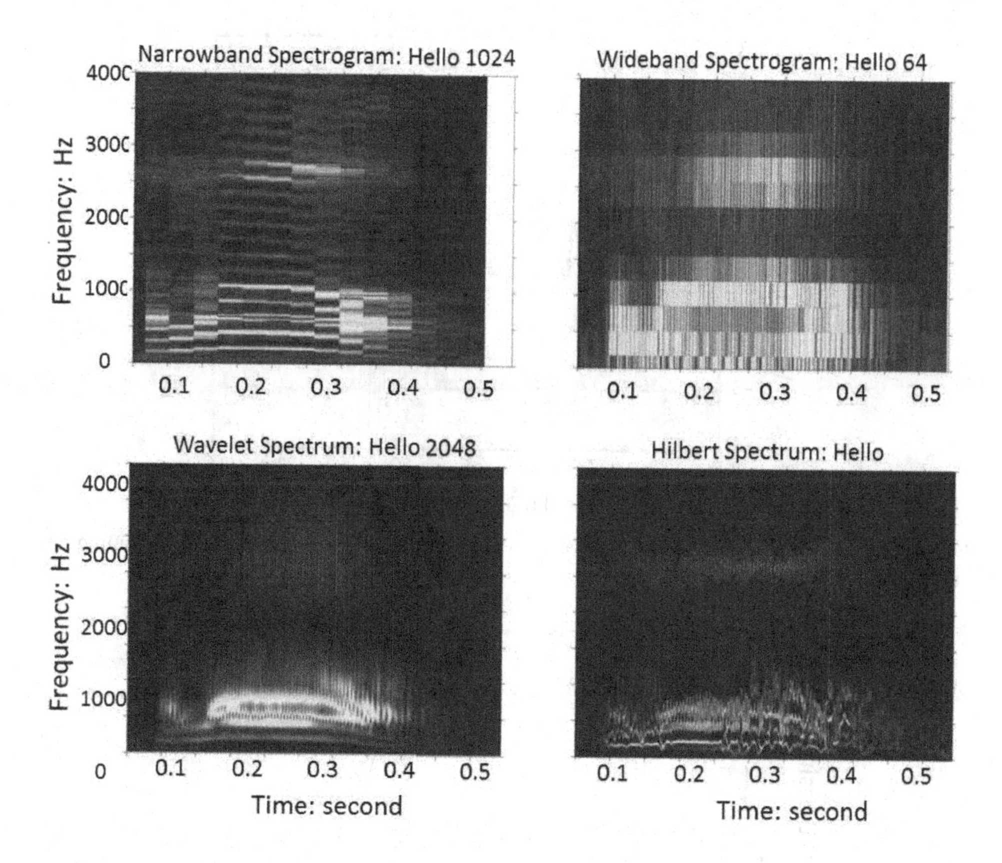

Fig. 3. (Color online) Fourier spectrograms (upper row: left panel for narrow band with window width 1024 points; right panel for wide band with window width 64 points), Morlet Wavelet and Hilbert spectral analyses for the 'Hello' signal.

Let us consider the case with two Gaussian modulate sine waves propagating in opposing directions independently and eventually meeting as shown in Figure 4.

As the Wigner-Ville is defined for the global domain Fourier transform,

$$WV(\omega,t) = \int_{-\infty}^{+\infty} R(\tau,t)\mathrm{e}^{-\mathrm{i}\omega\tau}\,\mathrm{d}\tau\,, \tag{8}$$

where $R(\tau,t)$ is the centrally valued correlation,

$$R(\tau,t) = \left\langle x\left(t+\frac{\tau}{2}\right)x\left(t-\frac{\tau}{2}\right)\right\rangle, \tag{9}$$

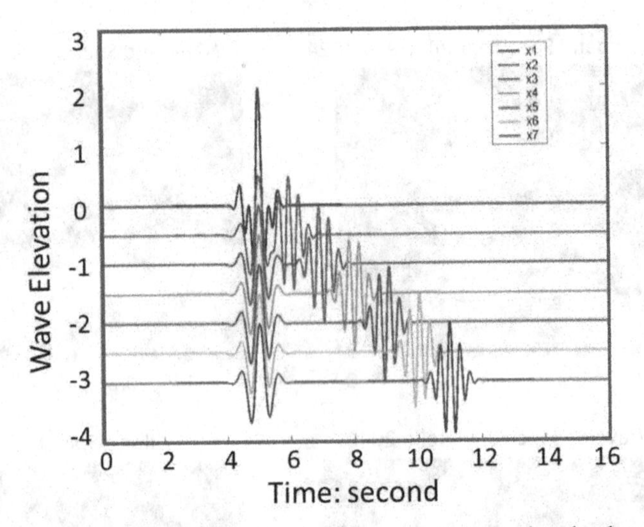

Fig. 4. Data of two Gaussian modulated wave packets propagate toward each other, (assuming one stationary).

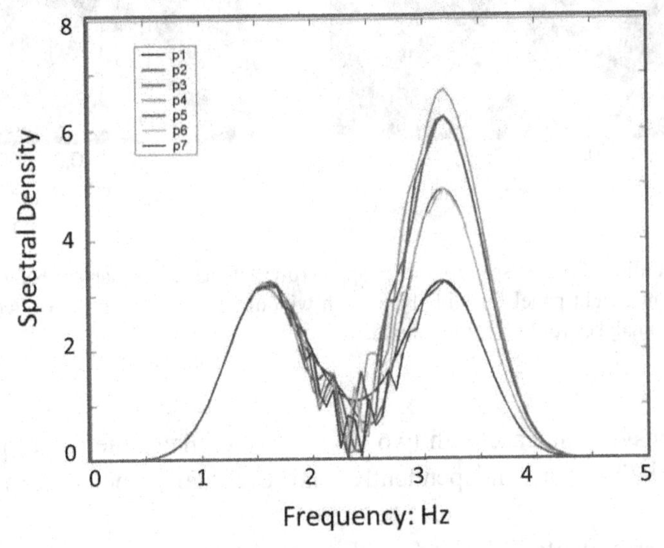

Fig. 5. Fourier spectra of the two waves at different locations along their paths.

the final Wigner-Ville distribution will be a global representation. The time variation came from the relative position of the waves. Furthermore, the marginal requirement would force the marginal distribution of the Wigner-Ville distribution to be exactly the same as the Fourier power spectrum shown in Figure 5, the two waves will have to be treated as a single event always depends on both the waves and their relative positions even if the waves are propagating independently.

As a result, the Wigner-Ville distribution suffers greatly in comparison to Wavelet and Hilbert spectrum as shown in Figure 6. Here, we have selected one particular position. The cross-talking effects are clearly seen, which renders the two waves are interconnected. Such relationship is not in the Wavelet result shown by contours or the Hilbert spectrum shown by the precise horizontal line segments giving the exact frequency values and the positions of the two waves, respectively. Although the Wavelet contours are more diffused because of the poor time-frequency resolution imposed by the uncertainty principle, the Wavelet results clearly separate the waves. There is no smear in either position or frequency value in HHT caused by the integration transform as required in either the Wavelet or Wigner-Ville distribution. The frequency in HHT is determined by differentiation. There are no limitations imposed by the uncertainty principle on the precision of time and frequency values.

Another serious flaw of Wigner-Ville distribution: the definition of instantaneous frequency is given by a mean value through moment method (Flandrin, 1995) as:

$$\omega(t) = \frac{\int_T \omega WV(\omega,t)\mathrm{d}t}{\int_T WV(\omega,t)\mathrm{d}t} \tag{10}$$

As a result, at any time, there could be only one single value designed as the instantaneous frequency. This is illogical and physically nonsensical. For example, in a symphony performance, one should be able to hear all kinds of sound from different instruments. Even in the application by Staszewski and Robertson (2006), they selected to eschew this definition and put the 'modal frequency' on the Wigner-Ville distribution to confirm the existence of such frequencies. To obtain such frequency values, one would have to make a priori selection subjectively. Thus, the use of Wigner-Ville is quite limited. Having defined the instantaneous frequency, we will proceed to define nonlinearity.

3. Degree of Nonlinearity

The term 'nonlinearity' has been loosely used, in most cases, not to clarify but as a fig leaf to hide our ignorance. As a result, any anomaly without obvious and ready explanations is labeled as 'nonlinear effect.' Such an approach could certainly be right for some cases, but, at the best, the answer is not complete; the nonlinear effects could have many causes. For example, nonlinear effects could be the results of the intrinsic properties of the system and also arise from the various initial conditions. Unfortunately, under the present state of our understanding, no better solution is available. The central of the problem is that the definition of 'nonlinearity' is only qualitative. The quantitative definition and method for computing the Degree of Nonlinearity was given by Huang *et al.* (2011) in a patent disclosure. Let us review the definition of nonlinearity.

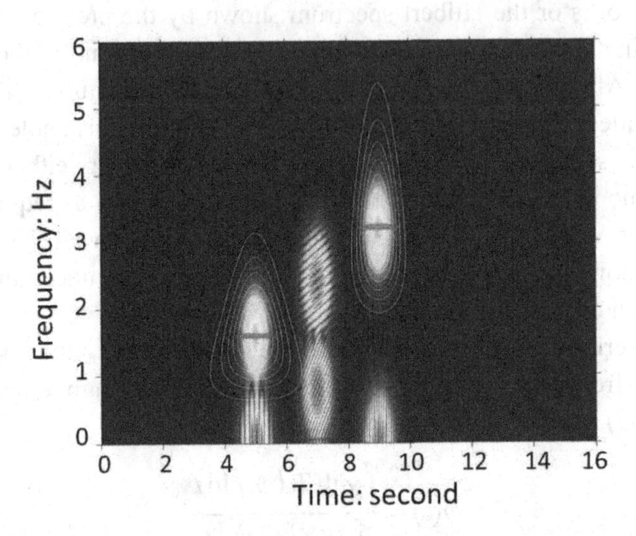

Fig.6. (Color online) The Hilbert spectrum (Pink horizontal bars), Morlet Wavelet (contours) andWigner-Ville distribution (colored contour) for the two waves propagating case. Note the clean distinct separation of the two waves using Hilbert spectrum and Wavelet. The Hilbert spectrum has a superb time-frequency resolution comparing to all the other methods.

Currently nonlinearity is defined based on a system point of view. And the definition is not directly on nonlinearity, but on linearity; any system that is not linear would be nonlinear. The linearity is based on linear algebra: For any system L is linear, if for inputs x_1 and x_2 we have outputs y_1 and y_2 respectively as:

$$L(x_1) = y_1; \\ L(x_2) = y_2; \tag{11}$$

we should also have

$$L(\alpha x_1 + \beta x_2) = \alpha y_1 + \beta y_2, \quad \text{for any } \alpha \text{ and } \beta \tag{12}$$

Any system that does not satisfy the superposition and scaling rules given in Equations (11) and (12) is a nonlinear system. This definition is rigorous but qualitative, for the answer offers 'yes' or 'no' without any quantitative measure. Furthermore, this system approach may not be very practical. For many phenomena, it might be difficult, if not impossible, to define the system in close form to test this input versus output approach. For example, many natural systems are so complicated that a closed system is hard to define. Even for those systems we could find explicit analytical expressions in closed form, the inputs and out puts could still be hard to define. Additionally, for the autonomous system, the state of the phenomena depends totally on the initial condition, which is hardly qualified as inputs. The motion or the flow of the system is also hard to be treated as output. Furthermore, for complicated systems, it may not be possible to track changes in a parameter as they progress from linear to nonlinear stage. All these difficulties made the discussions of nonlinearity hard to quantify. Without quantification, however, it would be impossible to discuss the nonlinearity more precisely and prevent its loose usage.

One of the unique characteristics of nonlinear processes, proposed by Huang *et al.* (1998), is the intra-wave frequency modulation, which indicates that the instantaneous frequency changes within one oscillation cycle from one instant to the next one and from one location to the neighboring one. Intra-wave frequency modulation is the hallmark of nonlinear system. Let us examine a very simple nonlinear system given by the non-dissipative Duffing type equation as:

$$\frac{d^2 x}{dt^2} + x + \varepsilon x^{n+1} = \gamma \cos \omega t \tag{13}$$

with n as an integer. This equation could be re-written as:

$$\frac{d^2 x}{dt^2} + x\left(1 + \varepsilon x^n\right) = \gamma \cos \omega t \tag{14}$$

We are not interested in the solution of this equation for the time being; rather, we will discuss the characteristics of the solution. For this purpose, we can treat the system given in Equation (14) as a virtual nonlinear spring or pendulum system. Then, the quantity within the parenthesis can be treated as the squared of the frequency in a linear system, which is similar to a variable spring constant, or a variable virtual pendulum length. According to simple pendulum law, the frequency of the pendulum should be inversely proportional to the square root of its length. With this view, we can see that the frequency of this virtual pendulum, with its length as a variable, should change from location to location, and time to time, even within one oscillation cycle. Indeed depending on the values of ε and n, the virtual pendulum could produce very different oscillatory modes. Without loss of generality, let us assume that ε is a positive constant. If $n = 0$, the virtual pendulum is linear; the oscillation will be simple sinusoidal. If $n = 1$ or any odd integer, the length of the virtual pendulum would be asymmetric during one swing: the length will be longer when x is on the positive side and shorter when x is on the negative side. As the frequency of a pendulum is inversely proportional to it length, the frequency will be high on the negative x value side comparing to the positive x value side. As a result, the resulting wave form is also asymmetric with sharp troughs and round peaks.

If $n = 2$ or any even integer, the virtual pendulum will be symmetric as shown in Figure 7. Depending on the value of ε, the final wave form could have sharp crest and trough for negative ε and round crest and trough for positive ε.

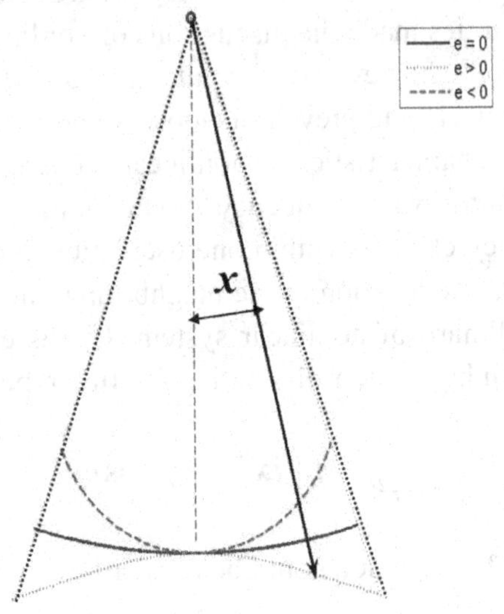

Fig.7. The schematic of a nonlinear pendulum with different signs of the nonlinear parameter.

The essence and the characteristics of the solution of the above equation can be captured by an intra-wave modulated wave form following the idea of perturbation analysis (Kevorkian and Cole, 1981) as:

$$x(t) = \cos(\omega t + \varepsilon \sin n\omega t) \quad (15)$$

The wave form is no longer sinusoidal as discussed by Huang *et al.* (1998, 2009). In the past, the best we could do with these distorted wave data was to represent it with harmonics, and hence the term, harmonics distortion. Mathematically, harmonic distortion is the result of perturbation analysis, which is obtained by imposing a linear structure on a nonlinear system. Those harmonics have only mathematic meaning: to fit the profiles of a distorted wave; they have no physical meaning, for they could not be any part of the real motion. This could be illustrated clearly using the water wave as an example. Deep water surface progress waves of permanent shape were solved by Stokes (1847) with perturbation analysis. The wave surface is asymmetric with sharp crest and rounded trough; hence, the wave profile has a full suite of harmonics. It is well known that deep water surface waves have a distinct property: they are dispersive. However, none of the harmonics has such property. All the harmonics are propagating phase locked at the phase velocity dictated by the fundamental wave frequency. As a result, the wave number and frequency of the harmonics cease to have any physical meaning; they are non-physical quantities. Furthermore, the form of the harmonics also depends on the basis used in the expansion. Based on perturbation analysis, all linear systems have a simple harmonic solution; sine and cosine functions become the natural choice of the building blocks. In general, the sine and cosine functions have been used only because they are the basis for Fourier analysis. Harmonics could also assume other functional form if wavelet expansion is employed. Thus, we can see that the harmonics are really mathematical artifacts to describe any wave form that is not intrinsically the basis used in the expansion. The physically meaningful way to describe the system should be in terms of the instantaneous frequency as discussed by Huang *et al.* (1998, 2009). To illustrate the effect of instantaneous frequency, let us consider the mathematical model used in Huang (1998):

$$x(t) = \cos(\omega t + \varepsilon \sin 2\omega t) \tag{16}$$

For ε set at 0.3, we would have the oscillation given in Figure 8. It has been shown by Huang et al (1998) that the wave form given in Equation (16) is equivalent to

$$x(t) \approx \left(1 - \frac{\varepsilon}{2}\right)\cos \omega t + \frac{\varepsilon}{2}\cos 3\omega t + \ldots, \tag{17}$$

which is exactly the Fourier expansion of the wave form given by Equation (16). Interestingly, this Fourier expansion would almost be the solution of the Duffing type cubic nonlinear equation, if one solves it with perturbation method (Kovorkian and Cole, 1981). Indeed, the form of solution from perturbation depends solely on the power of the nonlinear term: for a cubic nonlinear term, the solution would have third harmonics as indicated in Equation (17). In terms of instantaneous frequency, the wave given in Equation (15) should have an instantaneous frequency with intra-wave modulation as:

$$\Omega = \omega(1 + \varepsilon n \cos n\omega t) \tag{18}$$

Instead of a series of spurious harmonics, the instantaneous frequency would have intra-wave modulation as given in Equation (18). Figures 8a-8d give examples to illustrate the relationship between the wave form and the instantaneous frequency with intra-wave modulation.

When n is an even number, the wave form should be symmetric as shown in Figure 7. The frequency of intra-wave modulation could clearly be counted from the wave form and instantaneous frequency values. For $n = 2$ or 4, the frequency of intra-wave modulation is exactly 2 or 4 times of the wave form frequency. When n is an odd number, the wave form should be asymmetric also shown in Figure 8a and b. For $n = 1$ or 3 as shown in Figure 8c and d, the frequency of intra-wave modulation is exactly 1 or 3 times of the wave form frequency. Therefore, the comparison between the frequency of the intra-wave frequency modulation and the frequency of the wave data is a good indicator of the property of the controlling differential equation: for the n-time intra-wave frequency modulation, the order (power) of the nonlinear term in a simple equivalent differential equation should be $n+1$, as comparison between Equations (13), (14), (15) and (18). Thus the degree and the order of the equivalent nonlinear differential equation could be determined by simply examining the modulation pattern of the instantaneous frequency. Additionally, the instantaneous frequency is defined through a derivative; it is very local. Now, it can be also be used to describe the detailed variation of frequency within an oscillation and an indicator of the properties of the controlling differential equation. These are the feats unattainable with any of the traditional methods. Consequently, as discussed by Huang *et al.* (1998), the instantaneous frequency is such a natural solution for the describing the solutions of nonlinear systems. With all these considerations, we believe the measure of nonlinearity based on the deviation of instantaneous frequency from the constant mean value is logical and practical.

4. Numerical Model

Here we will present a numerical model on bilinear system, when the modulus of elasticity is represented by two segments of straight lines as shown in Figure 9. The reason this model is selected was that Liu (1977) had derived a close form solution, which make the computation mush easier. Basically, the equations governing the system have two phases: In the elastic phase, we have

$$m\frac{\mathrm{d}^2 x(t)}{\mathrm{d}t^2}+c\frac{\mathrm{d}x(t)}{\mathrm{d}t}+k_e x(t)=F(t)-r(t_i)-k_e x(t_i);$$
$$r(t)=r(t_i)+k_e\left[x(t)-x(t_i)\right] \tag{19}$$

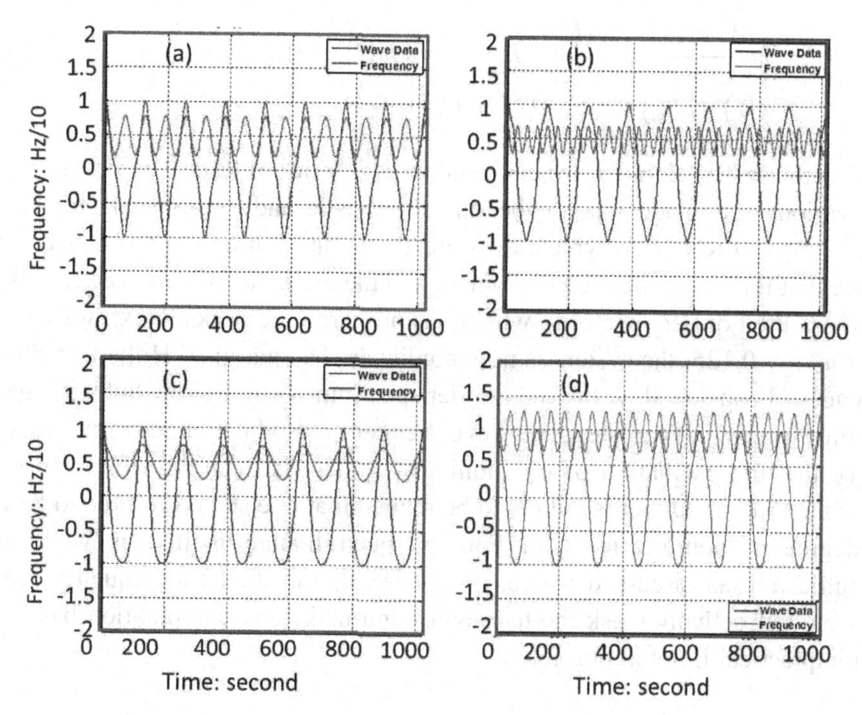

Fig. 8. The wave form and instantaneous frequency of a simple Duffing type of model wave. (a) When n=2, the intra-wave modulation is exactly twice the wave frequency; (b) When n=4, the intra-wave modulation is exactly four times the wave frequency; (c) When n=1, the intra-wave modulation is exactly the same as the wave frequency; (d) When n=3, the intra-wave modulation is exactly three times the wave frequency.

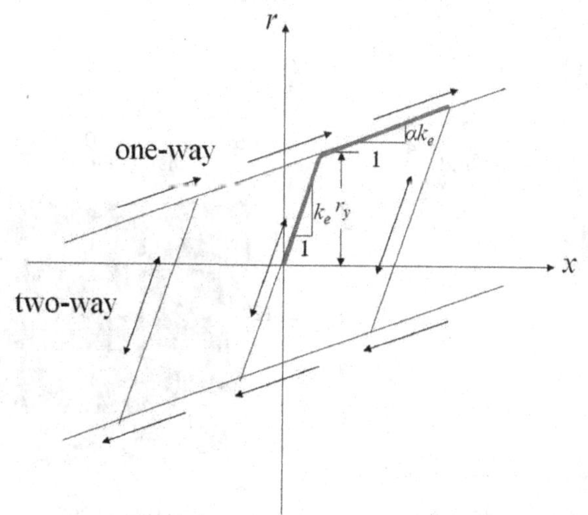

Fig. 9. A model on bi-linear system.

In the elastoplastic phase, we have

$$m\frac{\mathrm{d}^2 x(t)}{\mathrm{d}t^2}+c\frac{\mathrm{d}x(t)}{\mathrm{d}t}+k_e\alpha x(t)=F(t)-r(t_i)-k_e\alpha x(t_i);\qquad(20)$$
$$r(t)=r(t_i)+k_e\alpha\left[x(t)-x(t_i)\right]$$

With the system defined, we can conduct two series of tests. The first series is with various a and the loading beyond the elastic and elastoplastic transition point set at 400 KN. Two typical tests are shown in Figures 10a and b. Here we can see that for $\alpha = 1$, the system is linear and the marginal Hilbert spectrum is as narrow as the Fourier spectrum, which has no harmonic peaks. Next, we try the case for $\alpha = 0.125$, the system is now nonlinear: The marginal Hilbert spectrum is broadened considerably, but the Fourier spectrum is still narrow but with many harmonic peaks. Using the intra-wave frequency modulation pattern, we can deduce that the oscillation of this nonlinear system is up-and-down symmetric and of the order 3. Of course, it could be argued that one should be able to define the degree of nonlinearity with Fourier spectral analysis just as well. The difficulties are as discussed above: noises are also in the high frequency range that would effectively mask the harmonics and make any computation based on them impractical if not impossible.

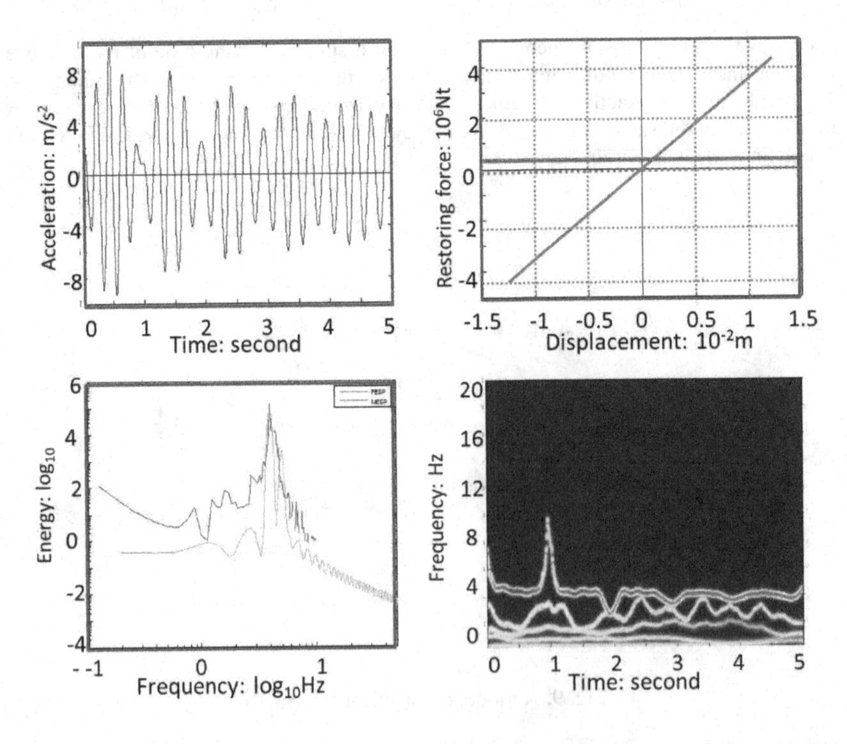

Fig. 10a. A test for loading beyond the elastic and elastoplastic transition point set at 400 KN. $\alpha=1$.

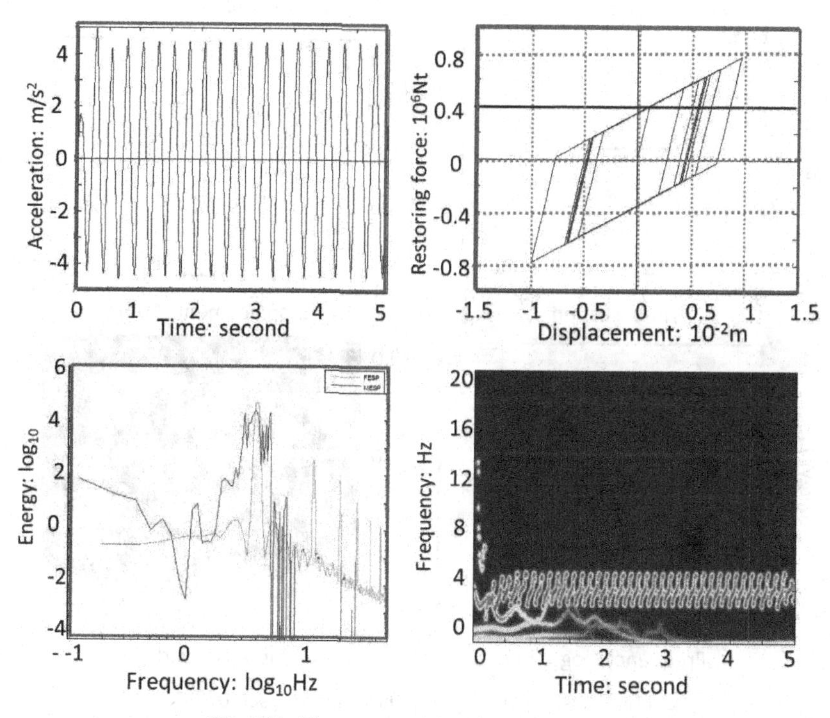

Fig. 10b. The same as Fig. 10a, but for α=1.

Now we will conduct the second test with a fixed a = 0.250, but changing the loading from 90 KN to 350 KN. The results are shown in Figures 11a and b. For the 90 KN loading, the system is linear. The response of the system is similar to that depicted in Figure 11: with narrow band for both Fourier and the marginal Hilbert spectra. If we increase the load to 350 KN, which is close to the plastic limit as shown in the phase diagram, the marginal Hilbert spectrum now has become broad band, while the Fourier spectrum exhibit the typical harmonic peaks. Thus we reaffirmed that nonlinearity would cause the broadening of the spectrum at the fundamental frequency range.

More importantly, this exercise illustrated the principle of using both light and heavy load to test the structure. A slightly damaged bridge should be considered as a deficient structure, but it might still be safe. The natural frequency would not change because of the damage; nevertheless, it might need some minor remedial action before the damage progresses and becomes irreversible. The sensitive test of nonlinearity with light and heavy loads would be an ideal tool for the first line of defense.

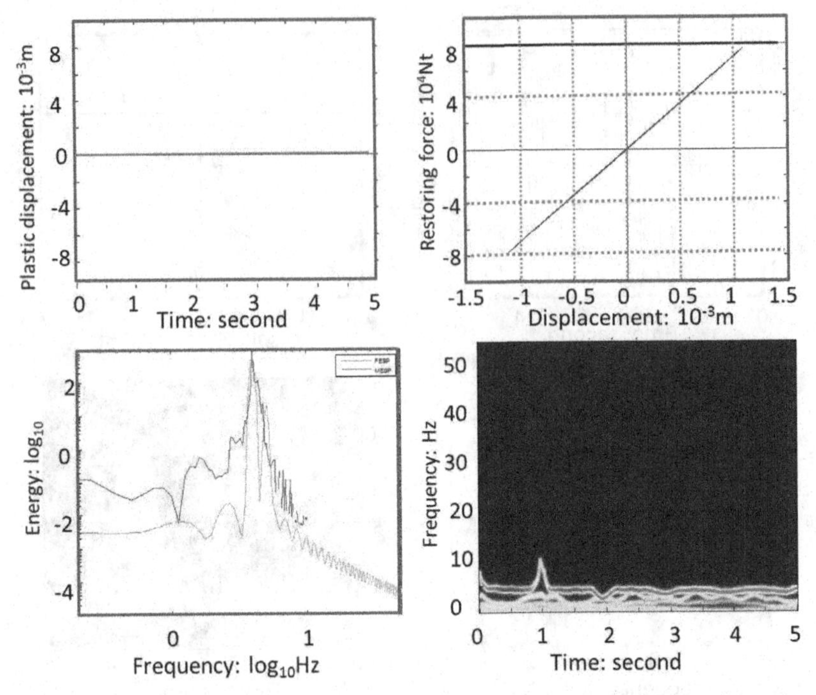

Fig. 11a. A test for fix a=0.25. The loading is set to 90 KN.

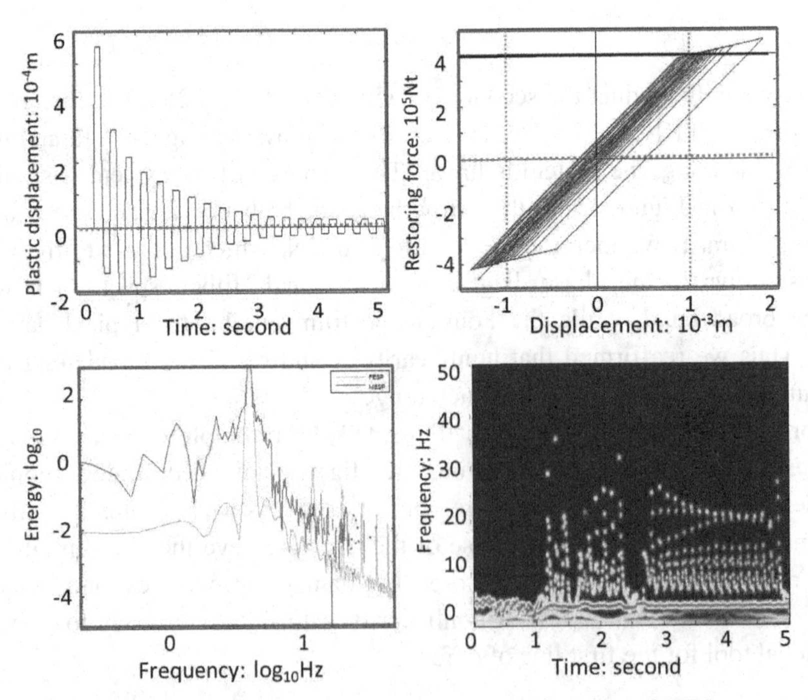

Fig. 11b. The same as Figure 11a, but for loading is set to 90 KN.

5. Bridge Structure Health Monitoring

Having verified the basic nonlinear behavior of a structure in the bilinear model, we can explore the application of this idea to a bridge structure. The basic idea was based on Huang (1998). A series of test was conducted on a bridge in Taiwan and reported by Huang *et al.* (2005), as discussed below. In that report, two novel approaches were introduced: compare the vibration characteristics under the normal traffic as moving exciting load and the light and heavy loads. The virtues of these new approaches are obvious.

For the use of normal traffic as the exciting load, we do not have to interrupt the traffic. The one of the reasons that the transient loading condition has not been utilized up to now is due to the lack of mathematical tools to analyze the nonstationary data so generated. For example, the main stream modal analysis requires the steady vibration to give the modal form. Transient load offers another advantage: the load will visit every point of the structure. Then the moving load would generate the displacement influence line that would make the detection of damage much easier and clearer as reported by Wang *et al.* (2011). The utilization of the normal traffic load also offers various loadings from light and heavy vehicles. Then the idea of using light and heavy load to reveal the nonlinear conditions to determine the structural health characteristics as in the bilinear model discussed earlier. The difference between the heavy and light load also made the requirement of comparisons between damaged and undamaged, new and used structure unnecessary. With these considerations in mind, we can proceed to the test case.

The structure of the bridge is of simply supported beam 30 m in span with continuous 15 cm reinforced concrete deck over three spans separated by construction joints. A single tri-axial force balance Kinematics EPI Sensor accelerometer is anchored at the middle span to measure the vibration of the bridge. The data, Hilbert marginal spectra and Fourier spectra for span 2 and span 3 are given in Figures 12a and b, 13a and b, 14a and b. As can be seen from the data, there are a variety of vehicles passing the bridge with acceleration from light cars to heavy trucks. The Hilbert marginal spectra for span 2 show a very narrow and constant frequency peaks at 4 Hz for all the loading conditions. For span 3, the Hilbert marginal spectra also show the peak frequency value at 4 Hz, but the spectral width broadened substantially for the heaviest load, an indication of nonlinearity in the wave forms.

The same data analyzed with Fourier spectral analysis did not yield any useful information. For span 2, the peaks are relatively narrow at a value slightly higher than 4 Hz, but the fluctuation is much larger than the corresponding Hilbert marginal spectra. For span 3, there is also a broadening of the spectral peak for the heaviest load, but the interpretation of the broadening is unclear physically or mathematically. From a mathematical point of view, the broadening of the spectral peak only indicated the unsteadiness, for an amplitude modulated signal would have sidebands (Huang *et al.* 1999). There is a suggestion of nonlinear behavior if one examines the second harmonics carefully. The harmonics at 8 Hz indeed showed some increase with the loading. As noise is also higher in the high frequency range, the magnitudes of the harmonics are hard to quantify either mathematically or physically.

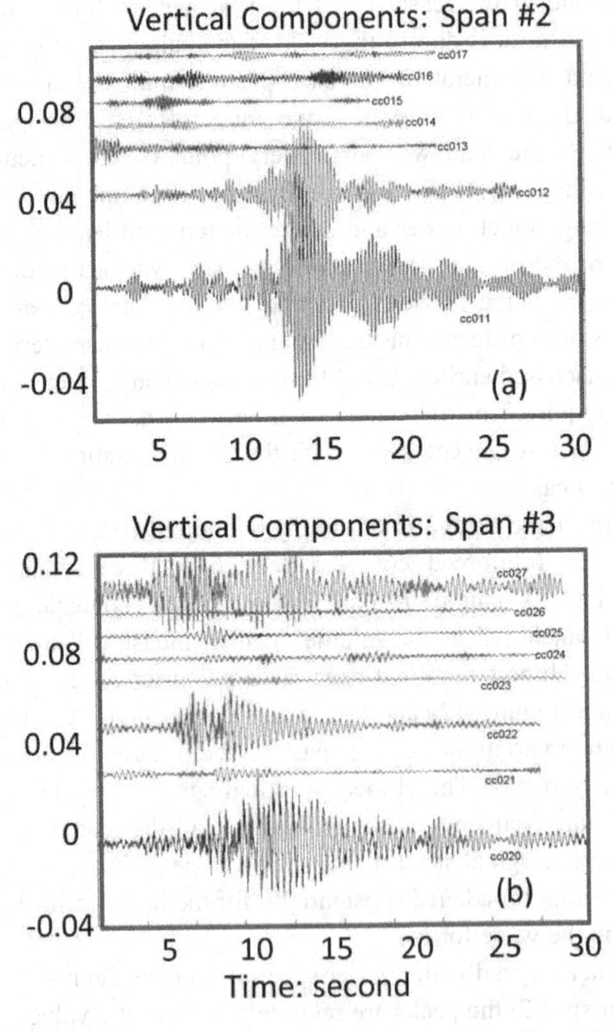

Fig. 12. Data of (a) Bridge Span 2 Vertical, and (b) Bridge Span 3 Vertical.

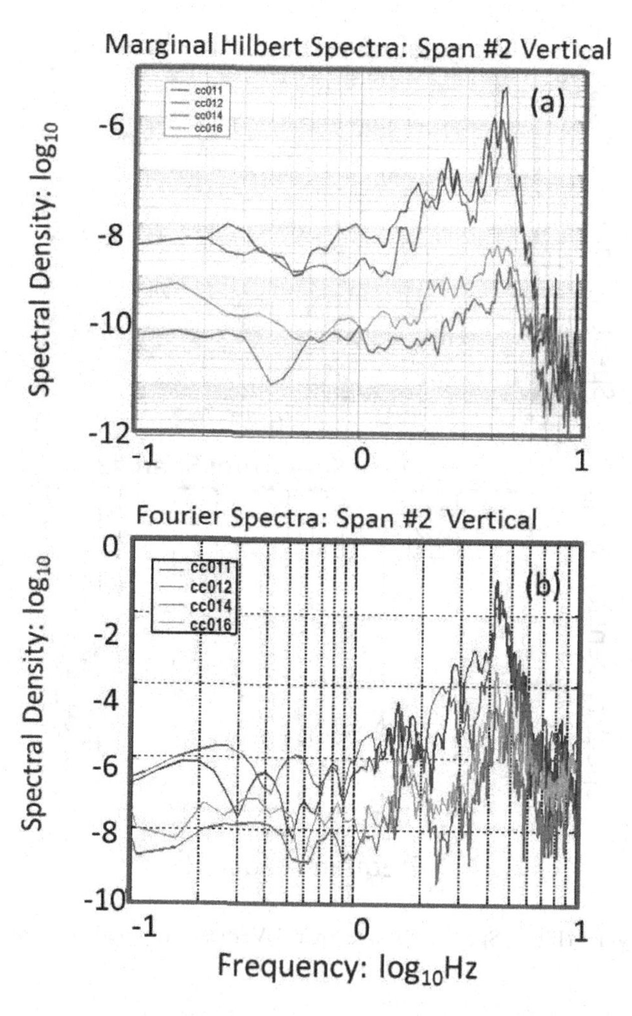

Fig. 13. (a) Marginal Hilbert Spectra : Bridge Span 2 Vertical; (b) Fourier Spectra : Bridge Span 2 Vertical.

Based on the numerical models discussed above, the results presented here indicate that the beams of both spans 2 and 3 did not sustained permanent damage, but span 3 has certainly entered the plastic range. This is an early warning sign generated by the plastic stage that any structure failure would have to go through. It is much easier to identify and once identified, it gives plenty of time for remedial actions. Therefore, we believe that the nonlinearity test should be the first line of defense for structural safety. It helps to set the safe loading conditions: limitations could be set for weight or speed of the vehicles over the bridge to guarantee that the bridge would not go beyond the linear elastic limit, used as the basis for the design of most of the structures.

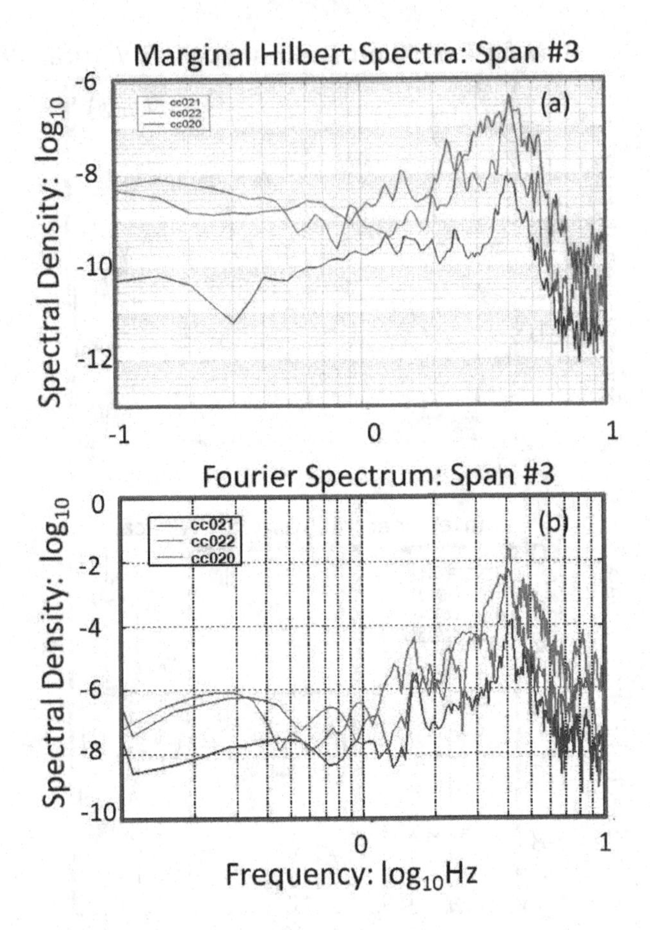

Fig. 14. (a) Marginal Hilbert Spectra : Bridge Span 3 Vertical; (b) Fourier Spectra : Bridge Span 3 Vertical.

Next, we will examine the piers. To monitor the piers, the accelerometer is anchored on the top of the pier next to the deck. Some typical data are shown is Figure 15a and the analysis results are given in Figures 15b and c.

Again the vibration of the piers under heavy load and light loads are drastically different as shown in Figure 15a for a heavy vehicle over piers 1 and 2. But for this case, the damage suffered by one of the piers is so severe that we could detect the downshift of the frequency already. The Marginal Hilbert and Fourier spectra shown in Figure 15b reveal that the Frequency of pier 2 has downshifted in the Hilbert spectra, but not in the Fourier spectra. The amount of frequency downshift is from 1.6 to 1.2 Hz at a ratio of 4 to 3. As the stiffness of the structure is proportional to the square of the frequency, the ratio of the stiffness of piers 1 to 2 is now 16 to 9, a loss of stiffness of almost 50%. But for such a severe damage, the Fourier spectral analysis could not pick up any clue. This frequency is not just a fluke. Figure 15c summarized the mean of many cases and some individual runs for piers 1 and 2. It is again clear that the frequency of pier 2 had shifted even under light loads in order to give the mean, but for the larger load, the downshift is clearer.

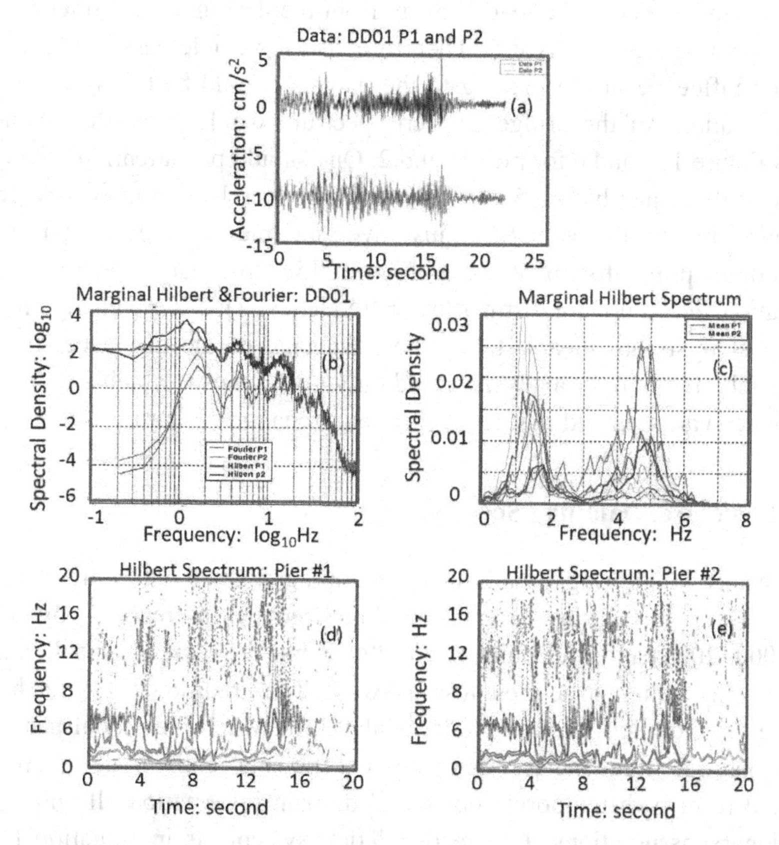

Fig. 15. (a) Data of a heavy vehicles over piers #1 and #2; (b) Marginal Hilbert and Fourier spectra for a heavy vehicles over piers #1 and #2; (c)Mean Marginal Hilbert spectra for a variety of vehicles over piers #1 and #2; (d) Hilbert spectra for the vehicle over piers #1; (e)Hilbert spectra for the vehicle over piers #2.

Here we have demonstrated the prowess of the HHT in health monitoring. The key is that HHT could be used to analyze data from nonlinear and non-stationary processes. As a result, we could use transient normal traffic load instead of artificially induced steady load. The normal traffic load is not only convenient, but the test condition is also the most logical for health monitoring. Most crucially, the instantaneous frequency employed in the HHT is much sharper in defining structural dynamic responses than the Fourier spectral analysis, which is based on the overall mean of the frequency throughout the whole domain.

The reason the instantaneous frequency could be more sensitive to damage detection is this: when the vehicle is over the damage part, it exerts more direct loading on the structure. This is the same idea of the influence line approach in structural analysis. When the load is away from the damage, the influence of the damage would be hard to discern. Therefore, as the vehicle passing through the bridge, the influence of the loading to the damage would be totally different at different locations of the bridge. Hilbert spectrum would show this change as shown in Figure 15d and e for piers 1 and 2. One should pay attention at the point near time scale at just before 5 second point. For pier 1, a large energy density spot occurs due to the vehicle is just over the pier in Figure 15d. At the corresponding point for pier 2 in Figure 15e, the large energy density concentration shows up at a distinct lower frequency. This sharp change in time-frequency analysis has clearly taken the advantage of the influence line idea. Unfortunately, the Fourier analysis would only give a mean, which would spread the frequency variation and render the frequency change obscure.

6. Ship Structure: Damping Spectral

In some health monitoring situations, the damping characteristics are important. The case we are discussing here is the ship structures as reported by Salvino (1999, 2000, 2001) and McDaniel *et al.* (2000), but the same approach could be applied to any other structures just as well. The traditional approach is to determine the damping through modal analysis, in which the modal parameters such as natural frequencies, damping coefficients and mode shapes are determined from both temporal and spatial domain information all under linear and stationary assumptions. For an oscillation system, as in Equation (19) or (20), we can normalized it as:

$$\frac{\mathrm{d}^2 x}{\mathrm{d}t^2} + 2\varsigma\omega_n \frac{\mathrm{d}x}{\mathrm{d}t} + \omega_n^2 x = f(t) \tag{21}$$

where ω_n is the natural frequency of the system,

$$\varsigma = \frac{c}{2m\omega_n} = \frac{c}{c_r} \tag{22}$$

is the damping ratio or fraction of critical damping coefficient, defined as:

$$c_r = 2m\omega_n = 2\sqrt{km} = \frac{2k}{\omega_n} \tag{23}$$

A complicated system could be treated as the sum of many vibration modes. Here, a collection of terms is defined as a mode with variable amplitude. Let x(t) be such a mode, then x(t) and its analytical pair, y(t), can be written as:

$$x(t) + \mathrm{i}\, y(t) = a(t)\mathrm{e}^{\mathrm{i}\theta(t)} = C\,\mathrm{e}^{-\phi(t)+\mathrm{i}\theta(t)} \tag{24}$$

Then, we can define a time dependent damping factor for this mode as:

$$\gamma(t) = 2\frac{\mathrm{d}\phi}{\mathrm{d}t} = -2\frac{1}{a}\frac{\mathrm{d}a}{\mathrm{d}t}. \tag{25}$$

Furthermore, we can also define a loss factor as:

$$\eta(t) = \frac{\gamma(t)}{\omega_0(t)} = \frac{2}{\omega_0(t)}\frac{1}{a}\frac{\mathrm{d}a}{\mathrm{d}t} \tag{26}$$

Finally, the time-dependent loss factor is defined as the positive valued RMS:

$$\left\langle \eta(x,t)^2 \right\rangle^{1/2} = \left\{ \frac{1}{T}\int_0^T \eta^2(x,t)\mathrm{d}t \right\}^{1/2} \tag{27}$$

As any signal could be decomposed through HHT into a finite number of orthogonal IMF modes, we can then compute the damping or the loss factor for each IMF and thus quantifying the damping and loss factor of the system.

Let us present a calibration and some specific examples of structure response studies. To calibrate the system, we select a know decay function given by

$$\begin{aligned} x(t) &= \sum_n C_n \mathrm{e}^{(\mathrm{i}2\pi f_n - \gamma/2)t}\, \mathrm{e}^{\mathrm{i}2\pi f_0 t} \\ &= I(t)\cos(2\pi f_0 t) - Q(t)\sin(2\pi f_0 t) \end{aligned} \tag{28}$$

The parameters are given as f_0=130 MHz; f=1/t>50 MHz; γ=10^6 sec-1 and n=340. With these parameter, the $I(t)$ and $Q(t)$ are given in Figure 16a.The loss factor obtained through HHT is given in Figure 16b, in which the dotted line indicates the theoretical value and the solid line indicates the experimental results from Hilbert spectral analysis. The HHT result compares very well with the theoretical values.

Now we will study a real test of the structure frames through two methods: the modal analysis and the HHT based analysis. To implement the modal analysis, one would have to limit the vibration to a single frequency and measure the response in temporal domain to obtain the loss factor of that particular frequency. To cover a range of frequency, the same test would have to be repeated for each frequency value. For HHT, the load could consist of a single bang on the structure. Figures 17a and b give the test results. While the HHT-based analysis, the results are continuous but all generated from a single band, which consists of all frequencies, the modal analysis would require repeated single continuous frequency vibrations each gives only one single point in the frequency space.

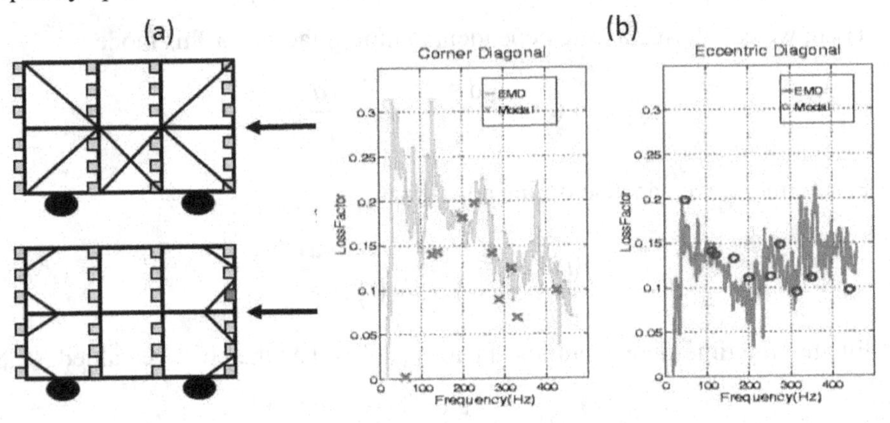

Fig. 16. (a) The calibration load; (b) The damping factor for the calibration load.

Clearly, the HHT-based method can give the damping or loss factor for all the frequency continuously with values comparable to but superior than the modal analysis. The reasons are simple: the modal analysis is Fourier based with the linear and stationary assumptions. It would work only for simple systems. The HHT-based analysis does not depend on a priori assumptions. With the specific capacity for nonstationary and nonlinear data analysis, HH- based method could be used for a shock load to define loss factor as function of time and frequency continuously. This method would work even if the structure is under extreme loading condition and the response goes beyond linear elastic limit.

In summary, the HHT method gives detailed local structural response behavior, and is a unique, valid, and important approach for shock data analysis. One of the advantages of this method is the effective use of the full complicated data in combined time-frequency representation. The full exploration of the physical significance of IMF components may offer new physical insight for analysis and modeling of shock impact and for identifying important structural failure mechanisms of large and complicated structures subjected to shock loading. The time-frequency dependent damping is the only method that is able to extract damping values from experimental measurements of complex systems, such as underwater shock test time series and express damping ratio as a function of both frequency and time.

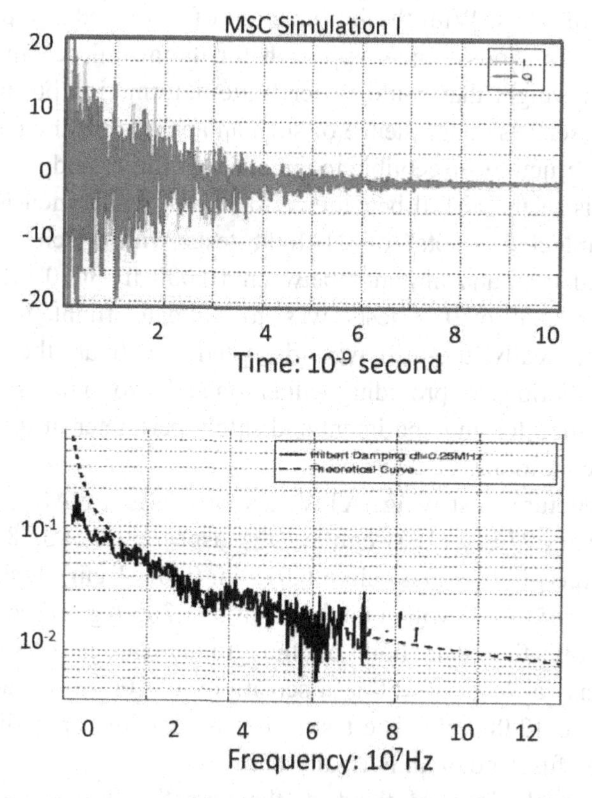

Fig. 17. (a) Two different structures: the top one is the corner diagonal braced the bottom one is the eccentric corner braced; (b) The damping factor for the different braced structures.

7. Aircraft Structure

Damping analysis is limited to the study of stable static structures. Certain structures such as machinery components, moving vehicles, and aircrafts could experience strong vibrations so much so that the vibration could experience negative damping or become instable. Therefore, the identification of structural damping is not a full description of all possible vibration behavior. To account for the full range of structural health monitoring, we have to include the possibility of instability and find a time dependent description of such behavior as reported by Huang *et al.* (2006). To define the stability is easy once we know how to determine damping. We only have to take the loss factor given in Equation (26) with the proper positive and negative signs, without computing the positive RMS value as given in Equation (27), and present it as a time-dependent stability spectrum.

We will present the stability spectral analysis as in Huang *et al.* (2006) to the aero-elastic airfoil tests. With the introduction of advanced composite materials and construction methods, a new approach for the airfoil design is to have the airfoil lighter in weight that could suffer finite deformation but having superior performance. A serious consequence of such an approach is the reduced stiffness could make the structure susceptible to the onset of flutter and instability. Thus, a stability analysis of the airfoil becomes critical. NASA Dryden Flight Research Center has conducted a total of 21 flight tests with Mach number ranging between 0.5 and 0.83 and altitudes between 10,000 ft (3050 m) and 20,000 ft (6100 m). The goal of the tests was to validate flight flutter prediction techniques. Specifically, the tests ware designed to validate the accuracy of in-flight flutter predictions by providing validation and also to investigate the ability of engineers to monitor an experiment and safely perimeter of the envelope near unstable flight conditions.

The aero-structures test wing (ATW) test article is a NACA 65A004 airfoil with a wing area of 97 in.2 (1271 cm2) and an aspect ratio of 3.28. The wing skin is made of a three-ply fiberglass cloth 0.015 in. (0.0381 cm) thick, and the wing core is made of rigid foam with a total weight of 1.205 Kg. Three accelerometers are used for collecting data both on the ground and in flight testing. Each accelerometer has a range of ±50 g, a sensitivity of 100 mV/g, and a frequency range from 0.3 to 12,000 Hz. The test airfoil is mounted on a flight-test-fixture (FTF), under the fuselage of an F-15 airplane.

During the final phase of the test, the objective is to explore the safety perimeter; therefore, the speed of the test plane is pushed to the limit of the airfoil could sustain to the point of structural failure. The data during the last 7 seconds of the final flight is shown in Figure 18. The HHT analysis produced the detailed Hilbert spectrum during the last 17 seconds is given in Figure 19. Here we can see the flutter frequency is confined in a broad band around 18 Hz. As the flutter amplitude increases in the last couple of seconds, the amplitude increases drastically as shown in the data as well as in the Hilbert spectrum. Coupled with the increase of flutter amplitude there were two interesting changes: the first is the decrease in the flutter frequency in the last second, an indication of the yielding of the airfoil. The second is the increase of the low frequency flutter with frequency in the range of a one Hz or lower near the baseline of the time-frequency of the Hilbert spectrum, as shown in Figure 19.

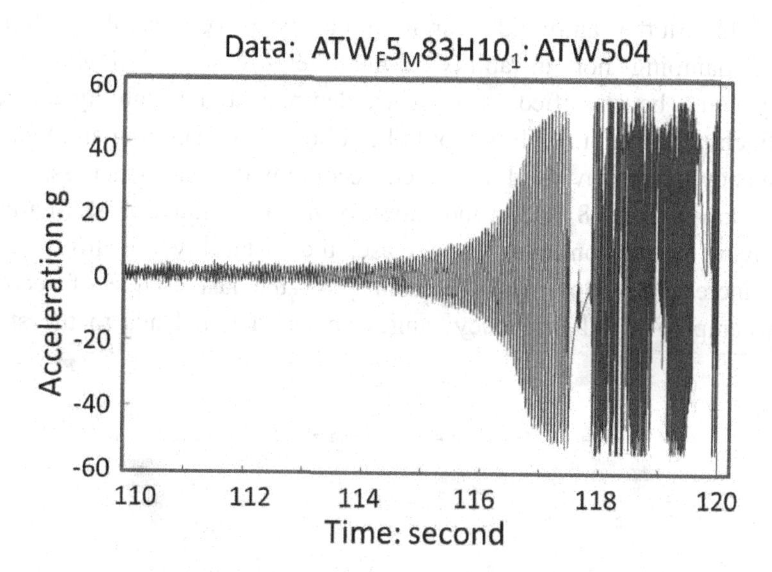

Fig. 18. The data from the accelerometer at the tip of the aero-elastic structure.

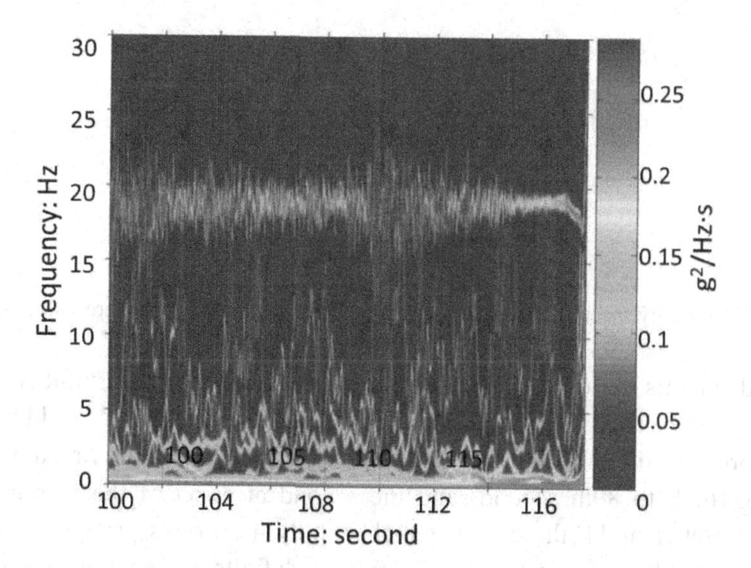

Fig. 19. The Hilbert time from the accelerometer data at the tip of the aero-elastic structure.

Now let us examine the stability spectrum given in Figure 20. Long before the failure, many visible anomalies could be detected. The first one is the stress stiffening: the stability spectrum exhibits a gradual increase in the flutter amplitude and frequency at around 60th to 80th second. Before this had happened, the flutter amplitude variation has been very stable with neither noticeable damping nor instability. After the stress stiffening, the flutter frequency could be classified as unstable: damping and instability all become visible. Each period of instability would be followed by damping and vice versa. The frequency ranges involved, however, become wider and wider, especially in the region lower than 18 Hz, an indication of the appearance of sub-harmonics. As the flying speed continues to increase, the instability (negative damping) becomes increasingly dominant. Eventually, at the last couple of second the negative damping at all frequency ranges appeared that leads to the structure failure.

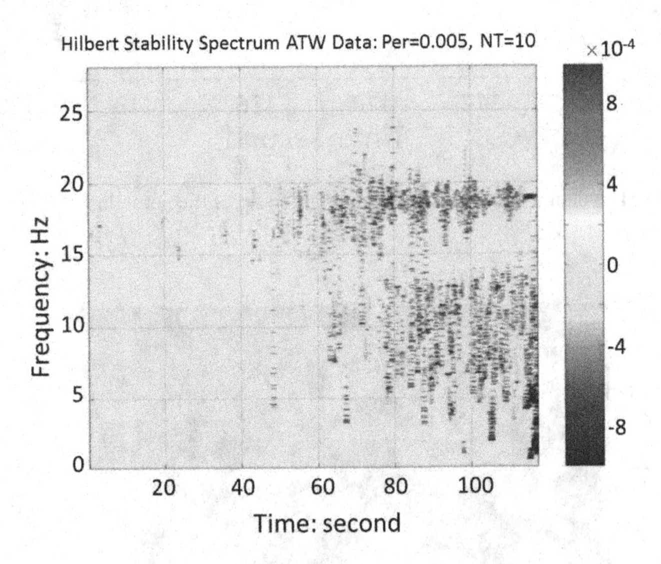

Fig. 20. The stability spectrum from the accelerometer data at the tip of the aero-elastic structure

Based on this series of events, we can conclude that the stability spectral analysis could be used as a health monitoring tool. Early warning could be issued long before the failure of the airfoil. Two obvious points are: the first at the stress stiffening (60th to 80th second) and the second at onset of the extremely low frequency flutter at 113th second, which is only 4 seconds prior to the failure. Of these possible warning points, the first is definite far enough ahead of the failure for us to take counter measure to prevent the failure from happening. The second warning point is only 4 seconds prior to the failure, which is still possible, but might be too short to be feasible, for us to do anything. At any rate, the HHT-based health monitoring, if used wisely, could be a tool to detect the precursor of the catastrophic events and prevent the structure failure.

8. Conclusions

We have introduced HHT for structural health monitoring. With the introduction of the time-frequency representation, one could monitor the dynamical behavior of structures in more detailed and quantitative way. Specifically, the time-frequency representation also enables us to quantify the nonlinearity, which by itself might not indicate damage, offers an early indicator of structural deterioration. Based on this approach, the HHT-based structural health monitoring has been applied to three types of structures: the civil infrastructures, the ship structures and the aircraft, all with laudable results.

For the civil infrastructures, the time-frequency method enables us three great advantages: the first one is to use transient loads from the normal traffic flow. This makes the health monitoring easier to implement without causing traffic stoppage and interruptions. The second and most innovative advancement proposed here is to use the heavy versus light loads. The comparisons could establish the safe loading condition for the structure. This will detect the potential damage long before anything drastic happened to the structure. Finally, the most important advantage of the time-frequency analysis from a moving load allows us to take advantage of the influence idea. When the moving load reaches the damaged spot, it would give the largest signal for damage detection. The example on ship structures introduced the damping spectral analysis in terms of continuous damping values for all frequencies. Damping characteristics is not limited to the health monitoring of ship structures. It could be used for other structures, especially the civil infrastructures for which damping characteristics can be monitored.

For the aircraft safety monitoring, the stability spectral analysis is introduced. This method can be used on other types of machinery with large range of movements or moving vehicles. An obvious target will be the high speed train systems, where uneven track or faulty suspension can cause the train to flutter, snake and even derailment. All these are dangerous operation conditions with different safety concerns. Presumably, the flutter and snaking would occur long before the derailment. Thus, continuous monitoring should enable us to detect anomalies long before anything drastic happen. Therefore, we should monitor and correct the anomalies for safety operations.

All the methods introduced here could be used for different types of structures. The critical additional research will be to establish the criteria for safety classifications. For example, how much nonlinearity is unsafe? How much permanent down shift of vibration should be the safety limit? In this aspect, this work is only the initial step toward a more powerful structural health/safety monitoring research.

Acknowledgments

N. E. Huang has been supported by a grant from Federal Highway Administration, DTFH61-08-00028, and the grants NSC 98-2627-B-008-004 (Biology); NSC 98-2611-M-008-004 (Geophysical) and support for the Center for Dynamical Biomarkers and Translational Medicine, National Central University, Taiwan (NSC 99-2911-I-008-100) from the National Science Council, Taiwan, and finally a grant from NCU 965941 that have made the conclusion of this study possible. He is also supported by a KT Lee endowed Chair at NCU.

References

Bedrosian, E. (1963). On the quadrature approximation to the Hilbert transform of modulated signals, Proc. IEEE, 51, pp.868–869.

Boashash, B. (1992a). Estimating and interpreting the instantaneous frequency of a signal. Part I: Fundamentals. Proc. IEEE, 80, pp. 520–538.

Boashash, B. (1992b). Estimating and interpreting the instantaneous frequency of a signal. Part II: Algorithms and applications. Proc. IEEE, 80, pp. 540–568.

Boashash, B. editor. (1992c). Time-Frequency Signal Analysis--Methods & pplications, Longman-Cheshire, Melbourne and John Wiley Halsted Press, New York.

Chen, J. (2009). Application of Empirical Mode Decomposition in structural health monitoring: some experience. Adv. Adapt. Data Analy.,1, pp. 601–621.

Chen, J., Xu Y. L. et al. (2004). Modal parameter identification of Tsing Ma suspension bridge under typhoon victor: EMD-HT method. J. Wind Eng. Ind. Aerod., 92(10), pp. 805–827.

Cohen, L., (1995). Time-frequency Analysis, Prentice Hall, Englewood Cliffs, NJ

Doebling, S. W., Farrar, C. R., Prime, M. B. & Shevitz D. W. (1996). Damage identification and health monitoring of structural and mechanical systems from changes in their vibration characteristics: a literature review. Los Alamos National Laboratory report LA-13070-MS.

Flandrin, P. (1999). Time-Frequency / Time-Scale Analysis, Academic Press, San Diego, CA.

Flandrin,P., Rilling G., and Gonçalvès P. (2004). Empirical Mode Decomposition as a filter bank. IEEE Signal Processing Letter,11, pp. 112–114.

Huang, K. (1998). A new instrumental method for bridge safety inspection based on a transient test load. US patent no. 6,192,758 B1.

Huang, N. E. (2003). Empirical Mode Decomposition for analyzing acoustic signal. US Patent 10-073857, August, 2003, Pending.

Huang, N. E., Shen, Z., Long, S. R., Wu, M. C., Shih, S. H., Zheng, Q., Tung, C. C. and Liu, H. H. (1998). The empirical mode decomposition method and the Hilbert spectrum for non-stationary time series analysis. Proc. Roy. Soc.London, A454, pp. 903–995.

Huang, N. E., Shen Z. and Long R. S. (1999). A new view of nonlinear water waves – the Hilbert spectrum, Ann. Rev. Fluid Mech. 31, pp. 417–457.

Huang, N. E., Huang K. and Chiang W. (2005). HHT-based bridge structural health-monitoring method. In Hilbert-Huang Transform and Its Applications, Ed. N. E. Huang and S. S. P. Shen, World Scientific, 2005, pp. 263–287.

Huang, N. E., Brenner M., Pak C.-G. and Salvion L. (2006). Application of Hilbert-Huang Transform to the stability study of airfoil flutter. AIAA,44, pp. 772–786.

Huang, N. E., Wu Z., Long S. R., Arnold K C., Chen X.and Blank K. (2009). On Instantaneous Frequency. Adv. Adaptive Data Anal., 2, pp. 177–229.

Huang, N. E., Wu Z., Lo M-T and Chen X. Y. (2011). On quantification of the degree of nonlinearity. US Patent Pending.

Infeld, E. and Rowlands G. (1990). Nonlinear Waves, Solitons and Chaos. Cambridge University Press, Cambridge.

Kevorkian, J. and Cole J. D. (1981). Perturbation methods in Applied Mathematics. Springer-Verlag, New York.

Liu, C.-S. (1997). Exact solutions and dynamic responses of SDOF bilinear elastoplastic structures. Journal of the Chinese Institute of Engineers, 20(5), pp. 511–525.

Kantz, H. and Schreiber T. (1997). Nonlinear Time Series Analysis. Cambridge University Press, Cambridge, UK.

Kootsookos, P. J., Lovell B. C. and Boashash B. (1992). A unified approach to the STFT, TFDs and Instantaneous Frequency. IEEE Trans. Signal Processing, 40(8), pp. 1971–1982.

Loughlin, P. J. and Tracer B. (1996). On the amplitude- and frequency- modulation decomposition of signals, J. Acoust. Soc. Am., 100, pp. 1594–1601.

Lovell, B. C., Williamson R. C. and Boashash B. (1993). The relationship between instantaneous frequency and time-frequency representations. IEEE Trans. Signal Processing, 41(3), pp. 1458–1461.

McDaniel, G., Dupont P. and Salvino L. W. (2000). A wave approach to estimating frequency-dependent damping under transient loading, J. Sound & Vibration, 231(2), pp. 433–499.

Nuttall, A. H. (1966). On the quadrature approximation to the Hilbert Transform of modulated signals, Proceedings of IEEE, 54, pp. 1458–1459.

Olhede, S and Walden A. T. (2004). The Hilbert spectrum via Wavelet projections, Proc. Roy. Soc. Lond., A460, pp. 955–975.

Picinbono, B. (1997). On instantaneous amplitude and phase signals, IEEE Trans. Signal Process, 45, pp. 552–560.

Rytter, A. (1993). Vibration based inspection of civil engineering structures. Ph.D. Dissertation, Department of Building Technology and Structural Engineering, Aalborg University, Denmark.

Salvino, L. W. (1999). Empirical mode analysis of structural response and damping, Proceedings of the 18th International Modal Analysis Conference, San Antonio, TX, 1999.

Salvino, L. W. (2000). Evaluation of structural response and damping using empirical mode and time-frequency analysis, Proceedings of the 71th Shock and Vibration Symposium, Arlington, VA, Nov. 2000.

Salvino, L. W. (2001). Evaluation of Structural Response and Damping using Empirical Mode Analysis and HHT, Proceedings of the 5th World Multi-conference on Systemics, Cybernetics and Informatics, Orlando, FL, July, 2001.

Staszewski, W. J. and Robertson A. N. (2006). Time-frequency and time-scale analysis for structural health monitoring, Phil. Trans. R. Soc. A, 365, pp. 449-477. doi: 10.1098/rsta.2006.1936

Stokes, G. (1847). On the theory of oscillatory waves, Trans. Camb. Phil. Soc., 8, pp. 441–455.

Vatchev, V. (2004). Intrinsic Mode Function and the Hilbert Transform, Ph. D. Dissertation, Industrial Mathematics Institute, Department of Mathematics, University of South Carolina, Columbia, SC.

Wang, C. Y., Huang C. K. and Chen C. S. (2011). Damage assessment of beam by a quasi-static moving vehicular load, Adv. Adaptive Data Anal., 3, (in press).

Whitham, G. B. (1975). Linear and Nonlinear Waves, New York, Wiley.

Wu Z. and Huang N. E. (2009). Ensemble Empirical Mode decomposition: A noise-assisted data analysis method, Adv. Adap. Data Analy. 1, pp. 1–42.

Xu, Y. L. and Chen J. (2003). Characterizing nonstationary wind speed using empirical mode decomposition, J. Struct. Eng. ASCE, 130(6): pp.912–920.

Xu, Y. L. and Chen J. (2004). Structural damage detection using empirical mods decomposition: Experimental investigation, J. Eng. Mech. ASCE, 130, pp. 1279–1288.

Yang, J. Lei N., Y. et al. (2003). System identification of linear structures based on Hilbert- Huang spectral analysis. Part 1: Normal modes, Earthquake Eng. Struct. Dynam., 32, pp. 1443–1467.

Yang, J. N., Lei Y. et al. (2004). Hilbert-Huang based approach for structural damage detection, J. Eng. Mech. ASCE, 130(1): pp. 85–95.

Yu, D. H. and Ren W. X. (2005). EMD-based stochastic subspace identification of structures from operational vibration measurements, Eng. Struct., 27(12), pp. 1741–1751.

Chapter 9

The Use of Genetic Algorithms for Structural Identification and Damage Assessment

C. G. Koh[1] and Z. Zhang[1]

[1]*Department of Civil and Environmental Engineering,*
National University of Singapore, 117576, Singapore
E-mail: cgkoh@nus.edu.sg; ceezzhen@nus.edu.sg

1. Introduction

Recent advances in sensor technology have generated strong interests in structural identification and health monitoring (Li *et al.*, 2004; Lynch and Loh, 2006). Much effort has been undertaken in applying system identification methods to practical applications. The process of system identification is to identify the unknown system parameters based on the analysis of measurement data of input and output (I/O). Many system identification methods focus on identifying the modal parameters (Cole, 1973; Ibrahim and Mikulcik, 1977; Beck and Jennings, 1980; Juang and Pappa, 1985; James *et al.*, 1995; Van Overschee and De Moor, 1996) as well as physical parameters (Koh *et al.*, 1991; Beck, 1979), which can then be used to quantify the structural health. Modal identification is relatively easy to apply but is rather insensitive to local damages due to difficulty in capturing higher modes accurately (Huang and Shen, 2005). Identifying the physical parameters such as stiffness directly in time domain from acceleration measurement provides a more promising way. Therefore, this study focuses on identification of physical parameters such that the numerically estimated response would match well with the measured response from field/lab testing. Among the methods to identify physical parameters, those with rigorous mathematical principles are referred to as classical methods (Caravani *et al.*, 1977; Shinozuka *et al.*, 1982; Hoshiya and Saito, 1984) while the others, known as genetic, are usually optimization or search methods established on heuristic rules such as natural selection and evolution. Compared to classical methods, one of the main strengths of non-classical methods in structural identification is the generally better robustness in achieving global optimum. Furthermore they do not require a good initial guess—unlike many classical methods.

The application of non-classical methods in structural identification is increasingly favored due to rapid advancement in computer power. Given the insensitive feature to noise and initial guess, genetic algorithm (GA) has shown great potential as an alternative for solving inverse problems as in structural identification. Recent research of GA-based structural identification has shown that damage as small as 4% can be successfully identified in the experimental study (Koh and Perry, 2007). Besides damage quantification, GA has also been adopted to many identification issues such as the identification of impact load location (Doyle, 1994), structural change (Chou and Ghaboussi, 2001) and identification of moving masses on a multi-span beam (Jiang et al., 2003). Most of the abovementioned references are, however, concerned with simple structures with few degrees of freedom or few unknowns. To reduce the computational cost in identifying large systems with more DOFs or unknowns, it is imperative to improve the search capability of GA. Recently considerable effort has been spent on the enhancement of identification strategies. It has been demonstrated that the identification process could be significantly expedited by using distributed computing (Koh *et al.*, 2002). Alternatively, in a sequential programming, a search space reduction method (SSRM) was developed to accurately and reliably identify the structural systems (Perry *et al.*, 2006). The strategy adaptively adjusts the search space to expedite the search and incorporates a modified GA based on migration and artificial selection (iGAMAS). The SSRM method was applied successfully into the identification of the system as well as the input force using only limited output observations (Perry and Koh, 2008).

Nevertheless, without better understanding of structural identification from the optimization perspective, it is hard to develop efficient GA-based strategy for this inverse problem. In this regards, this chapter will firstly present the investigation of fitness surface features of structural identification. Based on these insights into the fitness surface, an enhanced GA-based method will be developed by conducting both global search and local search. The global search is implemented by a uniformly sampled genetic algorithm (USGA). It is a dual-layer GA strategy, with an inner iGAMAS loop and a sampling test carrying out-layer solution space exploration. The local search is established by a gradient based BFGS method. Extensive parametric study will be carried out to determine the USGA parameters as well as the switch point from global search to local search. Finally, the proposed USGA with BFGS search will be verified by numerical and experimental studies.

2. Definition of the Problem: System Identification Using Genetic Algorithms

Treating the identification of physical parameters as an optimization problem, traditionally referred as output-error issue (Bekey, 1970; Bowles and Straeter, 1972), the statement of problem can be written typically in terms of maximization as

$$\max \phi(\mathbf{x}) \text{ subject to } \mathbf{x} \in \mathbf{S}, \qquad \text{where } \mathbf{S} = \{\mathbf{x} \in \Re^k \mid a_i \leq x_i \leq b_i, \ \ \forall i = 1,2,\cdots k\} \tag{1}$$

Here ϕ is a scalar-valued function based on the variable vector $\mathbf{x}$ which contains the unknowns to be identified, e.g. stiffness, mass and/or damping parameters. $\mathbf{S}$ is the search limit from a k-dimensional space $\Re^k$. The values of a_i and b_i are separately the lower and upper bounds of individual parameter.

In terms of global optimization, GA is a good choice owing to several desired characteristics including robustness with respect to measurement noise and initial guess, ease of adding constraints of optimization, and high concurrency for distributed computing. Using GA in structural identification, the idea can be illustrated in a block diagram in Fig. 1. The process typically incorporates a forward analysis to evaluate fitness function and a backward analysis to search new possibilities. The forward analysis is essentially a finite element procedure to obtain the estimated response using a trial population of unknown parameters. The backward analysis is an optimization process to obtain better candidate for the parameters to be identified.

The identifiability of the system is implied by the assumption that unique mapping is available between the measurement time signals and the unknown parameters. Although few researchers provide rigorous proof on the identification uniqueness, it is reasonable to avoid non-uniqueness by putting the measurement as distributed and sufficient as possible, based on the available results of limited research works (Udwadia and Sharma, 1978; Franco *et al.*, 2006). The second assumption is that the mathematical model used in the forward analysis is good enough to capture the physical behavior of the structures. This means that good understanding and modeling of the structural behavior is important.

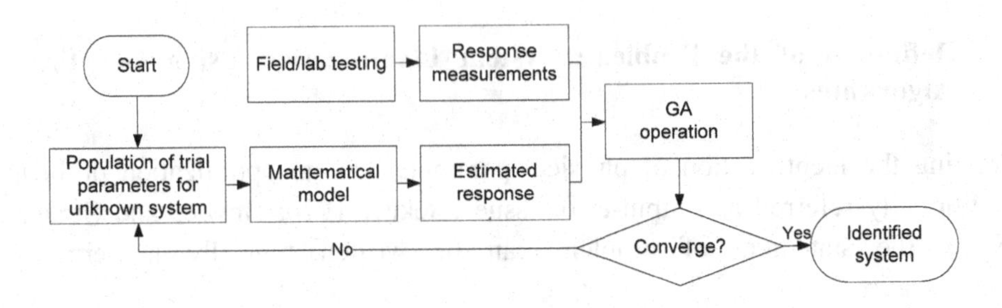

Fig. 1. System identification using genetic algorithms.

3. Characteristics of Structural Identification as an Optimization Problem

This section is to investigate the characteristics of structural identification as an optimization problem so as to develop efficient identification strategies. For numerical optimization, a good knowledge on the shape of fitness surface will help to choose a problem-suitable strategy. Identifying practical structural systems often involves a large number of unknown parameters, which makes it difficult to visualize the fitness surface via two- or three-dimensional plots. Through small scale testing with not more than two parameters, however, the nature of structural identification can be discovered from an optimization perspective.

To gain some insight, an investigation of the fitness surface is conducted for 1-DOF and 2-DOF lumped mass systems. Let K_1 and M_1 be the stiffness and mass of the 1-DOF system, respectively, as shown in the insert of Fig. 2. The 1-DOF system is considered with 600 kN/m stiffness, 500 kg mass and a damping ratio of 1% in the numerical study. The acceleration is measured at the only mass point which is excited by a random but known force. The shape of fitness surface is presented with respect to K_1 in Fig. 2 for three noise levels. For the 2-DOF system, the stiffness terms are K_1 and K_2 and mass terms are M_1 and M_2 as shown in the insert of Fig. 3 and Fig. 4. The stiffness and mass are 600 kN/m and 500 kg for the first (lower) level, and 350 kN/m and 300 kg for the second level. Rayleigh damping of 1% is considered in both vibration modes. Acceleration measurements at the two levels are numerically simulated corresponding to a known random force at the top. The mass is assumed to be known. For the ease of illustration, the identified parameter values are normalized with respect to the corresponding exact values in all these figures.

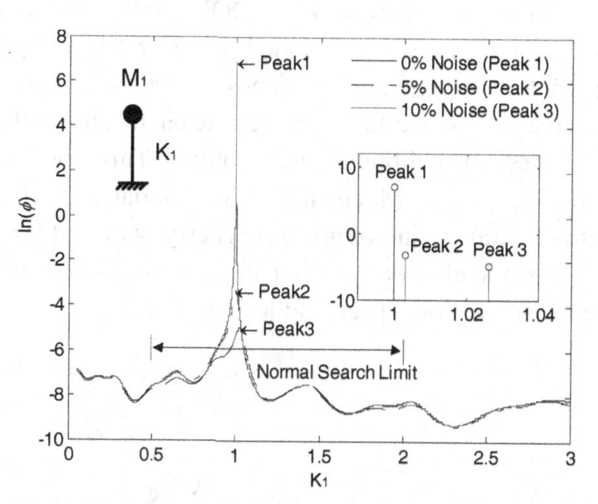

Fig. 2. Typical peak shifting of 1-DOF known mass system.

The investigation covers known mass and unknown mass systems in the [0.5-2.0] search space, i.e. search from half to twice of the baseline value. The search space is meshed evenly in a grid manner, as the contour of fitness surface is unknown in advance. The method is to evaluate the fitness function at each point on the regular grid of parameters. This is called an *s*-level problem if *s* possibilities of each variable are tried in the respective search range. Obviously the accuracy of the resultant surface inevitably depends on the grid density. The resolutions of sampling mesh are taken in the above way so that the fitness surface can be representatively described without unnecessarily additional fitness evaluations. In addition, a further 10-time finer mesh is found to make no difference in the shape of fitness function.

In this study, the fitness function ϕ is given as follows:

$$\phi = \frac{1}{c + \sum_{i=1}^{p} \sum_{j=1}^{q} (\ddot{u}_{ij}^{m} - \ddot{u}_{ij}^{e})^2} \tag{2}$$

where $\ddot{u}^m$ and $\ddot{u}^e$ represent the measured and estimated accelerations, respectively, at p measurement points with q data points in each measurement signal, and c is a constant to set an upper bound for the best fitness and to avoid a potential zero denominator. The value of c is chosen to be 0.001 which has the same order of magnitude as the error term defined in the double summation.

3.1. *Effect of measurement noise*

The measurement noise is assumed to be of Gaussian white noise distribution. Fig. 2 gives the distribution of fitness on a 1-DOF system. In the noise free case, there is one extremely high peak surrounded by several local optima within the search range of 0.5 to 2.0. The peak of fitness function decreases and shifts as well, with the increase of noise level. A significant feature is that noise in the measurements induces no additional local optima. This observation is herein referred to as "peak shifting" which in fact substantiates the idea of applying reduced search space strategy in optimization (Perry *et al.*, 2006). The reason is that SSRM cannot work efficiently when there is no significant difference in fitness height between the global peak and local optima.

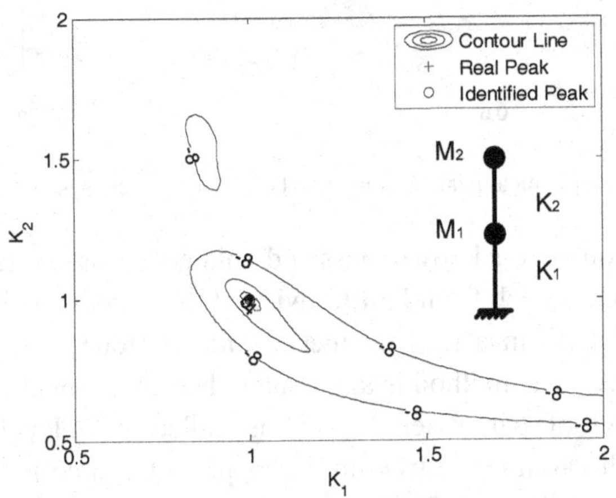

Fig. 3. Typical peak shifting of 2-DOF known mass system under 0% noise: contour line.

Further investigations on a 2-DOF system of known mass reproduce the "peak shifting". This is demonstrated in Fig. 3 and Fig. 4. Of great importance is that the extension (or flattening) of "peak ridge" indicates that noise in the measurement will have more influence to the lower-level stiffness K_1 than the upper-level stiffness K_2. This observation concurs with the previous finding (Koh *et al.*, 2003). That is, it is usually more difficult to identify the stiffness at lower-level of a lumped mass system. The reason is that the response excited is usually smaller at the lower level than at the upper level. Identification of unknown mass system is rarely reported in the literature as the identification becomes much more challenging than known mass system. The reason lies not only in the increase in search dimensions due to doubling the number of unknowns, but also in the fact that the simultaneous change in stiffness and mass parameters can produce the same eigen solutions which are the intrinsic properties of the system. An insight to the lower-dimension fitness shape of unknown mass case will be constructive to understand the physical meaning of higher-dimension identification. To this end, the 1-DOF system is again investigated but assuming that both the stiffness and mass are unknown (Zhang et al., 2010b). It is found from Fig. 5 and Fig. 6 that the shape of fitness function is strongly affected by the additional involvement of unknown mass to the unknown stiffness. Besides the noise-induced peak shifting, a rather long ridge is discovered along the direction of 45° in the K_1-M_1 space while the slope in the perpendicular direction is rather steep. This is because the eigen values and eigen modes will not be affected by simultaneously scaling the stiffness and mass by the same factor. The fitness surface analysis shows that more local optima and ridges are in the fitness surface than the known mass case.

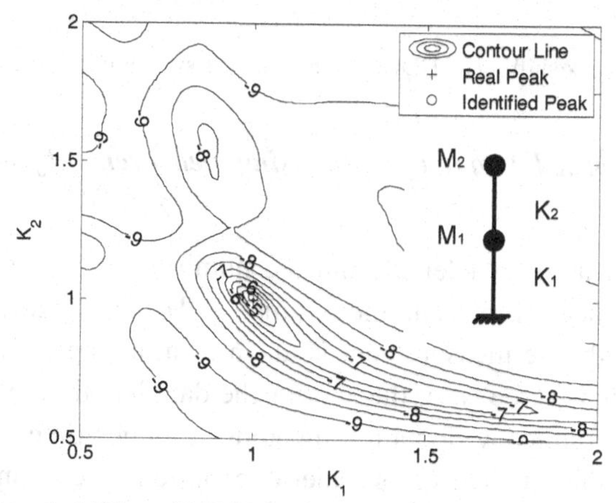

Fig. 4. Typical peak shifting of 2-DOF known mass system under 10% noise: contour line.

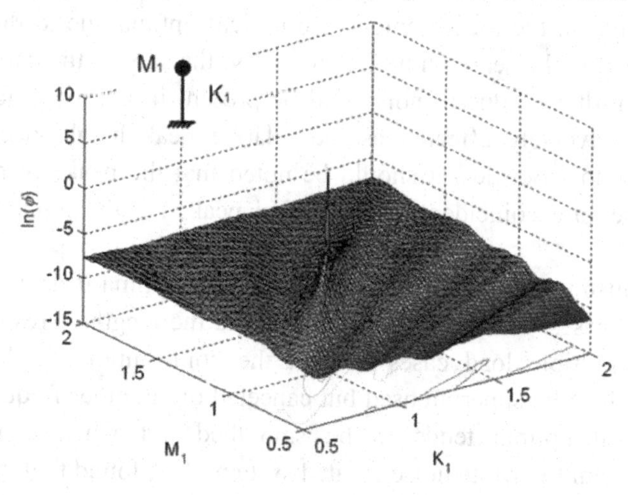

Fig. 5. Typical peak shifting of 1-DOF unknown mass system under 0% noise: 3D.

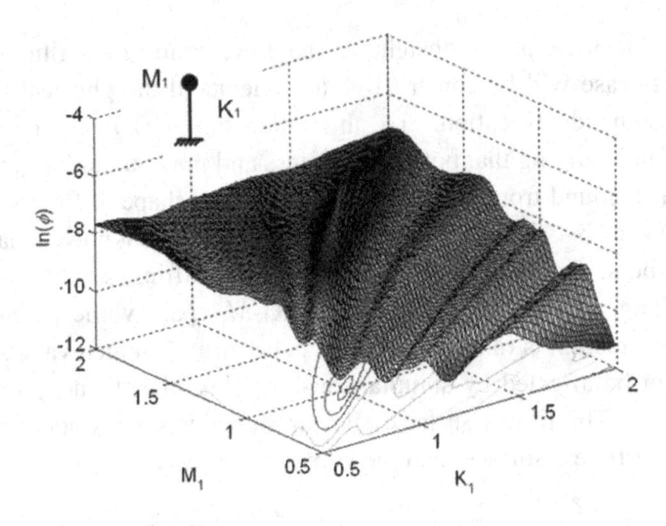

Fig. 6. Typical peak shifting of 1-DOF unknown mass system under 10% noise: 3D.

3.2. *Effects of recorded data length and using measurement from multiple load cases*

The effects of measured data length and the number of load cases are studied herein in the absence of measurement noise. To investigate the effect of measured data length, the measurement is taken from a single load case. Fig. 7 shows that, in the noise free case, the shorter the data length, the fewer the local optima. From the frequency domain viewpoint, a longer time history provides finer frequency resolution. Thus it is easier to capture the resonance between the system and external forces than using a shorter time history. Hence, by using longer data length in the measurement, more local optima due to the resonance will be found within the search limit. Conversely, the shorter the data length, the more negative influence due to noise will be present in the identification. The signal will be adversely affected as the relative peak height decreases with shorter data length. Besides, it should be noted that the peaks of all the three data-length cases are coincident at the global peak, i.e. the point with fitness value 1,000.

Figure 8 illustrates that there will be fewer local optima if data are collected from more load cases. This is reasonable because the length of response data is the same for multiple load cases so that the contribution on the dominant frequencies tends to be superimposed but canceled out at other frequencies. The number of local optima tends to be smoothed out when combining the measurements from the white noise excited system. It is found that, however, the use of more load case measurements will not necessarily benefit the identification. More load cases make the global peak sharper and taller, but this actually makes the identification more difficult through optimization as the chance of missing a narrow peak is higher than a broad peak.

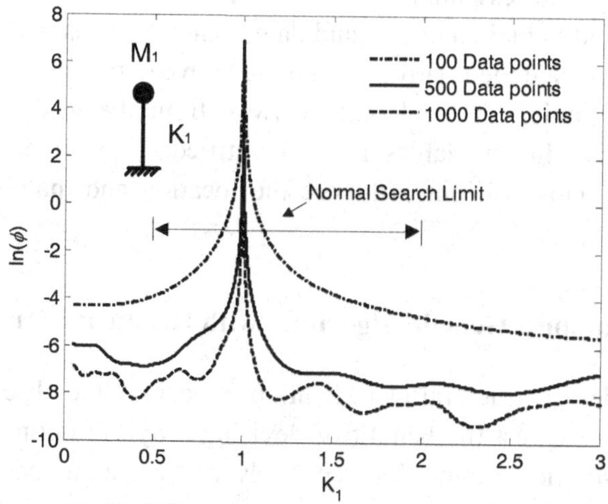

Fig. 7. Effect of data length on fitness function.

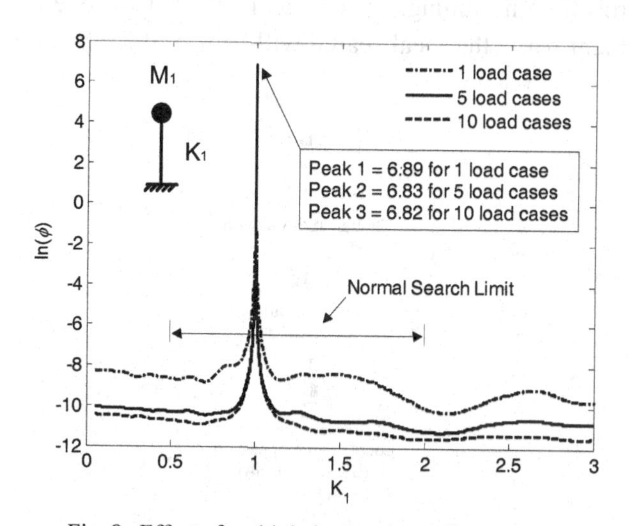

Fig. 8. Effect of multiple load cases on fitness function

It should be noted that the shape of fitness surface depends on the formulation of fitness function, i.e. Eq. (2), and thus on the measurement locations as well as data length. This is because the layout of sensor network will determine the richness of modal information contained in the measurements. Since modal participation will reflect the stiffness and mass distribution, identification of structures will be accordingly affected by measuring the accelerations from different parts of the structures. Of great importance is that peak shifting is a significant optimization characteristic of the global peak. The observation of peak shifting has nothing to do with specific formulation of the fitness function. The essential use of fitness function is to evaluate the performance of individual solution candidate. Therefore, using different fitness functions might change the relative magnitude between the global peak and local optima, the distribution of local optima away from the global peak, and the sensitivity of individual variables to be identified. Nevertheless, these three mainly affected items will not influence the location and magnitude of global peak.

4. Uniformly Sampled Genetic Algorithm with Gradient Search

The characteristics of structural identification, especially the observation of peak shifting, strongly suggest the benefit of devising a hybrid optimization strategy for performing solution exploration separately in the search space. The overall strategy is shown inFig. 9. This strategy thus comprises a global USGA search and a local search for fine tuning. The global search is to overcome the local optima trap problem while the local search will be useful for fine tuning near the peak.

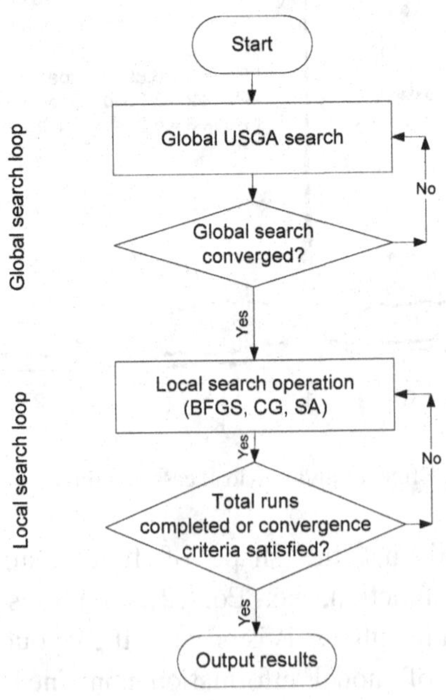

Fig. 9. USGA with local search.

In the following sections, the original idea of USGA is first introduced for global search. The second part will be on suggesting a local search method from gradient based conjugated gradient method and BFGS method as well as non-gradient based simulated annealing method. The GA parameters for USGA and switch point from global search to local search will be determined by parametric study. Finally the proposed USGA method with gradient search will be verified by numerical and experimental studies.

4.1. *Global search by USGA method*

The application of GA in structural identification began for relatively simple problems in early 1990s (Yao *et al.*, 1993; Doyle, 1994) and larger systems in 2000s (Koh *et al.*, 2000; Koh *et al.*, 2002; Koh *et al.*, 2003; Koh and Htun, 2004; Perry *et al.*, 2006). In these research works, GA has been investigated in direct application, distributed computing, GA-compatible local searcher, GA-associated identification uncertainty, and architectural modification. To tap the full potential of GA in large-scale structural identification, a crucial task has been to improve the search capability in high-dimensional space. The most recent improvement of sequential GA in physical domain identification is the SSRM, with versatile applications in structural parameter identification, damage detection, and offshore applications (Perry, 2006). The inherent strengths of SSRM have inspired its further advancement in the following sections. To formulate the proposed identification strategy, the essential ideas of search space reduction method (SSRM) (Perry *et al.*, 2006) are introduced.

The SSRM is a dual-layer evolutionary GA strategy with multi-species search. The outer layer conducts search space reduction. The purpose is to adjust the inner layer search to focus on the most promising solution space. The philosophy is that, when some of the parameters have converged, it is desirable that the computational effort can adaptively focus on other parameters in the exploration process. This is achieved by making use of the mean value and standard deviation of the identified parameters to obtain the new search range, since they can trace the convergence history of the candidate solution. The standard deviation decreases as the corresponding parameter is in the process of convergence, thereby reducing the search space adaptively. The rate of search space reduction is governed by the window size w (Perry *et al.*, 2006). The inner layer is a iGAMAS algorithm (Perry *et al.*, 2006) implementing detailed search in the parameter domain. The basic concept is to use multiple species to enhance diversity of the evolving population. One of the species is used mainly to keep the best solutions from all species, while the others are used to search for new solutions with different mutation and crossover strategies. In this regards, the evolution via single population in SGA is improved by multi-species exploration.

The adjustment of search window by the outer layer of the strategy is achieved by making use of the mean value μ_i^h and standard deviation σ_i^h of the identified parameters to obtain the new search range as follows:

$$\begin{cases} \hat{a}_i^h = \mu_i^h - w \times \sigma_i^h \\ \hat{b}_i^h = \mu_i^h + w \times \sigma_i^h \end{cases} \qquad (i = 1, 2 \cdots k) \tag{3}$$

where $\hat{a}_i^h$ and $\hat{b}_i^h$ are the lower and upper bounds of the current search limit. w is the window size, k is the number of variables and h represents the present run number. Convergence history is traced by the mean μ_i^h and standard deviation σ_i^h of the parameters. The standard deviation decreases as the corresponding parameter is in the process of convergence, thereby reducing the search space adaptively.

A drawback of SSRM is that the outer layer of search space reduction cannot be activated right from the beginning. The reason is that using SSRM too early may result in the so-called "jump out" problem, namely, the scaled search space may be outside the solution domain. Sufficient initial runs are necessary to make the means and standard deviations statistically meaningful. Nevertheless, these initial iGAMAS runs can cost up to 30% or 40% of the total search effort (Perry *et al.*, 2006). Hence there is a need to shorten this "learning curve". Furthermore, the initial population may not cover the search space as uniformly as it should – unless some good initial sampling strategy is devised.

This section presents a uniformly sampled GA specifically addresses these two issues, by providing a rational and optimal way to sample in the search space and defining the search limits according to the fitness of these initial samples. At the same time, the original mechanism of defining the search space by SSRM is retained as a "fall back" should the sampling method yield an inappropriate search limit. The flowchart of USGA is given in Fig. 10.

4.1.1. Sampling methods

The identification of multi-DOF (MDOF) system, as a multi-dimensional optimization problem, involves a substantially huge space of possible solutions. For stochastic optimization methods like GA, there is no specifically defined search direction unlike in gradient-based methods. The use of sampling test is thus significantly constructive to obtain a rough contour of the fitness surface before GA search so as to define an appropriate search limit. Usually a uniform sampling is desired to have unbiased samples in the multi-dimensional solution space. Nevertheless, sampling in a uniform grid manner is impractical as the computational effort will increase exponentially with the identification dimensions. For example, with k parameters and s points per parameter, a total of s^k fitness evaluations are required. An increase in grid density will of course provide more information of fitness surface but will also pose an expensive burden on computer time. Therefore, a cost-effective way is needed to take samples, and this can be achieved by conducting small scale logical sampling experiments for hyper-plane search. These samples could be much sparser than the s-k grid sampling manner, but remains as uniformly as possible to cover the whole solution space. In the present study, four sampling methods are investigated, incorporating orthogonal arrays (OA) (Hedayat *et al.*, 1999), randomly uniform distribution, Latin hypercube (McKay *et al.*, 1979), and Hammersley sequence (Hammersley, 1960). Distributions of these four sampling methods are given in Fig. 11 for comparison, and 289 samples are considered herein as an illustration.

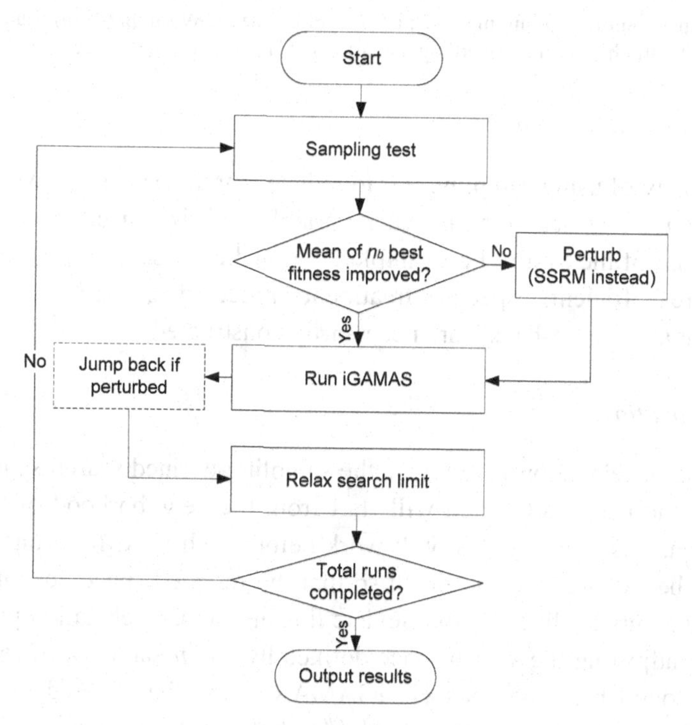

Fig. 10. Flowchart of USGA method.

Within the search space, it is possible to make trials by uniform samples. For randomly uniform distribution, Latin hypercube, and Hammersley sequence, if originally sampled between 0 and 1, an element of the *j*-th trial sample $(x_1,\cdots,x_i,\cdots,x_k)_j$, $(j=1,2,\cdots n)$ in the new search space is

$$x_i = \hat{a}_i^h + U(0,1)\times(\hat{b}_i^h - \hat{a}_i^h) \tag{4}$$

where *U*(0,1) stands for uniform samples taken by random uniform distribution, Latin hypercube and Hammersley sequence. The samples by OA are different from those by the other three methods, due to the formulation of OAs (Zhang *et al.*, 2010a).

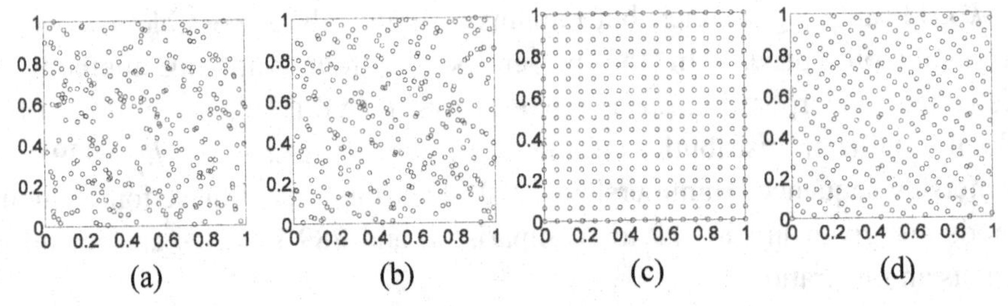

Fig. 11. Comparison of sampling methods: an illustration with 289 samples: (a) Random uniform distribution (b) Latin hypercube (c) Orthogonal array (d) Hammersley sequence.

4.1.2. *Treatment after sampling*

The philosophy of using sampling test to reduce search space is shown in Fig. 10. Sampling in the new search range, each variable limit is resized by the minimum and maximum of the first n_b best samples. To make the sampling-defined search space more efficient, special treatment procedures, namely, relaxation, perturbation and "jump-back", are necessarily constructed.

4.1.2.1. *Relaxation*

The purpose of relaxation is to revise the sampling-defined search space so as to ensure that the next exploration will start from the neighborhood of the present best solution. As sampling test will work before each iGAMAS run, the search space may be over-reduced so much so that the rescaled space does not contain the optimal solution. In order to alleviate this problem, a relaxation procedure is essential in adjusting the search space defined by the n_b samples. It requires that the newly found best solutions via iGAMAS should be covered by the larger radius of the present search range $[a_i^h, b_i^h]$, but should not exceed the original search range $[a_i^0, b_i^0]$. This is expressed mathematically as

$$\begin{cases} a_i^{h+1} = \max[x_i^h - |x_i^h - b_i^h|, a_i^0] \\ b_i^{h+1} = b_i^{h+1} \end{cases} \quad \text{if } |x_i^h - a_i^h| < |x_i^h - b_i^h| \tag{5a}$$

$$\begin{cases} a_i^{h+1} = a_i^{h+1} \\ b_i^{h+1} = \min[x_i^h + |x_i^h - a_i^h|, b_i^0] \end{cases} \quad \text{if } |x_i^h - a_i^h| > |x_i^h - b_i^h| \tag{5b}$$

where $i = 1,2\cdots k$. h represents the current number of runs, x_i^h is one of the variables in the current best solution $(x_1^h,\cdots,x_i^h,\cdots,x_k^h)$, and $[a_i^{h+1}, b_i^{h+1}]$ is the search space for the next sampling. This relaxation procedure considerably improves the balance between reducing the search space and ensuring the solution is within the new space.

On the other hand, if the mean fitness of n_b best samples is not improved, the sampling-defined search space may risk overshooting and missing the proper search space. Therefore a restoring mechanism is expected to recover the search range, which leads to the proposed perturbation procedure and jump-back procedure.

4.1.2.2. Perturbation

Perturbation is to discover the best candidate by a decoupled SSRM search. This automatic correction of search limits is essential in exploring the parameter space. The perturbation procedure requires running the current iGAMAS loop in an independent search limit defined by SSRM rather than those defined by samples. The search space is described in Eq. (3). The mean and variance are based on all the best chromosomes up to the present run. In fact, the SSRM is able to restore the search range to a reasonable limit, even if misled by the search limit defined by sampling test. The reason is that SSRM-defined search space provides a robust alternative by tracing the mean and standard deviation of the best parameter in each run. The standard deviation provides an indication of uncertainty of the identified parameter and the search space can be adjusted accordingly. If sufficient runs are carried out, the statistically defined search limit will become significant in suggesting a promising area containing optimal solution. More importantly, this SSRM-defined search space requires no additional evaluation of fitness function.

4.1.2.3. Jump-back

If perturbation is activated, a jump-back procedure then follows to make the next run start from one of the previous search ranges. The historical search range of each run has been recorded as the algorithm continues. This is shown in Fig. 10. Because the best candidate explored by SSRM in the perturbation procedure suppose to be inside the search range, the current range will keep jumping back until it covers the newly found best solution. Finally, the jump back procedure will keep the search range as $[a_i^{h-l}, b_i^{h-l}]$, which satisfies

$$x_i^h \in [a_i^{h-l}, b_i^{h-l}] \qquad (i = 1, 2 \cdots k) \tag{6}$$

where $l \in [1, h]$, denotes the number of runs that has jumped back to satisfy Eq. (5). When the jump-back is finished, it will step into the usual relaxation procedure. Then a new search space is modified by Eq. (7) on basis of the jumped search range $[a_i^{h-l}, b_i^{h-l}]$. That is

$$\begin{cases} a_i^{h+1} = \max[x_i^h - |x_i^h - b_i^{h-l}|, a_i^0] \\ \\ b_i^{h+1} = b_i^{h+1} \end{cases} \qquad \text{if } |x_i^h - a_i^{h-l}| < |x_i^h - b_i^{h-l}| \tag{7a}$$

$$\begin{cases} a_i^{h+1} = a_i^{h+1} \\ \\ b_i^{h+1} = \min[x_i^h + |x_i^h - a_i^{h-l}|, b_i^0] \end{cases} \qquad \text{if } |x_i^h - a_i^{h-l}| > |x_i^h - b_i^{h-l}| \tag{7b}$$

It is noted that the search space will be reset in three cases including sampling test, relaxation and jump-back procedures. The relaxation will be active regardless of the activation of perturbation procedure, and helps to enlarge the search limit if over-reduction is discovered by the samples. However, jump-back will not work if there is no perturbation, as the search space defined by the samples proves to be capable of producing progressive improvement with better solutions. Using these search space resizing processes, the balance could be considerably improved between the reduction of search space and ensuring a new search space that incorporates the optima.

In summary, the proposed USGA method with sampling test carries out an inner iGAMAS loop and an outer sampling exploration. The inner iGAMAS is to find the best solution in each run, which is set as the center of a new search limit. The outer sampling test is to define a radius of the search limit for the next run. Whether a new search limit defined in this way contains the real solution is judged by the improvement of the measure, i.e. the mean fitness value of the first n_b samples. If the measure is not improved compared to the previous sampling run, a wrong search limit is implied. In order to restore the search limit, changes are proposed to conduct in redefining both center and radius in this study. The center, i.e. the best solution, is then to be determined by the perturbation procedure. The radius is enlarged by jump-back procedure and the relax procedure.

4.2. *Local search by gradient based and non-gradient based methods*

The second part of the proposed identification method is the development of local search for fine tuning. The reason to adopt local search is well illustrated in Fig. 12. It is seen that local optima are frequently encountered in the first 20% of fitness evaluation as compared to the latter 80% evaluations. Thus, an appropriate local search method is expected to accelerate the convergence history of the latter portion.

The local searchers considered herein include gradient based conjugate gradient (CG) method, Broyden-Fletcher-Goldfarb-Shannon (BFGS) method, and non-gradient based simulated annealing (SA) method. The non-gradient local searcher will serve as a reference to check whether the peak found by gradient based methods is the global peak, by judging the achievement of the same identification accuracy. This is because simulated annealing starts a random search from the neighborhood of best estimates, and thus has better chance for locating the global peak than GA, which carries out species evolution within the whole search range. More importantly, simulated annealing is much better in capturing the global peak than the other two local searchers, i.e. CG and BFGS, as they are based on the gradients and easily converge to local optima. Detailed procedures can be referred to references (Zhang *et al.*, 2010ab) on the definition of objective function and convergence criteria for the proposed hybrid identification method.

A comprehensive parametric study has been conducted for the proposed identification method (Zhang *et al.*, 2010ab). The objective involves exploring a robust uniform sampling method for USGA, and the determination of the "switch point" between the global USGA search and the local search. With the recommended parameters from parametric study, the presented USGA method demonstrates better performance over the recent researches. The results are given in Table 1. Overall performance of the hybrid identification method is evaluated in the numerical studies of the next section. The parameters suggested for the use of proposed method are summarized in references (Zhang *et al.*, 2010ab).

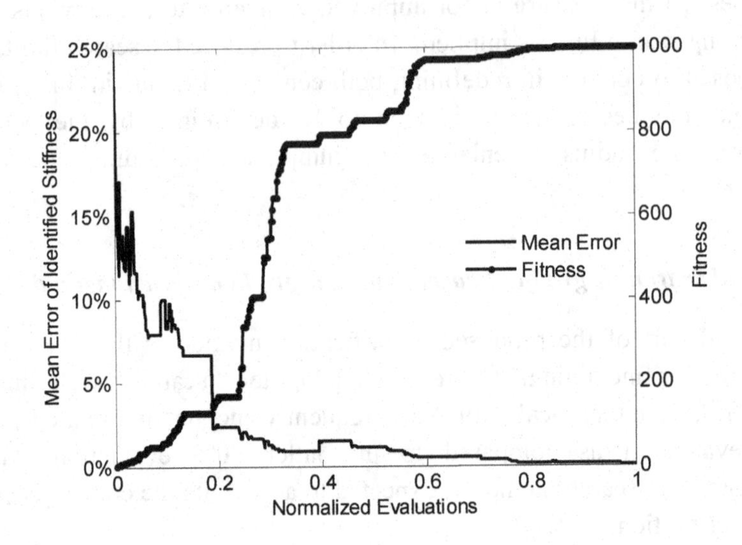

Fig. 12. Typical USGA convergence of 20-DOF known mass system under 0% noise: 40,000 total evaluations.

Table 1. Performance comparison for SSRM and USGA methods.

Results	Known-mass systems		Unknown-mass systems	
	10-DOF	20-DOF	10-DOF	20-DOF
Time (h:m:s) [a]	00:00:03	00:00:26	00:01:05	00:17:18
Evaluation ratio [b]	0.5	0.5	0.08	0.3
Mean error - k (%)	0.07(0.43) [c]	0.15(0.52)	2.05(2.98)	2.43(2.78)
Max error - k (%)	0.19(1.21)	0.38(1.60)	5.08(6.62)	6.78(8.64)
Mean error - m (%)	-	-	1.98 (3.00)	2.60(3.00)
Max error - m (%)	-	-	4.69(6.81)	7.54(10.40)

[a] Time is reported only for USGA, by a workstation with two 3.0GHz-CPUs.
[b] Evaluation ratio is obtained by current total evaluations divided by total evaluations by SSRM (Perry *et al.*, 2006).

5. Numerical Examples

To verify the proposed hybrid optimization strategy which stems from the "peak shifting" observation, three numerical examples are investigated. They are a lumped mass system and a truss structure shown in Fig. 13. These two examples are presented in the order of difficulty in terms of increasing number of unknown parameters – 12 and 31 respectively. The search space is set from 0.5 to 2.0. That is, the lower and upper bounds are half and twice of the actual values.

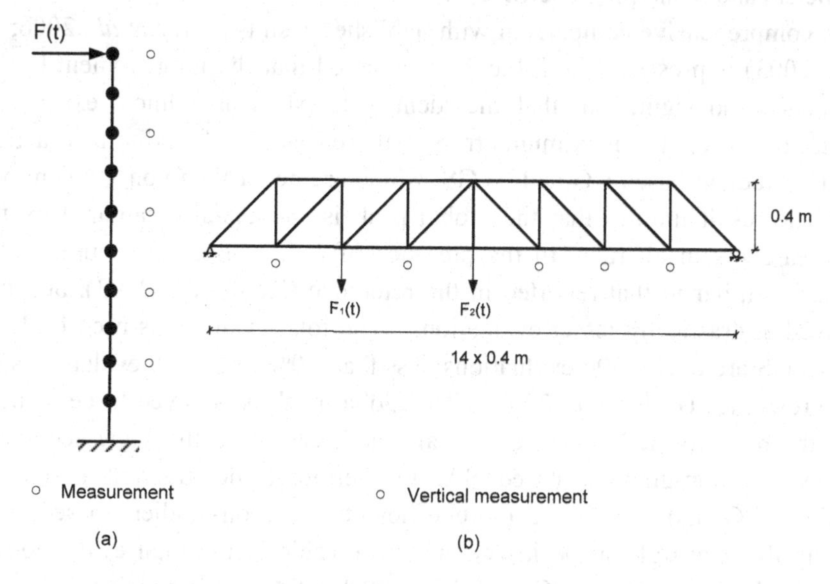

Fig. 13. (a) 10-DOF lumped mass system; (b) 29-element truss.

In the identification, mass distribution of the structures is assumed as known while the stiffness and two damping coefficient are to be identified. Due to random excitations, accelerations are generated by Newmark's constant-average-acceleration method. Both the input and output are assumed to be contaminated by Gaussian white noise with zero mean and variance adjusted to give a pre-defined noise level (0%, 5%, and 10%). The parametric study has been carried out for the proposed USGA and the local search methods, leading to the values used in this study. Considering the effect of measurement noise, the results reported in the following are the average from 25 runs. Computer time is reported based on a dual-processor workstation with 3.0-GHz CPU.

5.1. *10-DOF Lumped mass system*

In order to compare the performance of proposed identification strategy, a 10-DOF lumped mass system is used (Koh *et al.*, 2003) representing a plane shear building. As shown in Fig. 13(a), the input random forces are applied at the 10th level, and accelerations are measured at levels 2, 4, 6, 8, and 10. Proportional damping is assumed with modal damping ratio of 5% for the first two modes. The total number of parameters to be identified is 12, including 10 stiffness parameters and 2 damping coefficients.

A comprehensive comparison with published results (Perry *et al.*, 2006; Koh *et al.*, 2003) is presented in Table 2. It is noted that the improvement by local searchers is so significant that the identified results are almost exact in the absence of noise. The maximum errors in these cases are found to be practically zero for gradient based CG and BFGS, which are better than non-gradient based SA. The result implies that the global peak is successfully captured by three local searchers in all runs. In the presence of 10% noise, the accuracy of the results is similar to that reported in the reference (Koh *et al.*, 2003), but this is achieved at drastically lesser evaluations. The total evaluations used by USGA with BFGS are only 2,590 evaluations, less than 10% of 30,000 evaluations used in the reference (Koh *et al.*, 2003). The global peak is believed to be identified since the result by BFGS and CG local search can reach the same accuracy as that by the non-gradient based SA. Furthermore, the gradient based local searchers CG and BFGS are more efficient than non-gradient based SA in locating the sharp global peak. As shown in Table 2, the total evaluations are 2,590 for BFGS, 2,766 for CG, and 21,980 for SA in the case of 10% noise. Besides, it is observed in Table 2 that the number of evaluations required by BFGS is less than that of CG. The reason is that the line minimization adopted by BFGS is more efficient than CG.

5.2. *Truss of 29 elements and 28 DOFs*

The proposed method is now tested for in a more challenging problem from the viewpoint of measurement availability. The measurements available in this truss example are much less than the previous two examples in relation to the number of unknown stiffness parameters, i.e. higher ratio of unknown parameters to available measurements. The structure of 29 elements resembles a small-scale laboratory model of steel truss bridge with Young's modulus of the material being 200 GPa, and density 7800 kg/m^3. The outer diameter of the steel tube is

1.55 cm and the inner diameter is 1.09 cm, resulting in a sectional area of 0.9538 $\times 10^{-4}$ m^2. The structure is excited by two random forces and accelerations are measured at 5 locations, as illustrated in Fig. 13(b). Damping ratio of 2% is assumed for the first two modes. In total, there are 31 unknown parameters to be identified—29 stiffness unknowns and 2 damping coefficients.

Table 2. Results for numerical example 1: 10-DOF lumped mass system.

Results	10-DOF lumped mass system						
	SGA[a]	SSRM	GA-SW[a]	GA-MV[a]	USGA -BFGS	USGA -CG	USGA -SA
			0% Noise				
Total Evaluations	30,000	30,000	30,000	30,000	2,295	2,951	21,980
CPU Time (m:s)	-	00:52	-	-	00:04	00:05	00:34
Mean error - *k* (%)	2.90	0.58	0.40	0.20	0.00	0.00	0.06
Max error - *k* (%)	7.80	2.13	1.20	1.00	0.00	0.00	0.17
			10% Noise				
Total Evaluations	30,000	30,000	30,000	30,000	2,590	2,766	21,980
CPU Time (m:s)	-	00:52	-	-	00:04	00:05	00:34
Mean error - *k* (%)	5.10	5.37	2.70	3.00	2.90	2.79	3.40
Max error - *k* (%)	15.00	14.26	12.20	7.60	7.63	7.03	9.20

[a]Results by Koh *et al.* (2003)

The results in Table 3 show that the performance for the proposed hybrid optimization strategy is consistently excellent in the noise free case. The global peak is believed to be found by CG and BFGS with and without noise, because their accuracies of the identification results are similar to those by SA. It is also noted that the mean identification error by USGA with BFGS is as small as 3.23% even under 10% noise. At the same time, the computer time is only around a quarter of that is used by SSRM. Local search by CG will generally require more fitness evaluations, and is hence less efficient than BFGS.

Table 3. Results for numerical example 2: 29-element truss

Results	29-element truss				
	SGA	SSRM	USGA-BFGS	USGA-CG	USGA-SA
0% Noise					
Total Evaluations	60,000	60,000	10,780	110,169	25,960
Times taken (m:s)	14:07	13:25	02:42	24:19	05:55
Mean error - *EA* (%)	18.84	2.91	0.00	0.00	0.76
Max error - *EA* (%)	49.96	11.32	0.00	0.02	5.08
10% Noise					
Total Evaluations	60,000	60,000	12,619	54,324	25,960
Times taken (m:s)	14:21	14:01	02:55	11:54	06:00
Mean error - *EA* (%)	20.87	3.63	3.23	3.22	4.15
Max error - *EA* (%)	54.81	14.60	15.69	16.13	20.93

6. Experimental Verification

To further validate the proposed USGA with gradient search method in structural identification, an experimental study is carried out on a 7-level small-scale steel frame. The set up of the tested frame, sensors and shakers are illustrated in Fig 14 (a). Random excitations are input at the free end, i.e. level 7 of the frame. Full measurement of accelerations is available for the testing.

The experimental study considers six damage scenarios listed in Table 4. Damage is created by physical cut (partial or full) of selected column members as shown in Fig 14 (b). "D0" in Table 4 are the undamaged scenarios and serves as the benchmark for comparison with damage scenarios. The small damage and large damage correspond to 4.1% and 16.7% stiffness reduction respectively. To reduce the effects of measurement noise, the identification result is reported in average of experiments by different load combinations. Five different random forces are used in each damage scenario, i.e. forces A, B, C, D, and E. Dynamic test is carried out three times for each of these five forces. In the identification, the same force will be used in comparing undamaged state and damaged state. For example, if force A is used in the undamaged case D0, then the same force A is used in the three tests of damage case D1. Thus there will be 9 combinations for each force, giving 45 combinations in total for all 5 different forces.

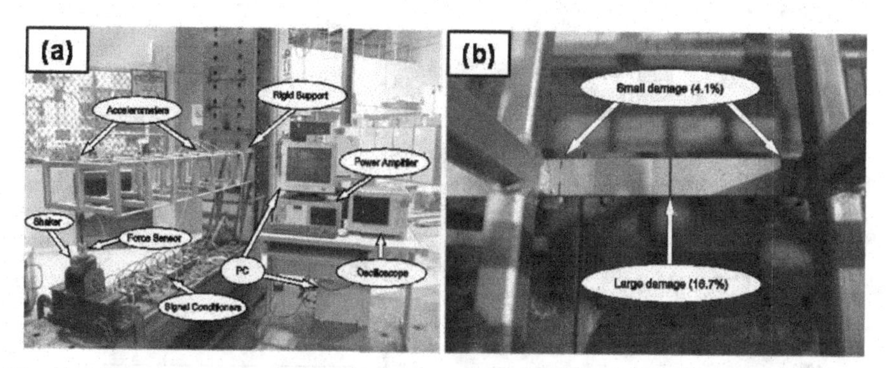

Fig. 14. (a) Experimental set up of dynamic testing; (b) Damages created by cut in a column.

Table 4. Damage scenarios in experimental study.

Cases	Small Damage (4.1%)	Large Damage (16.7%)
D0	-	-
D1	Level 4	-
D2	-	Level 4
D3	Level 6	Level 4
D4	Levels 3 and 6	Level 4
D5	Level 3	Levels 4 and 6
D6	-	Levels 3, 4 and 6

Typical results are demonstrated from Fig. 15. As shown in Fig. 15(a) for Case D2 (as an example of illustration), the identified damage is 17.5%. The difference is less than 1% when compared to the exact damage of 16.7%. Figure 15 (b) gives an excellent prediction of even small damage 4.1% at level 6 in the presence of large damage 16.7% at level 4, i.e. 4.2%. In Figure 15 (c), the two small damages (each 4.1%) are successfully identified at levels 3 and 6. The identified damages are 5.0% and 4.2% at levels 3 and 6, respectively. Figure 15 (d) gives good identification of three large damages (16.7%) at levels 3, 4, and 6, i.e. 18.9%, 16.5% and 19.2%, respectively. These results show that the proposed method successfully identifies not only single large damage, but also small damage of around 4% in the presence of multiple damages.

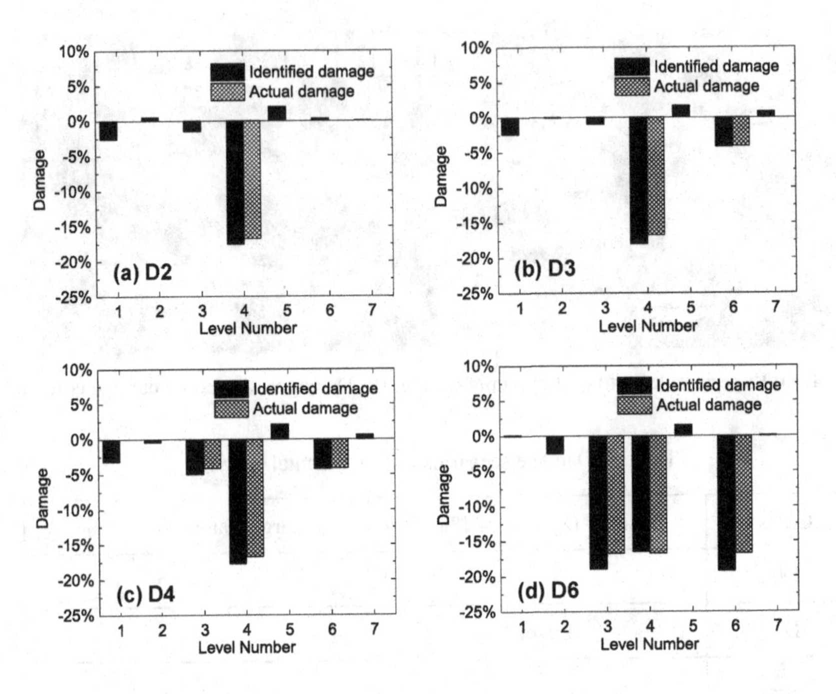

Fig. 15. Damage scenario (a) D2: damage of 16.7% at level 4; (b) D3: damage of 16.7% at level 4 and 4.1% at level 6; (c) D4: damage of 16.7% at level 4 and 4.1% at levels 3 and 6; (d) D6: damage of 16.7% at levels 3, 4 and 6.

7. Conclusions

A hybrid identification strategy for structural identification has been proposed with two complementary and necessary enhancements, namely global search by a uniformly sampled genetic algorithm and local search by gradient based methods. The dramatic improvement of structural identification efficiency is derived from insight of the optimization characteristics in the fitness surface analysis. The proposed USGA global search is a dual layer strategy with an inner layer to find the best solution in each run as the center of new search limit and an outer layer of sampling test to define the radius of search limit. Sampling testing is done by using the Hammersley sequence, which is shown to have strong scalability and uniformity within the solution space. Consequently, the USGA global search achieves significant reduction in computational time, and the local search by BFGS further improves the computational efficiency and accuracy of identification. The superior performance of the proposed hybrid identification is demonstrated by numerical examples including lumped mass system and truss structure. Experimental studies further demonstrate that the proposed strategy is effective in identifying challenging damage cases including small damage of 4% and multiple damages with different severities.

References

Beck, J.L. (1979). Determining models of structures from earthquake records, Technical Report EERL-78-01, California Institute of Technology.

Beck, J.L., and Jennings, P.C. (1980). Structural identification using linear models and earthquake records, *Earthquake Engineering and Structural Dynamics*, 8, pp. 145-160.

Bekey, G.A. (1970). System identification - an introduction and a survey, *Simulation*, 15, pp. 151-166.

Bowles, R.L., and Straeter, T.A. (1972). System identification computational considerations. System identification of vibrating structures: Mathematical models from test data; Proceedings of the Winter Annual Meeting, *American Society of Mechanical Engineers*, New York, pp. 23-43.

Caravani, P., Watson, M.L., and Thomson, W.T. (1977). Recursive least-square time domain identification of structural parameters, *ASME Journal of Applied Mechanics*, 44, pp. 135-140.

Chou, J.H., and Ghaboussi, J. (2001). Genetic algorithm in structural damage detection, *Computers and Structures*, 79, pp. 1335-1353.

Cole, H.A., Jr. (1973). On-line failure detection and damping measurement of aerospace structures by random decrement signatures, Technical Report NASA-CR-2205, NASA.

Doyle, J.F. (1994). A genetic algorithm for determining the location of structural impacts, *Experimental Mechanics*, 34, pp. 37-44.

Franco, G., Betti, R., and Longman, R.W. (2006). On the uniqueness of solutions for the identification of linear structural systems, *ASME Journal of Applied Mechanics*, 73, pp. 153-162.

Hammersley, J.M. (1960). Monte Carlo methods for solving multivariable problems, *Annals of the New York Academy of Sciences*, 86, pp. 844-874.

Hedayat, A.S., Sloane, N.J.A., and Stufken, J. (1999). Orthogonal Arrays: Theory and Applications, New York, Springer-Verlag.

Hoshiya, M., and Saito, E. (1984). Structural identification by Extended Kalman Filter, *Journal of Engineering Mechanics*, 110, pp. 1757-1770.

Huang, N.E., and Shen, S.S. (2005). Hilbert-Huang Transform and Its Applications. London: World Scientific.

Ibrahim, S.R., and Mikulcik, E.C. (1977). A method for the direct identification of vibration parameters from the free response, *The Shock and Vibration Bulletin*, 47, pp.183-198.

James, G. H., Carne, T.G., and Lauffer J.P. (1995). The natural excitation technique (NExT) for modal parameter extraction from operating structures, *Modal Analysis-the International Journal of Analytical and Experimental Modal Analysis*, 10, pp. 260-277.

Jiang, R.J., Au, F.T.K., and Cheung, Y.K. (2003). Identification of masses moving on multi-span beams based on a genetic algorithm, *Computers and Structures*, 81, pp. 2137-2148.

Juang, J.N., and Pappa, R.S. (1985). An eigensystem realization algorithm for modal parameter identification and model reduction. *Journal of Guidance Control and Dynamics*, 8, pp. 620-627.

Koh, C.G., and Htun, S. (2004). Adaptive search genetic algorithm for structural system identification, In: Mufti A, Ansari F, editors. *Proceedings of the Second International Workshop on Structural Health Monitoring of Innovative Civil Engineering Structures*. Winnipeg: ISI Canada Corporation. pp. 393-403.

Koh, C.G., and Perry, M.J. (2007). Structural damage quantification by system identification, *Journal of Earthquake and Tsunami*, 1, pp. 211-231.

Koh, C.G., Chen, Y.F., and Liaw, C.Y. (2003). A hybrid computational strategy for identification of structural parameters, *Computers and Structures*, 81, pp. 107-117.

Koh, C.G., Hong, B., and Liaw, C.Y. (2000). Parameter identification of large structural systems in time domain, *ASCE Journal of Structural Engineering*, 126, pp. 957-963.

Koh, C.G., See, L.M., and Balendra, T. (1991). Estimation of structural parameters in time domain - a substructure approach, *Earthquake Engineering and Structural Dynamics*, 20, pp. 787-801.

Koh, C.G., Wu, L.P., and Liaw, C.Y. (2002). Distributed GA for large system identification problems, *Conference on Smart Nondestructive Evaluation for Health Monitoring of Structural and Biological Systems*, Newport Beach, CA, pp. 438-445.

Koh, C.G., Wu, L.P., and Liaw, C.Y. (2002). Distributed GA for large system identification problems, In: Kundu T, editor. Smart Nondestructive Evaluation for Health Monitoring of Structural and Biological Systems, pp. 438-445.

Li, H.N., Li D.S., and Song, G.B. (2004). Recent applications of fiber optic sensors to health monitoring in civil engineering, *Engineering Structures*, 26, pp.1647-1657.

Lynch, J.P., Loh, K.J. (2006). A summary review of wireless sensors and sensor networks for structural health monitoring, *The Shock and vibration digest*, 38, pp. 91-128.

McKay, M.D., Beckman, R.J., and Conover, W.J. (1979). A comparison of three methods for selecting values of input variables in the analysis of output from a computer code, *Technometrics*, 21, pp. 239-245.

Perry, M.J. (2006). Modified GA approach to system identification with structural and offshore application, PhD Thesis, Department of Civil Engineering, National University of Singapore, Singapore.

Perry, M.J., and Koh, C.G. (2008). Output-only structural identification in time domain: Numerical and experimental studies, *Earthquake Engineering and Structural Dynamics*, 37, pp. 517-533.

Perry, M.J., Koh, C.G., and Choo, Y.S. (2006). Modified genetic algorithm strategy for structural identification, *Computers and Structures*, 84, pp. 529-540.

Shinozuka, M., Yun, C. B., and Imai, H. (1982). Identification of Linear Structural Dynamic Systems, *ASCE Journal of the Engineering Mechanics Division*, 108, pp. 1371-1390.

Udwadia, F.E., and Sharma, D.K. (1978). Some uniqueness results related to building structural identification, *SIAM Journal on Applied Mathematics*, 34, pp. 104-118.

Van Overschee, P., and De Moor, B. (1996). Subspace Identification for Linear Systems: Theory, Implementation, Applications. Dordretch, Kluwer Academic Publishers.

Yao, L., Sethares, W.A., and Kammer, D.C. (1993). Sensor placement for on-orbit modal identification via a genetic algorithm, *AIAA Journal*, 31, pp. 1922-1928.

Zhang, Z., Koh, C.G., and Duan, W.H. (2010b). Uniformly sampled genetic algorithm with gradient search for structural identification - Part II: Local search, *Computers and Structures*, 88, pp. 1149-1161.

Zhang, Z., Koh, C.G., and Duan, W.H. (2010a). Uniformly sampled genetic algorithm with gradient search for structural identification - Part I: Global search, *Computers and Structures*, 88, pp. 949-962.

Chapter 10

Health Diagnostics of Highway Bridges Using Vibration Response Data

Maria Q. Feng[1], Hugo C. Gomez[2] and Andrea Zampieri[3]

[1, 3]*Dept. of Civil Engineering and Engineering Mechanics, Columbia University, 500 W. 120th St. 610 Mudd, New York, NY 10027, USA*
[2]*MMI Engineering, 1111 Broadway Street, 6th Floor, Oakland, CA 94607*
E-mails: [1]mqf2101@columbia.edu & [2]hgomez@mmiengineering.com & [3]az2326@columbia.edu

1. Introduction

This chapter explores the potential of sensor-based structural health monitoring for objective and quantitative assessments of bridge structural integrity (i.e. health), which would enable priorities and targets to be set for timely, cost-effective maintenance interventions and post-event emergency responses. As a bridge structure ages, its present condition largely depends on previous performance history. This fact necessitates the availability of the past performance or response record of the bridge. If the record is available, it will significantly enhance the health condition assessment of a specific bridge. Response records may include displacement, inclination, settlement, strain, acceleration and velocity. In the context of long-term vibration monitoring, bridge condition assessment can be executed as new data become available. Any change in the structural condition is embedded in the measurements. Hence, engineers and decision makers can monitor bridge health on a periodic basis by analyzing the data. The information may come in the form of either ambient/ traffic or seismic-induced vibration response data.

Among all of the measurements used, acceleration data are frequently used for structural long-term vibration monitoring of highway bridges because accelerations are generally easier and less expensive to measure compared to other quantities. Furthermore, acceleration response can be easily incorporated into dynamic analysis of the system and several System Identification (SI) techniques described later in this chapter.

Although traffic-induced acceleration response provides valuable information for assessing the health status of the superstructure, it is less sensitive to possible damage in the substructure (columns). The reason is that traffic-induced vibrations tend to excite the vertical modes of the structure, so the collected signals do not carry much information about the behavior of the substructure. On the other hand, columns are excited and often damaged by lateral earthquake ground motion. Thus, to assess the column condition or detect seismic damage, one may analyze seismic response of the bridge. Therefore, seismic response data can be integrated with traffic-induced ambient vibration data for a comprehensive assessment of a bridge structure. In the following sections, ambient/traffic and seismic-induced vibration data will be discussed separately considering their different characteristics.

2. Methods for Structural Health Diagnostics

The structural health condition of a bridge is diagnosed by monitoring changes in structural parameters based on measured vibration data. Such parameters as stiffness of structural elements change as the bridge ages or suffers damage caused by a destructive event. The procedure of vibration-based structural health diagnostics can be generally described by the flowchart in Fig. 1. Once a sensor network is installed on the bridge, data collection can be triggered manually or automatically. Each time new data are acquired, system identification is carried out, using the acquired data, to estimate the current values of the selected structural parameters and track their changes with respect to a baseline, which is desirably established when the bridge structure is healthy. If the change exceeds a preset threshold, the bridge is considered to suffer unusual structural deterioration or damage, and thus maintenance intervention is suggested. Obviously it requires long-term monitoring data in order to determine a reliable threshold.

The following more detailed discussions on structural health diagnostic methods are based on the monitoring of change in structural parameters expressed as a change in stiffness matrix ΔK. This change directly results in the change of natural frequencies of and mode shapes of the system, according to the well-known dynamic characteristic equation:

$$|K - \omega^2 M| = 0 \tag{1}$$

Where K is the stiffness matrix, M is the mass matrix, and ω^2 represents the squared natural frequencies.

As modal parameters, i.e., the natural frequencies and mode shapes, carry the information about the stiffness of the structure, the first step of the identification procedure is often the extraction of the modal parameters of the bridge from vibration measurement data through experimental modal analysis. Depending on the type of data available, i.e. ambient/traffic or seismic records, modal properties are extracted using either output-only or input-output identification techniques. Section 2.1 provides a review of such methods.

It must be noted that the modal analysis techniques are based on the theory of linear systems. Modal properties are stable if the structure is linear during the measurement. It is reasonable to assume that bridge structural responses to ambient/traffic and moderate seismic excitations are linear. However, if damage (such as seismic damage) occurred to the structure causing a nonlinear behavior during the vibration measurement, the conventional modal analysis techniques are no longer valid. This problem can be overcome by approximating the nonlinear behavior of the structure as a sequence of linear states, and for each state the modal identification techniques are applied. As a result, the identified modal parameters change during the vibration measurement period.

While the modal parameters provide information on global structural conditions, structural parameters such as element stiffness, will enable the identification of location and extent of structural changes. While different approaches can be employed to estimate structural parameters, one basic idea is to operate through Finite Element (FE) model updating, by searching a set of values of the structural parameters that minimizes the discrepancy between measured modal parameters or measured response time histories or both and those predicted by the FE model. Emphasis must be placed in the careful selection of the structural parameters to be updated. It is impossible, in fact, to update all of the variables of the structure, since this would result in not only the increase in computational efforts, but also the ill-conditioning of the identification algorithm. Sensitivity analysis is beneficial in selecting the proper updating parameters.

A collection of N structural parameters to be updated is grouped into a vector $\boldsymbol{\theta}$ as:

$$\boldsymbol{\theta} = (p_1, p_2, \ldots, p_N)^T \tag{2}$$

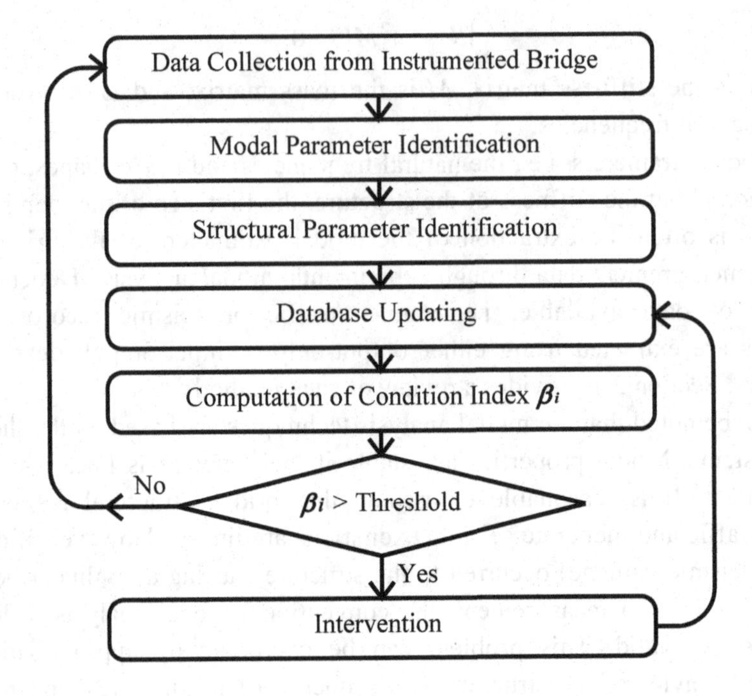

Fig. 1. Flowchart for bridge health assessment using long-term monitoring data.

The current values of the structural parameters identified are then stored on a database, and the corresponding elemental stiffness values are compared with the baseline values. While different algorithms are being developed to assess the structural health condition, a simple condition index β_i as defined blow, is used here to illustrate the concept of structural health diagnostics.

$$\beta_i = \frac{K_i^{UP}}{K_i^{BAS}} \tag{3}$$

Where K_i^{up} = updated stiffness of the i-th structural element, K_i^{BAS} = baseline element stiffness at a previous healthy state of the bridge, stored in a database. It is important to note that condition indexes can be calculated with respect to either the structural state just preceding the current one or the pristine health condition, or at least at the structural state at the begin of the monitoring program that can be considered healthy (Feng, *et al,* 2004). When condition indexes are calculated as in the former manner, they are particularly useful to assess the effect of a damaging event such as an earthquake, for they provide a direct comparison between the post- and pre-event structural health condition. On the other hand, condition indexes calculated with respect to the pristine state are important to track structural degradation due to aging.

If the condition index exceeds a certain threshold, maintenance intervention of the bridge is suggested. After the intervention, the database will need to be updated based on new measurements. A long-term bridge structural monitoring database is crucial for determining a reasonable and reliable threshold. Such a database must consider a number of factors, such as environmental effects, that can also affect the identified elemental stiffness.

Based on the values of the condition indexes, structural deficiencies due to aging can be detected at an early stage, which will enable more cost-effective maintenance interventions, ultimately leading to enhanced safety of the travelling public with limited resources. In addition, in the aftermath of a major event, the sensor-based structural health diagnostics will enable a rapid and remote bridge condition assessment to improve emergency response operations and decisions on cost-effective restoration.

2.1. *Modal identification*

While abrupt changes in the dynamic properties of bridges are usually caused by a destructive event such as an earthquake, previous studies on long-term monitoring showed the natural frequencies of bridge structures follow a gradual and approximately linear decrease over the years (Soyoz and Feng 2009; Gomez, *et al.* 2011). Such effects cannot be disregarded in an accurate structural health condition assessment. As mentioned earlier, acceleration data are possibly the most frequently used for structural long-term vibration monitoring of highway bridges. Usually, acceleration response data are used to generate modal models by means of experimental modal analysis. In this case a modal model consists of eigenfrequencies, mode shapes and, if needed, damping ratios. In the case of ambient and traffic-induced vibration it is impractical, if not impossible, to measure the excitation forces. As a consequence, in a long-term monitoring program the available data is mainly composed by structural response (output) at sensor locations. Consequently, the SI techniques applied are called output-only techniques. On the other hand, strong motion excitations can be more easily measured, and thus input-output SI techniques are generally applied. Both output-only and input-output SI methodologies will be discussed as follows.

2.1.1. *Output-only modal identification*

The most straightforward output-only modal identification approach is the so-called peak-picking technique, where the structural response, measured in time domain, is transformed to frequency domain through either the Discrete Fourier Transform (DFT) or the Power Spectral Density (PSD) function. Afterwards, the system eigenfrequencies are identified as the peaks of the frequency spectrum. For dealing with multiple outputs, other methods have been adopted, including the complex mode indication function (Shih, *et al.* 1988) and the maximum likelihood identification (Pintelon, *et al.* 1994). An extension of the idea behind the peak-picking technique resulted in the singular value decomposition of the spectral density matrix of the response using multiple outputs. This approach was referred to as the Frequency Domain Decomposition (FDD) (Brincker, *et al.* 2000). A description of other output-only techniques can be found in (Peeters and De Roeck 2001).

The FDD technique takes the Singular Value Decomposition (SVD) of the spectral density matrix of the response, which is decomposed into a set of auto spectral density functions, each corresponding to a Single Degree Of Freedom (SDOF) system. Results of the FDD are exact under three assumptions: (1) the input excitation is white noise, (2) the structure is lightly damped, and (3) the mode shapes of close modes are geometrically orthogonal.

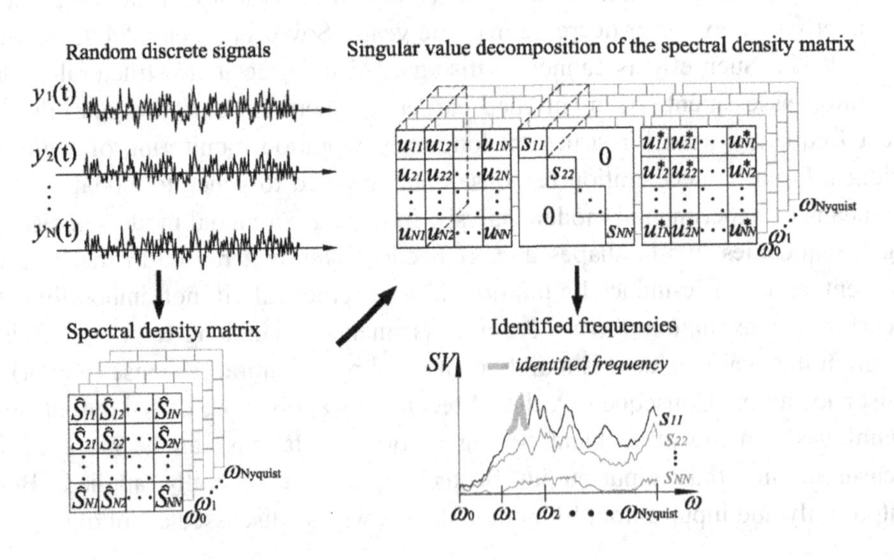

Fig. 2. FDD algorithm applied to discrete-time signals.

Figure 2 shows how the modal parameters of a structure are extracted from measured discrete-time response signals by using the FDD technique. The first step is to compute the response spectral density matrix, which is defined as the distribution of power per unit frequency and is given by:

$$\hat{S}_{rs} = \hat{S}_{y_r y_s}(\omega) = \sum_{\tau=-\infty}^{\infty} R_{y_r y_s}(\omega) e^{-j\omega\tau} \tag{4}$$

The response spectral density matrix $\hat{\boldsymbol{S}}_{YY}$ is estimated at discrete frequencies ω_i and decomposed by taking the SVD of the matrix at each discrete frequency:

$$\hat{\boldsymbol{S}}_{YY}(\omega_i) = \mathbf{U}_i \mathbf{S}_i \mathbf{U}_i^H \tag{5}$$

where $\mathbf{U}_i(\omega_i)$ is a unitary complex matrix containing the singular vectors u_{ik} and $\mathbf{S}_i(\omega_i)$ is a diagonal matrix containing nonnegative scalar numbers or singular values s_{ik}. In the proximity of a peak related to the *i-th* discrete frequency in the Singular Value (SV) spectrum in Fig. 2, this frequency, or maybe a possible close one, will be a dominating mode. If only the *i-th* frequency is a dominating mode, the corresponding first singular vector u_{i1} is an estimate of the mode shape associated to that frequency, and the corresponding singular value is an estimate of the auto spectral density function of the corresponding mode. In the vicinity of this mode (ω_i), the associated spectral density function is identified around the peak by approximating the mode shape estimate $\hat{\boldsymbol{\phi}}$ with the singular vectors for the discrete frequencies around the peak as shown in Eq. (6) A detailed explanation of the FDD algorithm can be found in (Brincker, *et al.* 2000).

$$\hat{\boldsymbol{\phi}} = u_{i1} \tag{6}$$

2.1.2. *Input-output modal identification*

When input motion can be measured such as during an earthquake, input-output modal identification techniques will be preferred for identifying dynamic properties of the structure. Post-event natural frequencies can be used to detect structural damage.

The post-event natural frequencies are estimated by means of input-output SI techniques, which are generally based on the state-space form of the equation of motion. For instance, the Eigensystem Realization Algorithm with Data Correlations (ERA/DC), a time-domain SI method, was developed for modal parameter identification using the impulse response of a system (Juang and Pappa 1985). The method was later extended by adding a procedure, called Observer/Kalman Filter Identification (OKID), to recover the impulse response from input-output data (Juang, *et al.* 1993). The method presents a unified procedure to build state-space models of linear time-invariant systems from the measured data. Applications of the method to civil engineering structures are reported in (Luş, *et al.* 1999; Luş, *et al.* 2002, and Ulusoy, *et al.* 2011.). In this procedure, a finite dimensional, linear, time-invariant system is described by:

$$\begin{aligned} \mathrm{x}(k+1) &= \mathbf{A}\mathrm{x}(k) + \mathbf{B}\mathrm{u}(k) \\ \mathrm{y}(k) &= \mathbf{C}\mathrm{x}(k) + \mathbf{D}\mathrm{u}(k) \end{aligned} \tag{7}$$

where $\mathrm{x} \in \mathbb{R}^n$ is the state vector, $\mathrm{y} \in \mathbb{R}^l$ is the measurement vector, $\mathrm{u} \in \mathbb{R}^m$ is the excitation vector, while $\mathbf{A} \in \mathbb{R}^{n\times n}$ is the state matrix, $\mathbf{B} \in \mathbb{R}^{l\times m}$ is the input matrix, $\mathbf{C} \in \mathbb{R}^{n\times l}$ is the output matrix, $\mathbf{D} \in \mathbb{R}^{n\times m}$ is the discrete transmission matrix, and k is the discrete time variable.

2.2. *Identification of structural parameters*

While the modal parameters provide information on the global behavior of a structure, structural parameters, such as element stiffness, can be further identified in order to access the location and extent of structural deterioration or damage. The identified structural parameters can also be used to update the FE model of the structure.

In this section some relevant techniques for structural parameter identification are presented. It must be noted that these inverse analysis techniques do not represent a unique solution to the problem of parameter identification. Furthermore, the FE model with the indentified (i.e. updated) structural parameters is only an approximation of the actual bridge structure, so discrepancies between measurements and analytical predictions are unavoidable.

2.2.1. *Bayesian updating*

A Bayesian framework was laid down by Beck (1989) for structural system identification that selects the most probable model from a class of models based on input/output measurement. It was later expanded to structural health monitoring and reliability analysis. The Bayesian approach, however, often results in extremely demanding computation in high dimensional spaces. Chen and Feng (2009) proposed to describe a structure using a probabilistic model, whose parameters and uncertainties are periodically updated using measured data in a Recursive Bayesian Filtering (RBF) approach. Operating in a recursive mode, RBF is suitable for continuous monitoring. Such a model of a structure, with gradually diminished uncertainties by the monitoring data, is essential in reliably evaluating its current condition and predicting its future performance. This section first presents the RBF formulation and then will briefly discuss two solution methods: The extended Kalman filter, and the central difference filter.

The Bayesian theorem provides a way to infer conditional probability $P(B_i|A)$ from conditional probability $P(A_i|B)$, where events $\{B_i\}, i = 1$ to n are a set of mutually exclusive and collectively exhaustive events, by

$$P(B_i|A) = \frac{P(A|B_i)P(B_i)}{\sum_{i=1}^{n} P(A|B_i)P(B_i)} \tag{8}$$

In a Bayesian point of view, conditional probability $P(B_i|A)$, or the *a posteriori* probability, conveys the entire knowledge about B_i after observing the occurrence of event A. Based on $P(B_i|A)$, one can obtain optimal estimation of B_i, as well as the uncertainty associated with the estimation. Similarly, $P(B_i)$ conveys the entire *a priori* knowledge about B_i, which is subjected to correction or refinement. Conditional probability $P(A|B_i)$ represents the procedure of deduction: how likely A is to occur as a result of B_i. Therefore, a Bayesian framework implements the deductive–inductive approach outlined previously. When recursively applied to a dynamic system represented by a state-space model, to estimate the current system states based on previous and current observations, the framework is referred to as RBF.

A state-space model is defined as

$$\begin{aligned} \mathbf{X}_k &= f(\mathbf{X}_{k-1}, \mathbf{U}_{k-1}, \mathbf{W}_{k-1}; \boldsymbol{\theta}) \\ \mathbf{Z}_k &= h(\mathbf{X}_k, \mathbf{U}_k, \mathbf{V}_k; \boldsymbol{\theta}) \end{aligned} \tag{9}$$

where $\mathbf{X}$ = hidden states; $\mathbf{Z}$ = observations (measurements); $\mathbf{U}$ = deterministic input; $\mathbf{W}$ = process noise; $\mathbf{V}$ = measurement noise; and k = time index. The state transfer function f and observation function h are argumented by parameters $\boldsymbol{\theta}$, which could be dependent on k.

For the structural health monitoring purpose, structural properties (e.g. element stiffness) are of concern, which usually are represented by parameter $\boldsymbol{\theta}$ in functions f and h. To trace structure changes or to update the model and its uncertainty, it suffices to estimate $\boldsymbol{\theta}$ and its probability distribution. To this end, $\boldsymbol{\theta}$ is regarded as part of the extended state $\{\mathbf{X}_k, \boldsymbol{\theta}_k\}^T$ and the state-space model is restructured as

$$\{\mathbf{X}_k, \boldsymbol{\theta}_k\}^T = F(\{\mathbf{X}_{k-1}, \boldsymbol{\theta}_{k-1}\}^T, \mathbf{U}_{k-1}, \mathbf{W}_{k-1}) \tag{10}$$

$$\mathbf{Z}_k = H(\{\mathbf{X}_k, \boldsymbol{\theta}_k\}^T, \mathbf{U}_k, \mathbf{V}_k) \tag{11}$$

where Eq. (10) is the result of incorporating the first of Eq. (9) with

$$\boldsymbol{\theta}_k = \boldsymbol{\phi}\boldsymbol{\theta}_{k-1} + \mathbf{W}_{k-1}^{\boldsymbol{\theta}} \tag{12}$$

and Eq. (11) is Eq. (9) undergoing necessary adjustment. When the structural system changes at a slow rate, usually, in Eq. (12), $\boldsymbol{\phi} = \mathbf{I}$, where $\mathbf{I}$ = unit matrix.

Unfortunately, even when f and h are linear with respect to **X**, F and H usually are no longer linear. Gaussian probability densities, when propagating through nonlinear functions, are no longer Gaussian or tractable, rendering the RBF not to have a close-form solution.

Central difference filter

The Central Difference Filter (CDF) algorithm is a new version of RBF. It proceeds by weighted statistical linear regression of functions F and H. Instead of linearizing the functions at one local point, the CDF selects a limited number of representative points, called the sigma points, spreading according to the *a priori* distribution, then propagates each individual point through the nonlinear functions, and obtains the new center and spread of the distribution by linear regression of the points cast forward. The implementation of the CDF is described by Van der Merwe (2004) and its application for SHM of highway bridges by Chen and Feng (2009).

Extended Kalman filter

The Extended Kalman Filter (EKF) method aims at identifying the time-varying structural parameter vector $\boldsymbol{\theta}$ (such as bridge column stiffness), which is treated as a part of the extended-state vector in Eq. (10). The EKF determines the optimal estimate of the extended state by minimizing the trace of the error covariance:

$$\boldsymbol{C}_{k|k} = E\left[(Y_k - \hat{Y}_{k|k})(Y_k - \hat{Y}_{k|k})^T\right] \tag{13}$$

Where $Y_k = \{X_k, \theta_k\}^T$ indicates the extended state vector. $Y_{k|k}$ is the optimal extended state, and the ^ denotes estimation.

There are mainly two conceptual phases in EKF, namely prediction and correction phases. In the prediction phase, state estimate $\hat{Y}_{k-1|k-1}$ and the error covariance $\mathbf{C}_{k-1|k-1}$ are projected ahead in time resulting in *a priori* estimates of $\hat{Y}_{k|k-1}$ and $\mathbf{C}_{k|k-1}$. In the correction phase, these *a priori* estimates are filtered using the information from the new measurements resulting in *a posteriori* estimates $\hat{Y}_{k|k}$ and $\mathbf{C}_{k|k}$. More details about the formulation of the EKF method and its application for structural health monitoring are provided by Grewal and Andrews (2001) and Soyoz and Feng (2008).

2.2.2. *Optimization-based FE model updating*

Structural parameters can also be identified and updated by formulating an optimization problem with the objective of minimizing the error between the measured and simulated responses of the structure. The updating parameters are those related to structural deterioration and damage, such as element stiffness, i.e., the Young's modulus, the sectional moment of inertia, the area of the section, the stiffness of abutments and foundations, and the Rayleigh damping coefficients. In this context, FE model updating can be formulated as a constrained optimization problem aimed to find the optimal set of model parameters $\boldsymbol{\theta}$ which minimizes the *objective function* F, the aforementioned measure of error. The optimization of the objective function can be achieved by means of sophisticated optimization algorithms, e.g. quasi-Newton (Polak 1997), genetic algorithm (Holland 1975; Friswell, *et al.* 1998) or direct search (Hooke and Jeeves 1961; Lewis, *et al.* 2000).

The following objective functions have been shown to produce satisfactory results for bridge health assessment (Baghaei and Feng 2010; Gomez 2011):

$$F(\boldsymbol{\theta}) = \sum_{i=1}^{n}\left\{w_i^f\left|\frac{f_i^e - f_i^a}{f_i^e}\right| + w_i^\phi(1 - MAC_i)\right\} \tag{14}$$

$$MAC_i = \frac{\left|\phi_i^{eT}\cdot\phi_i^a\right|^2}{\left(\phi_i^{eT}\cdot\phi_i^e\right)\left(\phi_i^{eT}\cdot\phi_i^e\right)} \tag{15}$$

where $\boldsymbol{\theta}$ is the set of updating parameters, n represents the number of dominant modes of vibration of the structure, the superscripts e, a denote experimental and analytical results respectively, f_i, ϕ_i =natural frequency and mode shape of the ith mode respectively, and MAC denotes the Modal Assurance Criterion (Allemang and Brown 1982). The inclusion of weighting factors (w_i^f, w_i^ϕ) is intended to improve the results of the optimization, but their manipulation depends on engineering judgment.

Many optimization procedures to minimize the objective function are available in the literature. The Genetic Algorithm (GA), introduced by Holland (1975), is one of the most frequently used. It consists of a stochastic search algorithm based on heuristic concepts of natural selection and genetic operations (crossover, mutation, elitism, new blood). Similar to genetic algorithms, Direct Search (DS) optimization algorithms do not require information about the objective function gradient, and for this reason they are often known as "derivative-free" or "zero-order methods" (Lewis, *et al.* 2000). One of the most popular DS algorithms is the generalized Pattern Search (PS) (Hooke and Jeeves 1961).

2.2.3. *Artificial neural networks*

Artificial Neural Networks (ANN) are also used to identify structural parameters based on measured structural response. Resembling biological nerves, a neural network has the ability of storing knowledge, which is later used to produce reasonable outputs for inputs not included in a previous learning process. The constitutive unit of an ANN is the *neuron,* an abstraction of a biological neuron reduced to a mathematical function transforming inputs into outputs. Neurons can be grouped in an entity called *layer*, and a network contains one or more intermediate (hidden) layers and *input* and *output* layers. A particular network has a specific *architecture* defined by the number of layers, the number of neurons in each layer, and the type of mathematical functions used by the neurons. Fig. 3 shows a common architecture referred to as the back-propagation. This ANN consists of three inputs, two hidden layers with twenty neurons and an output layer with three neurons. The hidden layers have a *logsig* transfer function whereas the output layer has *purelin* transfer functions. A more detailed description of back-propagation and other types of networks can be found in (Hecht–Nielsen 1989; Hagan and Menhaj 1994).

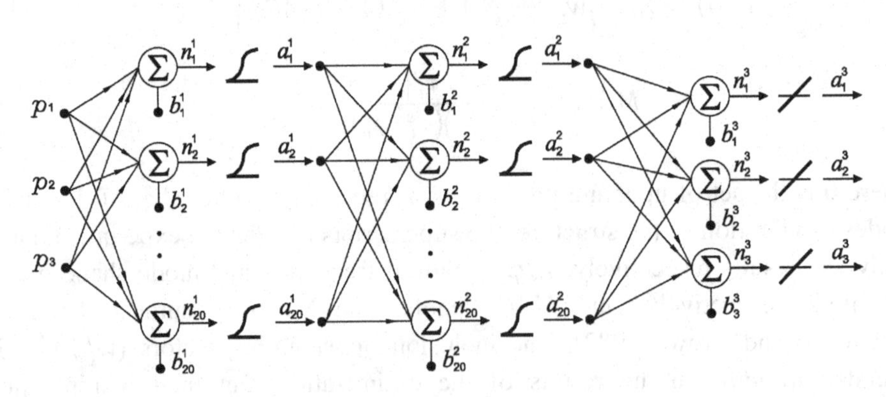

Fig. 3. Architecture of a feed-forward back-propagation neural network.

The procedure of the ANN-based identification involves the following steps: (1) determining the types of input and output patterns; (2) preparing the training and testing patterns through FE analyses; (3) training the neural network; and (4) estimating the structural parameters of the baseline FE model by inputting the measured natural frequencies and mode shapes to the well-trained neural network. Applications of ANN for SHM of highway bridges can be found in (Feng, *et al.* 2004; Soyoz and Feng 2009).

3. Validation of Health Diagnostics Methods through Large—Scale Seismic Shaking Table Tests.

Some experimental studies have been carried out to validate the efficacy of the structural health diagnostics methods. A notable one is the seismic shaking table tests using a large concrete bridge model (Chen, *et. al.*, 2008, Soyoz and Feng, 2008). Shaken by a sequence of simulated seismic ground motions with increasing intensity, the bridge suffered damage of different severities, providing invaluable opportunity to verify the vibration-based structural health diagnostic methods presented above, by comparing analytical results with visual inspection observations and experimental measurements obtained by strain gauges embedded on the reinforced concrete and displacement sensors.

3.1. *Test specimen, instrumentation and procedure*

Figure 4 shows the bridge test specimen built at the university of Nevada, Reno. Each of the three bents is supported by an individual shaking table. The bents are linked by the bridge deck, with total length 18.29 m (720 in). The bents are of different heights, 1.83 m (72 in), 2.44 m (96 in) and 1.52 m (60 in) for Bents 1, 2, and 3, respectively, so that they present different stiffness. To resemble the inertia of other parts of the superstructure not built into this specimen, compensative masses were added. The shake tables were driven by input acceleration signals in the transverse direction of the bridge, and were shaking in unison. Eleven accelerometers were installed on the specimen to obtain the acceleration inputs and responses of the bridge.

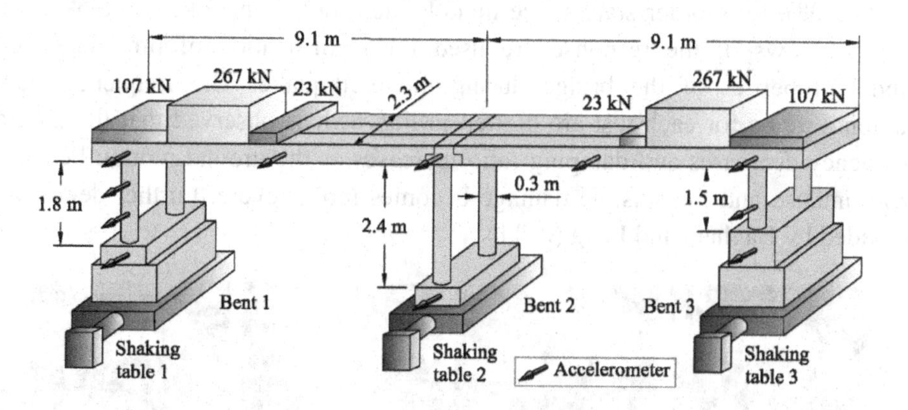

Fig. 4 Design and sensor layout of bridge specimen.

The bridge specimen was subjected to a sequence of earthquake excitations in the traverse direction based on the 1994 Northridge earthquake ground motion record, applied with increasing intensity from a pre-yield demand (0.07g PGA) to bent failure (1.66g PGA), where PGA is the Peak Ground Acceleration of the seismic motion.

The strong motions induced progressively increasing damage to the structure. White noise excitations were input to the bridge in between the earthquake input motions, for system identification purposes. White noise tests were low amplitude coherent motions that were not large enough to induce any damage to the bridge structure. Indicating as "WN#" the white noise excitations, and "T#" the earthquake excitation tests, the experiment protocol is outlined as: WN1 (0.1)→T12 (0.075)→T13 (0.15)→T14 (0.25)→WN2 (0.1)→T15 (0.5)→T16 (0.75)→T17 (1.0)→WN3 (0.1)→T18 (1.33)→WN4 (0.1)→T19 (1.66)→WN5 (0.1). The numbers in parenthesis are the PGA of each excitation in *g*. According to visual inspections conducted after each test, damage sequence can be outlined as follows: T13 (Bentl yielded) →T14 (Bent3 yielded) →T15 (Bent2 yielded) →T19 (Bent3 steel buckled). Fig. 5 shows the damage state after two different tests for bents 2 and 3.

3.2. *Modal identification*

The modal properties of the bridge are identified from the earthquake and intermediate white noise excitations using the aforementioned SI techniques. As stated earlier, identification of modal properties of the bridge during a damaging earthquake requires a special treatment due to the system nonlinearities (Chen, *et al.* 2008). Low-order state space models identified from successive 5-second long windows of the response are used for identification of time-dependent modal properties of the bridge during the earthquake. The frequencies and damping ratios for each test are plotted in Fig. 6. It is observed that the natural frequency decreases and damping ratio increases as the ground motion becomes more intense and the seismic damage becomes more severe. Further details are provided by Baghaei and Feng (2011).

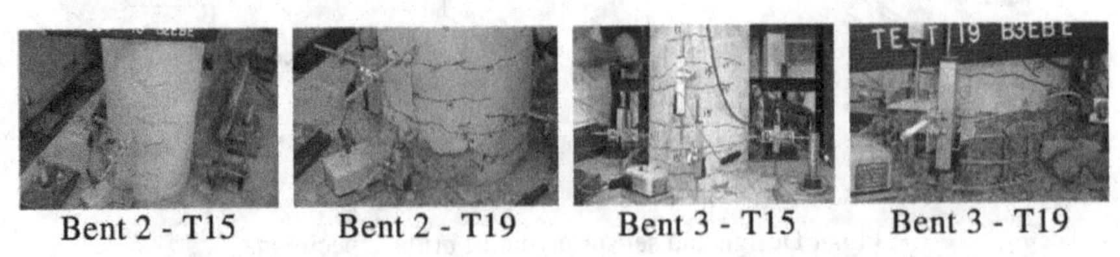

Bent 2 - T15 | Bent 2 - T19 | Bent 3 - T15 | Bent 3 - T19

Fig. 5 Damage state at different tests for bents 2 and 3.

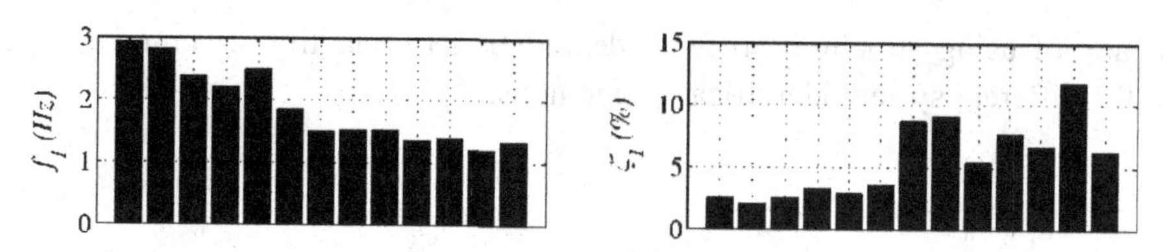

Fig. 6 Shaking table test identified modal parameters.

3.3. *Damage assessment*

Fig. 7 presents the progressive deterioration of the stiffness of each bent as a function of time during each of the seismic excitations estimated using the EKF technique (Soyoz and Feng 2008). The results are well correlated to the observed damage sequence previously outlined.

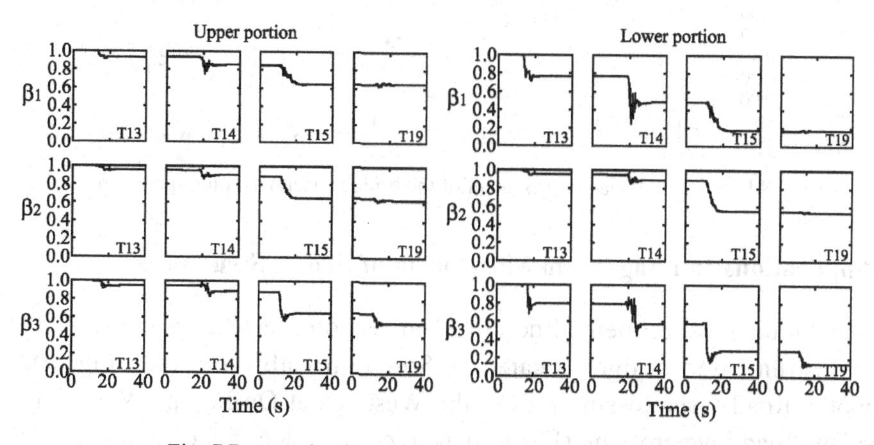

Fig. 7 Progressive identification of instantaneous stiffness

Additionally, structural parameter identification has been carried out by using other techniques, and the results are compared in Fig. 8. Chen (2006) focused on using pre- and post-event low amplitude vibration to estimate the location and the extent of seismic damage to this bridge model using the Bayesian updating framework. Chen and Feng (2009) applied the CDF algorithm to implement the RBF. Baghaei and Feng (2011) identified the seismic damage of the same bridge model based on the optimization approach using GA. The stiffness degradations associated with the seismic damage, which were identified by the different SI methods, are all consistent among each other. More importantly, they agree with the visually observed structural damage and the measurements from displacement sensors installed on the structure. The condition index at each of the three bents decreases as the ground motion becomes more intense and the damage becomes more severe. The seismic shaking table tests validated the efficacy of using structural stiffness degradation to quantify structural damage and the different system identification methods.

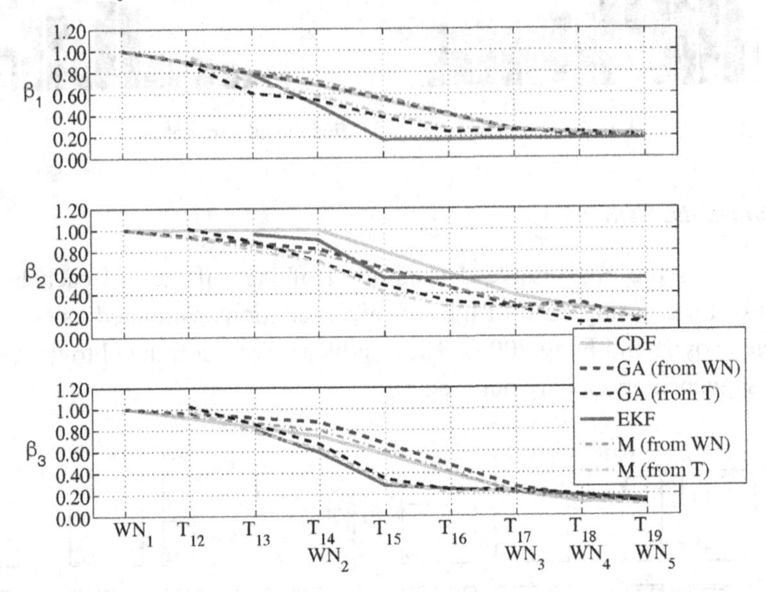

Fig. 8 Post-event damage assessment of the bridge specimen column elements.

4. Applications In Long-Term Monitoring of Bridge Structures

The health diagnostics method described in section 2 are being implemented in three instrumented bridges located in Southern California (see Fig. 9), the Jamboree Road Overcrossing (JRO), the West Street On-Ramp (WSO) and the Fairview Road Overcrossing (FRO), (Chen, *et. al.*, 2006, Soyoz and Feng, 2009, Gomez, 2011, Gomez, *et al.*, 2011). The bridges have been monitored since 2002, providing a unique opportunity to update FE models using long-term monitoring data and to evaluate the current state of each bridge. Details of the

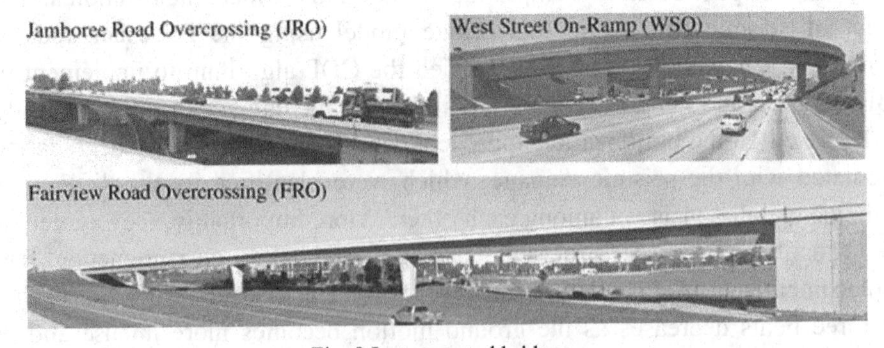

Fig. 9 Instrumented bridges.

bridges are given in Fig. 10. The three selected bridges have similar deck systems which consist of a pre-stressed and post-tensioned box girder supported on single column bents and seat-type abutments. Some of the long-term monitoring results on these bridges are dicussed.

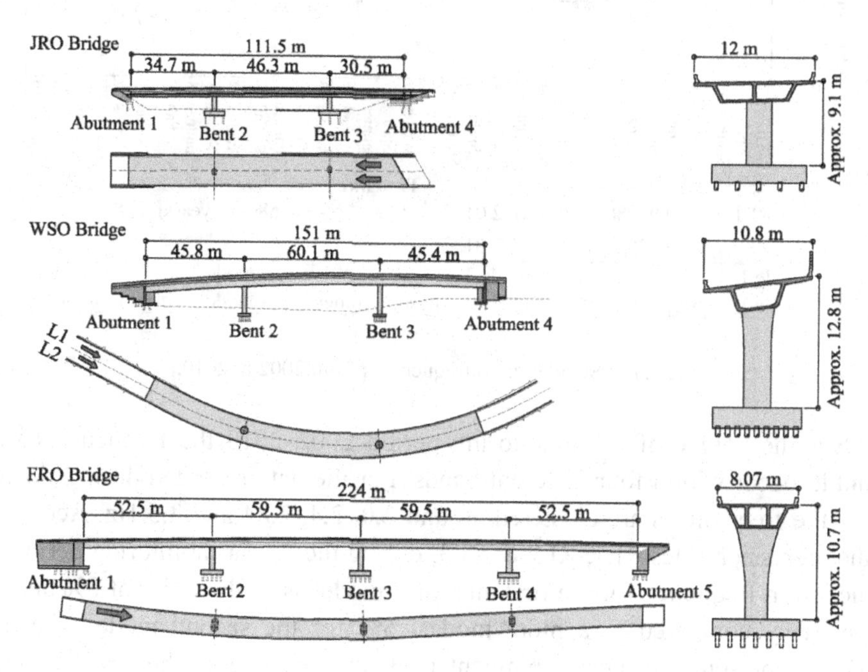

Fig. 10 Details of the three studied bridges.

4.1. *Use of ambient and traffic-induced vibration data*

Starting in 2001, the first author and her team, in collaboration with the California Department of Transportation (Caltrans), have installed permanent monitoring systems including accelerometers, strain gauges, pressure sensors, and displacement sensors on these highway bridges. Approximately 1,600 sets of data were collected at the JRO Bridge from 2002 to 2006, 1,350 sets at the WSO Bridge from 2002 to 2010 and 50 sets at the FRO Bridge during the years 2006 and 2010.

4.1.1. *Monitoring of natural frequencies*

The first four WSO Bridge frequencies identified from singular value plots, constructed for all ambient vibration records, are plotted in Fig. 11 (Gomez *et al.*, 2011). The frequencies were obtained by means of the FDD SI technique.

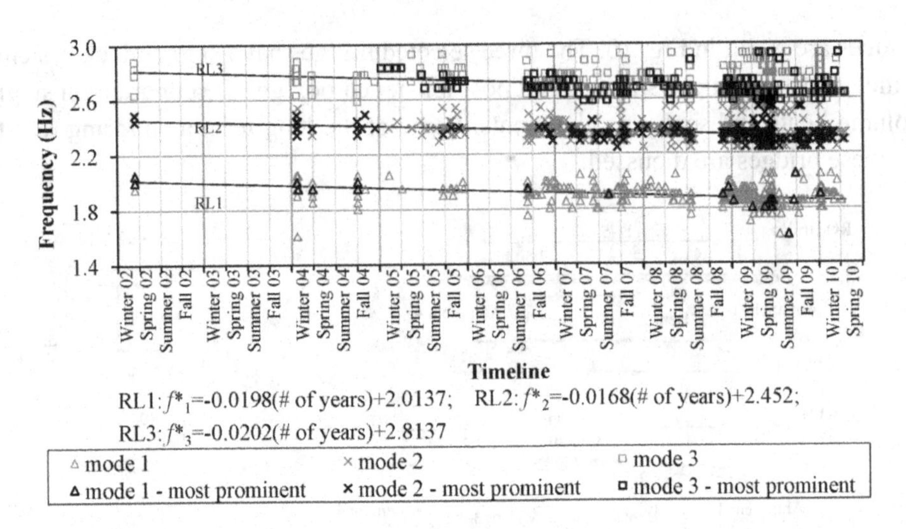

Fig. 11 WSO identified frequencies from 2002 to 2010.

Over the course of the monitoring period 2002–2010 the frequencies were found to occur within four different bands. For the data recorded during 'Winter 2002' the frequencies are clustered around 2.0, 2.4, 2.84 and 3.5 Hz. According to the regression lines RL1, RL2 & RL3, during the 8 year monitoring period the reduction in frequency for the first and third modes is of the order of 8% and 7% respectively compared to a more modest 5% for the second mode. The dark markers represent the most prominent peak in each one of the 1,350 singular value plots. The prominent peak represents the mode with larger contribution in the bridge response. It is observed the first mode at WSO is not always prominent. Here, the second mode has the most important contribution in the response.

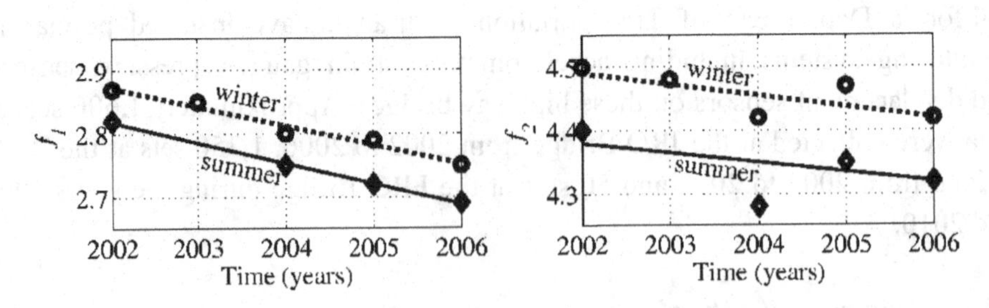

Fig. 12 Variation of JRO Bridge natural frequencies due to season change.

Similarly, a linear decrease in the natural frequencies was observed in the JRO Bridge respectively for the winter and summer seasons. Fig. 12 (Soyoz and

Feng 2009) shows the identified frequencies during winter are slightly larger than those identified during summer. This difference is commonly attributed to environmental effects including different temperature and moisture, which is a topic of ongoing research.

4.1.2. *Monitoring of mode shapes*

In order to study whether the gradual reduction in frequencies presented in the previous section is associated with a change in the mode shapes, long-term monitoring of these mode shapes is presented based on the (MAC) criterion. Similar to frequencies, mode shapes were obtained by means of the FDD SI technique. Using the MAC value, two identical mode shapes yield a MAC value of unity while truly orthogonal mode shapes result in a MAC value of zero. Recall the MAC value between two modes (ϕ_1, ϕ_2) can be computed using Eq. (15). The variation in the resulting MAC matrix terms over the duration of the monitoring period of the WSO Bridge is plotted in Fig. 13. The modes are substantially consistent (MAC values of unity for diagonal terms) and orthogonal (MAC values of zero for off-diagonal terms). Unlike the eigenfrequencies, the mode shapes show no change as suggested by the MAC values along the monitoring period.

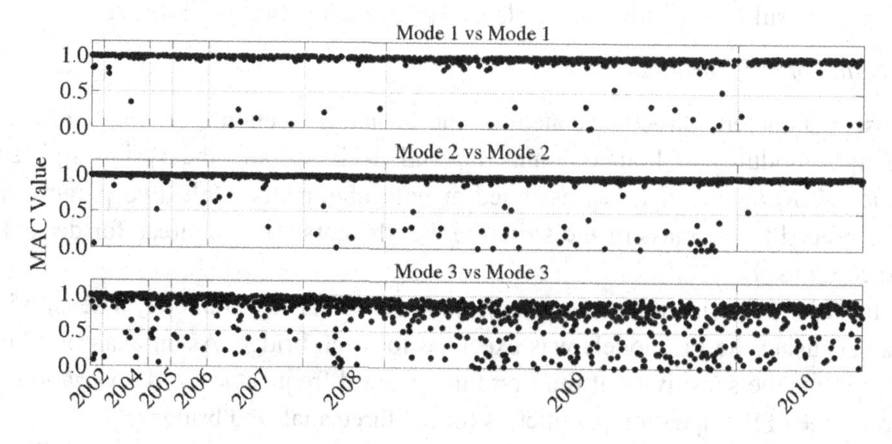

Fig. 13 Variation in MAC values from 2002 to 2010 of the WSO Bridge.

4.1.3. *Monitoring of structural stiffness*

Three-dimensional linear FE models were defined for the three bridges and the updating of the models was executed annually in order to assess the condition of the bridge structures. The FE models were defined using linear elastic beam elements. Fig. 14 shows, as an example, the FE model for WSO.

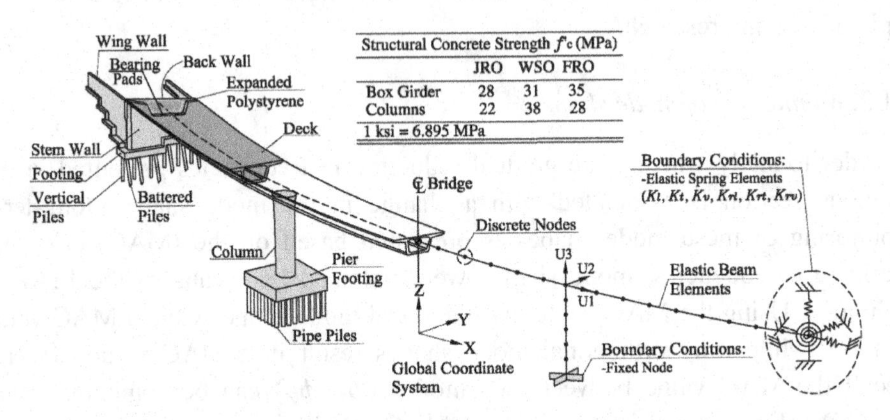

Fig. 14 WSO Bridge finite element model definition (the geometry varies for each bridge).

The boundary (support) conditions were defined as fixed at the base of each bent and linear springs were used to model abutment behavior. Although different abutment models are available, a single spring was assigned to each degree of freedom, i.e. transverse, longitudinal, vertical, and their respective rotations, resulting in a total of 12 elastic springs in the bridge FE models.

Selection of structural parameters

Those parameters directly related to the structural bending stiffness are the Young's modulus of both columns E_d and deck E_c, and the spring stiffness values $K_l, K_t, K_v, K_{rl}, K_{rt}, K_{rv}$ assigned at both abutments. Only two parameters were related to the mass of the structure, i.e. the corss section areas for deck A_d and columns A_c.

In order to define adequate search domains for each parameter, a sensitivity analysis using the FE models was executed for each bridge. As an example, Fig. 15 presents the sensitivity of the first three natural frequencies to the variation in percentage of the updating parameters for the three analyzed bridges.

An abscissa-value of zero represents the values assigned in the preliminary model based on drawings and design aids. During the sensitivity analysis the updating parameters are incrementally changing one at a time. The results indicate the most sensitive parameters are the Young's modulus of both deck and columns (E_d, E_c) along with the spring stiffnesses in the transverse direction (K_t). The natural frequencies decrease as these three parameter values decrease. The frequencies show negligible sensitivity to the variation of other parameters.

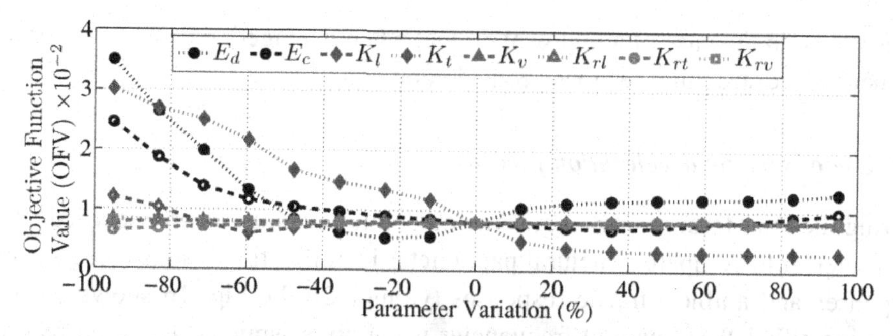

Fig. 15 Sensitivity analysis of the WSO Bridge FE model.

4.1.4. *Health diagnostics*

Table 1 lists the model updating results using three different structural identification techniques for the health assessment of the JRO Bridge. The change between preliminary and baseline superstructure stiffness is larger when a direct search optimization algorithm is used. Nevertheless, the three compared techniques (Backpropagation ANN, RBF ANN and DS optimization) show the largest variation corresponds to the spring stiffnesses at abutments. Similar results were found for the FRO Bridge (Gomez, *et al.* 2011).

In the case of the WSO Bridge, the results showed the elastic modulus of both the superstructure and the columns was gradually decreasing in an approximately linear fashion. A reduction of approximately 9% occurred in a period of five years. After the eight years of monitoring, the resulting state indexes β, given by Eq. 16), for E_d, E_c, K_{1t}, and K_{4t} are approximately 12, 16, 1, and 1% respectively. The reduction in stiffness is in accordance to the formation of a cracked section and it justifies the use of an effective moment of inertia for the columns in design practice.

Table 1 JRO Bridge structural parameters identification results

Updating parameter	Correction factor after updating the preliminary model		
	Backpropagation ANN	RBF ANN	Direct Search
Superstructure bending stiffness	1.00	0.96	0.66
Superstructure bending stiffness	0.99	0.87	0.50
Abutment 1 lateral stiffness	0.16	0.87	1.94
Abutment 4 lateral stiffness	0.67	0.87	1.94

Reduction of the structural stiffness may be regarded as an indication of aging and deterioration of the structure due to numerous factors that are extremely difficult to detect and quantify. These aging factors include creep and shrinkage of concrete, and relaxation of tendons, which eventually leads to pre-stress losses

(Lin 1963). In the case of a highway bridge, when damaged, the pavement profile influences the structural response considerably.

4.2. *Use of seismic acceleration records*

During the long-term monitoring, some moderate earthquakes are recorded and it is of interest to compare structural paprameter identification results using seismic responses and ambient/traffic responses (Gomez 2011). Fig. 16 shows the first three identified WSO natural frequencies using six seismic acceleration records. For comparison purposes, the black bars represent the average frequencies identified from the long-term ambient/traffic response monitoring (2002 to 2010) using FDD. The grey scale bars represent those frequencies identified from the seismic records using the OKID-ERA/DC SI technique. For the strongest earthquake (Chino–Hills) the identified frequencies are nearly 20% smaller than for all other records. An interesting observation is the frequencies before (2008) and after (2009) the Chino–Hills earthquake, practically have the same values meaning the system did not suffer any damage because of the earthquake. One possible explanation for the change of the bridge frequencies is the changing boundary conditions such as soil properties at footings and abutments.

Once the eigenfrequencies were estimated, the structural identification results showed an interesting relation between structural stiffness and earthquake peak ground acceleration in the transverse direction (PGA_T). The deck and columns modulus of elasticity decrease as the earthquake PGA_T increase. Also, the first two modal damping ratios increase as the earthquake PGA_T increase. The updated modal damping ratios were approximately 20% of the critical which is a much higher value than the conventional assumption of 5%. This is in accordance with previous research on evaluating the equivalent modal damping ratio of short-span bridges where damping ratios where found as high as 25% due to an earthquake of $PGA_T = 0.55g$ (Lee, *et al.* 2011).

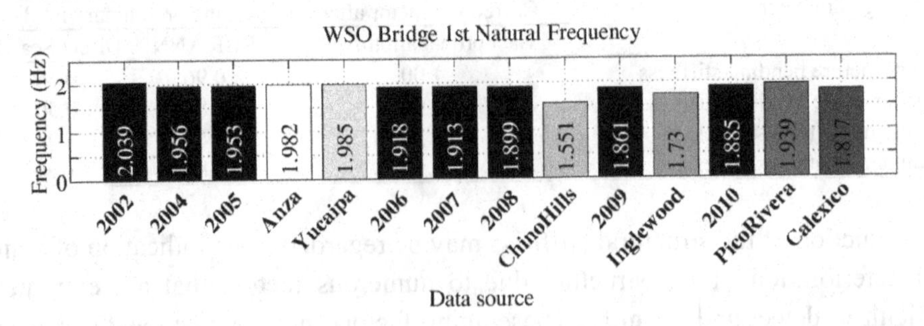

Fig. 16 First three WSO identified natural frequencies using six seismic records and ambient/traffic induced vibration records from 2002 to 2010.

5. Conclusions

This chapter discussed how structural vibration monitoring data can be processed and used for diagnosis of structural health conditions, through the presentation of some fundamental and emerging system identification techniques, experimental validation using seismic shaking table tests, and field applications in long-term monitoring of actual highway bridges. Considering certain structural parameters, such as element stiffness, change as the structure ages, deteriorates or suffers from damage, the structural health is defined here as the change of such parameters. Therefore, the bridge structural health monitoring problem becomes a system identification problem to detect changes of these structural parameters. Seismic shaking table tests of a large-scale three-bent concrete bridge validated the efficacy of using the change in structural element stiffness to locate and quantify seismic damage on the bridge, as well as the various system identification methods. The application of these methods to three instrumented concrete bridges has produced valuable database regarding the change of modal and structural parameters over an eight-year period. As more long-term monitoring data and analysis results from actual bridges become available in the future, more reliable diagnostics can be achieved regarding structural health conditions. This will eventually enable timely detection of damage/deterioration, scientific prioritization, and cost-effective intervention to ensure structural reliability and public safety.

References

Allemang, R., and Brown, D. (1982). "A correlation coefficient for modal vector analysis." *Proc., 1st Int. Modal Analysis Conf.*, Society for Experimental Mechanics, Orlando, FL, U.S.A, 110–116.

Baghaei, R., and Feng, M.Q. (2010) "Damage assessment of the bridge structures using a hybrid optimization strategy." SPIE, San Diego, CA, USA, 76491N-76411.

Baghaei, R., Feng, M.Q. (2011). "Utilization of Strong Motion Data for Damage Assessment of Reinforced and Concrete Bridges." Proceedings of SPIE, San Diego, CA, USA, 7983–92.

Beck, J.L. (1989). "Statistical System Identification of Structures." Proceedings of the 5th Internbational Conference on Structural Safety and Reliability, San Francisco, CA.

Brincker, R., Zhang, L., and Andersen, P. (2000). "Modal Identification from Ambient Response Using Frequency Domain Decomposition." *Proc., 18th Int. Modal Analysis Conf.*, Society for Experimental Mechanics, San Antonio, TX, U.S.A, 625–630.

Chen, Y., Feng, M.Q., and Tan, C-A. (2006), "Modeling of Traffic Loads for System Identification of Bridge Structures.", *Computer-Aided Civil and Infrastructure Engineering*, Vo. 21, pp. 57–66.

Chen, Y., Feng, M.Q., Soyoz, S. (2008), “Large-Scale Shaking Table Test Verification of Bridge Condition Assessment Methods” *ASCE Journal of Structural Engineering*, Vol. 134, No. 7., pp. 1235–1245.

Chen, Y. (2006). "Methodology for Vibration-Based Highway Bridge Structural Health Monitoring." Ph.D. Dissertation, University of California, Irvine, CA, USA.

Chen, Y., and Feng, M.Q. (2009). "Structural Health Monitoring by Recursive Bayesian Filtering." *J. Eng. Mech.*, 135(4), 231–242.

Feng, M.Q., Kim, D.K., Yi, J.-H., and Chen, Y. (2004). "Baseline Models for Bridge Performance Monitoring." *J. Eng. Mech.*, 130(5), 562–569.

Friswell, M.I., Penny, J.E.T., and Garvey, S.D. (1998). "A combined genetic and eigensensitivity algorithm for the location of damage in structures." *Comput. Struct.*, 69(5), 547–556.

Gomez, H.C. (2011). "System Identification of Highway Bridges Using Long-term Vibration Monitoring Data." Ph.D. Dissertation, University of California, Irvine, Irvine, CA.

Gomez, H.C., Fanning, P.J., Feng, M.Q., and Lee, S. (2011). "Testing and long-term monitoring of a curved concrete box girder bridge." *Eng. Struct.*, 33(10), 2861–2869.

Grewal, M.S., and Andrews, A.P. (2001). Kalman Filtering: Theory and Practice Using Matlab John Wiley & Sons, Inc., New York.

Hagan, M.T., and Menhaj, M.B. (1994). "Training feedforward networks with the Marquardt algorithm." *Neural Networks, IEEE Transactions on*, 5(6), 989–993.

Hecht-Nielsen, R. "Theory of the backpropagation neural network." *Proc., Neural Networks, 1989. IJCNN., International Joint Conference on*, 593–605 vol.591.

Holland, J. (1975). "Adaptation in Natural and Artificial Systems."University of Michigan, Press.

Hooke, R., and Jeeves, T.A. (1961). ""Direct Search" Solution of Numerical and Statistical Problems." *J. ACM*, 8(2), 212–229.

Juang, J.-N., and Pappa, R.S. (1985). "An eigensystem realization algorithm for modal parameter identification and model reduction." *J. Guid. Control Dyn.*, 8(5), 620–627.

Juang, J.-N., Phan, M., Horta, L.G., and Longman, R.W. (1993). "Identification of observer/Kalman filter Markov parameters - Theory and experiments." *J. Guid. Control Dyn.*, 16(2), 320–329.

Lee, S., Feng, M.Q., Kwon, S.-J., and Hong, S.-H. (2011). "Equivalent Modal Damping of Short-Span Bridges Subjected to Strong Motion." *J. Bridge Eng.*, 16(2), 316–323.

Lewis, R.M., Torczon, V., and Trosset, M.W. (2000). "Direct search methods: then and now." *J. Comput. Appl. Math.*, 124(1-2), 191–207.

Lin, T.Y. (1963). Design of prestressed concrete structures, Wiley, New York.

Luş, H., Betti, R., and Longman, R.W. (1999). "Identification of linear structural systems using earthquake-induced vibration data." *Earthquake Eng. Struct. Dyn.*, 28(11), 1449–1467.

Luş, H., Betti, R., and Longman, R.W. (2002). "Obtaining refined first-order predictive models of linear structural systems." *Earthquake Eng. Struct. Dyn.*, 31(7), 1413–1440.

Peeters, B., and De Roeck, G. (2001). "Stochastic System Identification for Operational Modal Analysis: A Review." *Journal of Dynamic Systems, Measurement, and Control*, 123(4), 659–667.

Pintelon, R., Guillaume, P., Rolain, Y., Schoukens, J., and Van Hamme, H. (1994). "Parametric identification of transfer functions in the frequency domain-a survey." *Automatic Control, IEEE Transactions on*, 39(11), 2245–2260.

Polak, E. (1997). Optimization: algorithms and consistent approximations, Springer-Verlag.

Shih, C.Y., Tsuei, Y.G., Allemang, R.J., and Brown, D.L. (1988). "Complex mode indication function and its applications to spatial domain parameter estimation." *Mech. Syst. Sig. Process.*, 2(4), 367–377.

Soyoz, S., and Feng, M.Q. (2008). "Instantaneous damage detection of bridge structures and experimental verification." *Structural Control and Health Monitoring*, 15(7), 958-973.

Soyoz, S., and Feng, M.Q. (2009). "Long-Term Monitoring and Identification of Bridge Structural Parameters." *Comput. Aided Civ. Infrastruct. Eng.*, 24(2), 82–92.

Ulusoy, H.S., Feng, M.Q., Fanning, P.J. (2011). "System Identification of a Building from Multiple Seismic Records." *Earthquake Engineering and Structural Dynamics*, 40(6), 661–674.

Van der Merwe, R. (2004). "Sigma-point Kalman filters for probabilistic inference in dynamic state-space models." Ph.D. Dissertation, Oregon Health & Science University, Portland, OR.

Chapter 11

Sensors Used in Structural Health Monitoring

Mehdi Modares[1] and Jamshid Mohammadi[2]

[1,2]Department of Civil, Architectural and Environmental Engineering, Illinois Institute of Technology, Chicago, Illinois 60616, USA
E-mails: [1]mmodares@iit.edu & [2]mohammadi@iit.edu

1. Introduction

In reality, the response of structures to load applications varies from the values computed during the design stage. This is owing to the assumptions made in structural modeling, differences between the design drawings and the actual built structure as well as the unpredictable and changing natures of service and environmental loads (Park *et al.*, 2007). Hence, structural health monitoring (SHM) is necessary to monitor structures by employing a system that can detect structural defects and provide early indication of problems before any catastrophic failure occurs. An early warning provided by the SHM system is then used to determine the best remedial solution. Not only is SHM used for existing structures, it is also used during the construction process to manage safety risks (Karbhari and Ansari, 2009). The system for health monitoring includes sensors, hardware and software of data acquisition, transmission and procession units, and data management system (Li *et al.*, 2006). An ideal SHM system requires that the sensors used are inexpensive, durable, easy to install and maintain, and can output information directly regarding the health status of the structure (Ou and Duan, 2006). This chapter provides an overview of sensor types that can be employed in SHM system applications in measuring a variety of structural parameters, such as displacements, strains, stresses, cracks, fatigue damage, corrosion, moisture, temperature, and applied loads.

2. Traditional Structural Health Monitoring

The traditional practice of SHM has been done by visual inspectors with the aid of simple instruments such as plumb lines and levels (Park *et al.*, 2007). Visual inspections can detect structural flaws due to aging, excessive vibrations or settlement (Kijewski-Correa, 2005; Feng, 2009). Although critical information is gathered from visual inspections, it is labor intensive yet cost effective. More significantly, through visual inspection internal damages may go unnoticed (Huston, 2010). Hence the need for more accurate measures leads to using modern sensors to monitor structural health conditions. Sensors can be used on a routine inspection and maintenance basis for inventory data gathering purposes. This type of application is referred to as "ongoing" data acquisition and monitoring in this chapter. The other application is corresponding to cases where a structure has gone through a known damage condition. Sensors for this application are used for operating data gathering and monitoring purposes. This type of application is referred to as "on-demand" monitoring in this chapter.

3. Strain Sensors

Many types of strain sensors have been developed to measure the maximum induced strain to evaluate the safety of a structural member. These sensors are generally installed at fixed locations and only cover a limited range of structural members. Hence, the reliability of, and the range of data compiled by, strain sensors depend on the number and locations of them, which are often influenced by the distribution of the actual induced stresses (Park *et al.*, 2007). There are different methods for mounting a strain gage onto a structure. Earlier sensors required grinding the surface of a structural member, cleaning the area and spot welding the gage onto the structure. Newer models can be applied by using industrial adhesives. However, generally, the process of mounting strain gages onto structure is labor and skill-intensive. Moreover, the type of adhesive used to mount the strain gage onto the structure would affect the sensor's precision and durability (Huston, 2010).

3.1. *Foil strain gage*

Foil strain gages are the most commonly used strain gages since they are inexpensive to produce and relatively easy to install. These gages consist of a pattern of very fine wire or foil conductors bonded to the structure. The most popular alloys used for foil strain gages include copper-nickel and nickel-chromium alloys. When the structure deforms, the foil deforms with it causing a change in electrical resistance that can be correlated to deformation in the form of strain (Huston, 2010; Evans *et al.*, 2009). To measure the individual component of strain, the strain gage is configured in rosette or spider designs. The 2-D rosette design uses three gages attached at angles of 0°, 45°, and 90°. The 3-D spider design is similar to the rosette design but this configuration requires six strain gages. Since the resistance-to-strain relationship is linear, the strain is then determined by measuring the change in resistance (Huston, 2010). However, the detected resistance is affected by temperature, material properties, environmental noise and the type of adhesive used (Feng, 2009; Yenilmez *et al.*, 2007).

3.2. *Semiconductor strain gage*

Although semiconductor strain gages are more sensitive to strain than foil gages, they tend to be more expensive, more sensitive to temperature changes, and are more fragile than foil gauges. Semiconductor strain gages use piezoelectric resistance of silicon to determine the change in resistance (Huston, 2010). Piezoelectric fibers are easy to install into small places and have good sensitivity. These fibers are durable and have a better signal to noise ratio than foil strain gages (Yenilmez *et al.*, 2007). Since the resistance-to-strain relationship is nonlinear in semiconductors, error is introduced. To overcome this shortcoming, software programs are employed (Huston, 2010).

4. Accelerometers

Accelerometers measure the fluctuation patterns in displacements or vibrations from the various levels of acceleration at given points of a structure. These sensors must be mounted onto the structure in order to correctly measure the vibrations. Accelerometers are based on the concept of a mechanical system made up of a damped mass on a spring. Upon experiencing vibrations, the mass is displaced causing the spring to accelerate proportionally to the actual acceleration (Huston, 2010; Bowling and Richy, 2000). Through a numerical integration, the displacement can be determined from the acceleration data. A disadvantage of this approach is that the resulting numerical integration will have errors, which can be minimized by adjusting and improving the time variation of displacements (Huston, 2010).

4.1. *Piezoelectric accelerometers*

Piezoelectric fibers generate electric voltages, when they are subject to tension or compression proportional to the load (Yenilmez *et al.*, 2007). These fibers are used in various sensors including accelerometers, pressure gages, and ultrasonic sensors. Piezoelectric accelerometers use piezoelectric materials such as quartz to convert the vibration into an electrical signal that is proportional to the applied acceleration (Feng, 2009; Huston, 2010). Piezoelectric accelerometers are durable, easy to install, have a large dynamic range and long life span (Huston, 2010; Bowling and Richy, 2000). Such devices have a wide range of frequencies measuring vibrations from 10–4 g to above 104 g. Due to its difficulty in measuring low-frequency vibrations, piezoelectric accelerometers cannot measure vibration of large structures (Karbhari and Ansari, 2009; Huston, 2010). Moreover, due to its sensitivity to excessive heat, the accuracy of piezoelectric accelerometers decreases at higher temperatures.

4.2. *Micro electro-mechanical systems (MEMS) accelerometers*

MEMS accelerometers use silicon and capacitive strain-gage displacement measurement to detect low-frequency vibrations (Huston, 2010; Park *et al.*, 2008). Capacitive measurements are based on the ability of a body to hold an electrical charge when its geometry changes. MEMS accelerometers are composed of movable proof mass with moveable plates. The displacement of proof mass is proportional to the capacitance difference (Bowling and Richy, 2000). The advantage of MEMS accelerometers includes their low cost, high sensitivity and low temperature sensitivity. This device is susceptible to electromagnetic interference; and as such, appropriate shielding has been developed to prevent this problem (Huston, 2010).

5. Displacement Sensors

Linear variable differential transformers (LVDTs) and dial gages are traditional structural displacement sensors used to measure linear displacement such as structural crack location, and length and rate of crack growth (Huston, 2010). These sensors can only measure the displacement at one point of a structure and require a stationary platform installed near the structure as a measurement reference (Fu and Moosa, 2002). However, the recent advancements in global positioning systems (GPS) offer a substitute and a device that can be used for continuous displacement measurement (Kijewski-Correa, 2005).

5.1. *Linear variable differential transformer (LVDT)*

LVDTs transduce displacements with a piston-line motion of a free moving soft magnetic cylindrical core inside three cylindrical wire coils. The two outer secondary coils are wrapped in the opposite direction of the center primary coil. Displacement is determined from measuring the change in voltage levels in thesecondary coils generated from the displacement of the core (Karbhari and Ansari, 2009; Huston, 2010). LVDTs must be attached on the structure which can be costly and impractical for applications in large structures (Park *et al.*, 2007).

5.2. *Global positioning system (GPS)*

A GPS is typically used to measure static and dynamic displacements as well as the displacement history of a structure. The deflection is measured by installing a GPS antenna to the structure which receives signals from satellites orbiting the Earth. Since the signals are received from satellites, displacements cannot be measured inside a structure. The transmission of the GPS signals are additionally vulnerable to various interferences, which can introduce errors (Kijewski-Correa, 2005; Psimoulis *et al.*, 2008). The deflection accuracy for health monitoring requires measurement accuracy in an order of millimeters. The precision of a GPS, however, is ±5mm horizontally and ±10mm vertically (Li *et al.*, 2006; Huston, 2010). A GPS can also be used to measure the frequency oscillation of flexible engineering structures, such as long span bridges, most accurately for up to 2 Hz (Psimoulis *et al.*, 2008).

6. Photographic and Video Image Devices

Modern digital cameras can capture and store images of structures at low cost due to their ease of operation and availability. Digital cameras are used to monitor both the deformed shape and maximum values of displacement (Huston, 2010; Evans *et al.*, 2009). Since these devices do not require contact with the structure, they are a desirable device to use for SHM (Evans *et al.*, 2009). A reference point, typically a highly reflective object, is placed onto a structure and monitored using a camera. The disadvantage of this monitoring technique is primarily due to its limited capability to capture robust images at night. Although the night vision technology has been developed, it has not been yet widely implemented for SHM applications (Park *et al.*, 2007). A video camera can continuously record data. Such a device can be used to monitor traffic crossing over a bridge to correlate each vehicle with corresponding data (Fraser et al., 2010). New software technologies allow for converting digitized images directly to displacements. However, the accuracy of the results depends on the number of pixels captured in each image.

6.1. *Charge-coupled-devices*

Charge-coupled-device (CCD) cameras have been employed to monitor structural displacements. These cameras use a CCD, which is an optical sensorthat converts the amount of light received to electric charges and creates a digitized image of a large number of pixels. Unlike traditional displacement sensors, a CCD camera does not need any transducers to be attached to the structure. The common approach to measure displacement with a CCD camera is to obtain and compare images before and after applying loads on the structure. In order to obtain 3D coordinate information, additional computation will be necessary (Park *et al.*, 2007; Evans *et al.*, 2009; Fu and Moosa, 2002).

7. Fiber-Optic Sensors

Fiber-optic sensors are used to monitor various parameters including strain, temperature, pressure, chemical composition, deformations and corrosion (Fraser *et al.*, 2010). These sensors are classified based on the intensity, phase or polarization of the light passing through the sensor (Huston, 2010; Rivera *et al.*, 2007).

Fiber-optic sensors can measure every point over the distance of the entire optical fiber unlike traditional strain gages that measure strain at a localized point (Park *et al.*, 2007). Thus, fiber-optic sensors can acquire a continuous distribution of strain or temperature variations; and as such they are useful in applications with elongated structures such as dams, long-span bridges and pipelines (Haus, 2010). However, fiber-optic sensors are not capable of measuring localized strains or temperatures of a structure (Kim *et al.*, 2003). Despite this, fiber-optic sensors are light weight and resistant to corrosion and possess immunity to electromagnetic fields, insensitivity to external disturbances and ability for remote monitoring (Ou and Duan, 2006; Matta *et al.*, 2008; Kim *et al.*, 2003). Although fiber-optic sensors are more expensive, their cost is justified considering their durability and long life (Rivera *et al.*, 2007).

7.1. *Fiberbragg grating sensors*

Fiber Bragg Grating (FBG) sensors have been developed to measure various parameters including strain, temperature and cracks. With these sensors, a light is emitted into the fiber and it is reflected back if its wavelength corresponds to the Bragg wavelength. The reflected wavelength is affected by changes in both the strain and temperature in the effective Bragg grating spacing (Huston, 2010; Inaudi and Glisic, 2006) Thus, it is imperative that both the strain and temperature be measured simultaneously and therefore, the thermally induced strain in static strain measurements can be accounted for and corrected properly (Evans *et al.*, 2009). FBG sensors do not need electrical wiring, thus, they are light weight and are less vulnerable to effects from electromagnetic radiations (Matta *et al.*, 2008). Two common methods used to manufacture these sensors are: (1) the holographic method; and (2) the phase mask method. The holographic method is considered to be an easier method for creating FBG sensors of different wavelengths. This is because of the simplicity of adjusting the angle between two beams to create different periods. However, this method requires a stable setup and a good light source (Zhou and Jinping, 2004). In general, FBG sensors are fragile and they should not be directly applied to structures without encapsulation (Li *et al.*, 2006; Fraser *et al.*, 2010). Different ways are available to encapsulate FBG sensors (Thursby *et al.*, 2008).

7.2. *Distributed Brillouin sensors*

Distributed Brillouin sensors (DBS) are used for strain and temperature measurements. The typical applications are for fire and crack detection. Brillouin sensors are based on brillouin scattering in which optical and sound waves interact. The scattering occurs when light emitted into the fiber is reflected back by thermally excited waves. The scattering is dependent on the frequency of the acoustic wave as well as the shifts in frequency of the light. These sensors provide a direct method to measure changes in both temperature and strain along the entire length of an optical fiber (Haus, 2010; Ravet *et al.*, 2007).

7.3. *Ramon distributed sensors*

Ramon distributed sensors have been developed to measure temperature. These sensors are based on Ramon scattering, in which there exists a non-linear interaction between the light traveling in a fiber and silica (Haus, 2010). Raman scattering occurs from propagating light pulses emitted into the fiber and reflected back by thermally excited wavesvibrations from the sensors. The vibration is associated with optical photons (Huston, 2010).

8. Ultrasound Waves

A method to detect macroscopic structural defect is to launch an ultrasound wave into the structure and examine the corresponding response. The high-frequency sound waves are affected when they encounter structural anomalies such as holes, corrosion spots, cracks and delimitations. The process of detecting flaws is achieved by comparing the response of the waves when the structure is in a known healthy stage to those corresponding to the case, where it has been in service for several years (Matta *et al.*, 2008; Thursby *et al.*, 2008; Iyer *et al.*, 2005). Ultrasound waves are generally used, as an "on-demand" basis. This is especially true when the failure of an individual part of a structure has been known to cause a severe consequence (Huston, 2010). Ultrasound waves can identify subsurface cracks and fatigue induced cracks in early stages of development as small as 100μm (Huston, 2010; Psimoulis *et al.*, 2008). Typically ultrasound waves are detected using FBGs (Matta *et al.*, 2008).

A disadvantage of ultrasonic detection is that only the distance to the first acoustic interference is measured, and the waves do not penetrate further into the structure (Iyer *et al.*, 2005).

9. Laser Scanning

Laser scanning measures the surface geometry and kinematics of a host structure using techniques such as triangulation, speckle pattern analysis and Doppler frequency shifts. From its collection of data of a structure, a laser scanner constructs a digital model, which is then used for further analysis (Huston, 2010). The advantages of laser scanning include the ability to detect to an entire structure without being restricted to a specific location or by environmental conditions and without adding additional weight to a structure (Park *et al.*, 2007).

9.1. *Terrestrial laser scanning*

Terrestrial laser scanning (TLS) is used to obtain three-dimensional displacement information of a structure as well as its static deformed shape. The laser scanner used for TLS is not airborne and it is located within 350m of the structure. The advantages of TLS are that (1) there is no restriction to reach structures, (2) the system is independent of the natural light, and (3) there is no wiring cost involved. This method, however, can have a maximum error of 10mm, which is insufficient for the purpose of most SHM applications. To overcome this shortcoming, a displacement measurement model has been presented. The 3D coordinate information is then determined by measuring the time for the laser pulse to travel back to its source as well as computing the distance based on the speed of light (Park *et al.*, 2007; Olsen *et al.*, 2010).

9.2. *Laser dopplervibrometer*

The laser dopplervibrometer (LDV) is a non-contact scanning vibration measurement system. This system employs lasers to measure the velocity and displacement of points on a vibrating object.

LDV is based on the doppler effect, where a frequency change between the radiation laser beam and the reflected laser beam is induced. The vibration of each component direction is determined using the laser beam angles. LDVs offer a suitable system for vibration data acquisition, since they do not add additional mass to the structure and can measure high frequency components of small vibrations over long spans (Ou and Duan, 2006).

10. Temperature Sensors

The variation of temperatures in a structure offers useful information about the underlying conditions such as fatigue, microcracking and yielding. Since many sensors used in SHM applications are affected by temperature that may lead to erroneous diagnostic results, temperature must be monitored to overcome this limitation(Huston, 2010; Haus, 2010). Temperature must also be monitored due to the deformation and strain that it induces (Karbhari and Ansari, 2009). However a full thermal characterization of structures may require data collection over a long period of time (Huston, 2010).

10.1. *Thermocouples*

Thermocouples are among the most common types of sensors for temperature measurement. They are inexpensive and offer a reasonably accurate temperature measurement. These devices combine two materials that produce a temperature-dependent voltage difference. The voltage is generated when the junction of the two metals experiences temperature changes. Although the temperature measurement is precise and accurate, the output relationship between temperature and voltage is nonlinear (Karbhari and Ansari, 2009; Huston, 2010).

10.2. *Resistance temperature detector*

A resistance temperature detector (RTD) measures temperature by employing the liner relationship of resistance to temperature. Platinum is preferred for RTDs due to its long-term stability; even though it has a small coefficient of resistance change with temperature. Generally, RTDs have a smaller temperature range, higher initial cost and sensitivity to high vibration environments (Karbhari and Ansari, 2009; Huston, 2010).

10.3. Thermography

Every object naturally emits infrared (IR) radiation proportional to its surface temperature. As the surface temperature increases, the amount of radiation emitted increases. Thermography utilizes this fact by using thermographic cameras to measure the emitted radiation and create a digitalized image of the surface temperature variations. Such images can be used to provide a record of surface temperatures. This no-contact method is typically used to monitor and detect delimitations, state of cure and quality control in concrete. Advantages of thermography include its speed of inspection, compared with the ultrasound system, and application for detection in large surface areas. However, thermography does not indicate the depth of a flaw and is affected by atmospheric radiation (Huston, 2010; Iyer *et al.*, 2005).

11. Load Cells

Various load cells have been developed to measure forces and other loads on structures by converting the force acting on it into a proportional electrical signal. These devices can only measure a single component of force, since load cells are not affected by components in other directions as well as by bending moments. Mounting configurations are most critical to successful use of load cells. Typically single-axis load cells are used with a large hole in its center for the concentric passage of a rod to be attached to the structure (Huston, 2010).

12. Anemoscopes

Wind load is a critical load on long-span bridges and high rise buildings. Anemoscopes have been developed to measure various wind load parameters such as wind velocity, angle, and orientation. These devices work properly under bad weather conditions and have a high temperature range. Though various types of anemoscopes exist, ultrasonic anemoscopes has shown to have the best performance among others (Li *et al.*, 2006; Ou and Duan, 2006).

13. Fatigue Sensors

Fatigue damage is due to a combination of microcracking and mechanical loading. It is not easily detected until significant strength reduction has occurred leading to a sudden and catastrophic failure (Huston, 2010). Although dynamic strain gages can accurately determine the cumulative fatigue damage in a structural component, fatigue sensors have been developed to directly measure the actual amount of cumulative fatigue damage and estimate the structure's remaining life. Such sensors are commonly used in structures under cyclic loading such as passage of vehicles over bridges (Karbhari and Ansari, 2009; Evans *et al.*, 2009).

14. Summary Table for Sensors

The table summarizes the information previously presented and includes various technical parameters typically found in manufacture data sheets such as sensor accuracy, measurement range, data acquisition rate, operating temperature and mean time between failures (MTBF). This information will be useful in choosing the right sensor for a given application, location and degree of accuracy needed for the compiled data. Note that the data acquisition rate can change with the type of hardware or software employed in the data acquisition. Also, it is important to point out that MTBF is a statistical measure of the average time the sensor will function before failing and should not be confused with the life of the sensor.

Acknowledgment

The authors would like to acknowledge the help provided by Ms. Natalie Waksmanski, civil engineering student at Illinois Institute of Technology, for her assistance in preparing this chapter.

	Sensor Name	Application	Accuracy	Measurement Range	Sensing Range	Data Acquisition Rate/Response Time	Operating Temperature	Reliability/ MTBF
	Foil Strain Gage	Static Strain Fatigue Stress	± 50000 μstrain (±0.1%)	Point	0.1 - 40,000 με	< 1 usec	-270°C to +370°C	10 years
	Semiconductor Strain Gage	Static Strain Dynamic Strain	± 1500 μstrain	Point	0.001 - 3,000 με	< 10 usec	-34°C to +66°C	75 years
	Piezoelectric accelerometer	Vibration Seismic loads	±1%	Point	-80000g to 80000g	< 10 usec	-20°C to +135°C	10 years
	MEMS Capacitive Accelerometer	Vibration Chemical Sensing	±0.5% to ±1%	Point	-20000g to 20000g	0.5 ms	-40°C to +185°C	2,000,000 hours
	Laser Doppler Vibrometer	Vibration	< ±0.05%	250 mm - 200m	0 ± 60m/s	3.5s	0°C to +40°C	100,000 hours
	Fiber Bragg Grating Sensors	Strain Temperature Pressure	± 1 μstrain ± 0.2 °C to ± 0.5 °C	up to 25 km	-270°C to +300°C	0.2 - 8.5s	0°C to 50°C	25 years
	Distributed Brillouin Sensor	Strain Corrosion Crack Temperature	± 2 μstrain ± 0.1 °C	up to 50 km	-270°C to +800°C -2% to 3%	>1s	0°C to 40°C	23 years
	Raman Distributed Sensor	Temperature	±1°C to ±2°C	5km -30 km	-180 to +300 °C	>10s	0°C to 40°C	28 years

	Sensor Name	Application	Accuracy	Measurement Range	Sensing Range	Data Acquisition Rate/Response Time	Operating Temperature	Reliability /MTBF
	Thermocouples	Temperature	± 1°C to ± 2°C	Point	-200°C to +2600°C	0.05s - 1.3s	-200°C to +2600°C	4,500 - 400,000 hours
	Resistance Temperature Detectors	Temperature	± 0.15°C to ± 0.3°C	Point	-200°C to +650°C	0.5s - 5s	-200°C to +650°C	20 - 100 years
	IR Thermographic Camera	Surface Temperature *Moisture Content*	± 2°C	10 cm to infinity	-40°C to +2000°C	490,000 points/sec	-20°C to +60°C	100,000 hours
	LDVT	Displacement	±0.25%	Point	0 to ±102mm	0.4ms	-50°C to +150°C	1,000,000 hrs
	GPS	Static Displacement Dynamic Displacement	±5mm horizontally ±10mm vertically (1.3 - 3.6%)	Infinite	0- 10 mm horizontally 0- 20 mm vertically	100 samples/sec	-55°C to +65°C	60,000 hours
	CCD Cameras	Displacement	± 2 μm	30 mm to infinity	0 -150mm	2.5 frames/sec to 10 frames/sec	-20°C to +70°C	80,000 hours - 600,000 hours
	Terrestrial Laser Scanning	Surface Geometry Displacement	±10mm	2m -1000m	2m -1000m	up to 50,000 points/sec	0°C to 40°C	70,0000 hours
	Load Cells	Force	±0.05% to ±1%	Point	0 – 90,700 kg	1ms	-60°C to 200°C	50 years
	Ultrasonic Anemoscopes	Wind Load	2% to 3%	Point	0- 75m/s	0.25s - 3s	- 50°C to +700°C	26 years

References

Bowling, S. and R. Richey. Two Approaches to Measuring Acceleration. *Sensors Magazine 2000.*

Evans, J.E., J.M. Dulieu-Barton and R.L. Burgueta. Modern Stress and Strain Analysis: A state of the art guide to measurement techniques. *Eureka Magazine*, 2009.

Feng, M. Q. Application of Structural Health Monitoring in Civil Infrastructure. *Smart Structures and Systems* 2009; **5.4**: 469-82.

Fraser, M., A. Elgamal, X. He, and J.P. Conte. Sensor Network for Structural Health Monitoring of a Highway Bridge.*Journal of Computing in Civil Engineering* 2010: 11-23.

Gongkang, F. and A. G. Moosa. An Optical Approach to Structural Displacement Measurement and Its Application. *Journal of Engineering Mechanics* 2002; **128.5**: 511-19.

H.S. Park, H.M Lee, H. Adeli, and I. Lee. A New Approach for Health Monitoring of Structures: Terrestrial Laser Scanning. *Computer-Aided Civil and Infrastrure Engineering* 2007; **22**: 19-30.

Haus, Jörg. *Optical Sensors Basics and Applications.*Weinheim: Wiley-VCH-Verl., 2010.

Huston, D. *Structural Sensing, Health Monitoring, and Performance Evaluation.* Boca Raton: CRC, 2010.

Inaudi, D. and B. Glisic.Distributed Fiber optic Strain and Temperature Sensing for Structural Health Monitoring.*The Third International Conference on Bridge Maintenance, Safety and Management,* Portugal: 2006.

Iyer, S. R., S. K. Sinha, and A. J. Schokker.Ultrasonic C-Scan Imaging of Post-Tensioned Concrete Bridge Structures for Detection of Corrosion and Voids.*Computer-Aided Civil and Infrastructure Engineering* 2005; **20.2**: 79-94.

Karbhari, V. M., and F. Ansari.*Structural Health Monitoring of Civil Infrastructure Systems.* Cambridge: Woodhead, 2009.

Kijewski-Correa, T. GPS: A New Tool for Structural Displacement Measurements. *Association for Preservation Technology International* 2005;**36.1**: 13-18.

Kim, S.H., J. J. Lee, D.C. Seo, and J.O. Lim. Application of Point and Distributed Optical Fiber Sensors to Health Monitoring of Smart Structures. *International Journal of Modern Physics* 2003; **17.8**: 1368-373.

Li, H., J. Ou, X. Zhao, W. Zhou, H. Li, Z. Zhou, and Y. Yang. Structural Health Monitoring System for the Shandong Binzhou Yellow River Highway Bridge.*Computer-Aided Civil and Infrastructure Engineering* 2006;**21.4**: 306-17.

Matta, F., F. Bastianini, N. Galati, P. Casadei, and A. Nanni.Distributed Strain Measurement in Steel Bridge with Fiber Optic Sensors: Validation through Diagnostic Load Test. *Journal of Performance of Constructed Facilities* 2008; **22.4**: 264-72.

Olsen, M. J., F. Kuester, B. J. Chang, and T. C. Hutchinson.Terrestrial Laser Scanning-Based Structural Damage Assessment.*Journal of Computing in Civil Engineering* 2010; **264**: 264-72.

Ou, Li and Duan.*Structural Health Monitoring* of *Intelligent Infrastructure. London: Taylor & Francis Group plc, 2006.*

Park, S., C.B. Yun and D.J. Inman. Structural Health Monitoring Using Electro-mechanical Impendence Sensors. *Fatigue & Fracture of Engineering Materials & Structures* 2008; **31**: 714-724.

Psimoulis, P., S. Pytharouli, D. Karambalis, and S. Stiros.Potential of Global Positioning System (GPS) to Measure Frequencies of Oscillations of Engineering Structures.*Journal of Sound and Vibration* 2008; **318.3**: 606-623.

Ravet, F. L. Zou, X. Bao, T. Ozbakkaloglu, M. Saatcioglu and J. Zhou.Distributed Brillouin Sensor for Structural Health Monitoring.*Canadian Journal of Civil Engineering* 2007; **34.3**: 291-297.

Rivera, E., A.A. Mufti, and D.J. Thomson.Civonics for Structural Health Monitoring.*Canadian Journal of Civil Engineering* 2007; **34**: 430-37.

Thursby, G., B. Culshaw, and D. C. Betz.Multifunctional Fibre Optic Sensors Monitoring Strain and Ultrasound.*Fatigue & Fracture of Engineering Materials & Structures* 2008; **31.8**: 660-73.

Yenilmez, A., A. Yapici, C. Velez, and I.N. Tansel.Development of Piezoelectric Strain Gages for Structural Health Monitoring Applications. *Review of Quantitative Nondestructive Evaluation* 2007; **26**: 902-09.

Zhou, Zhi, and Jinping OU.Development of FBG Sensors for Structural Health Monitoring. In: Farhad Ansari ed. *Civil Infrastructure in Sensing Issues in Civil Structural Health Monitoring*. Netherlands: Springer, 2004: 197-206.

Chapter 12

Sensor Data Wireless Communication, Sensor Power Needs, and Energy Harvesting

Erdal Oruklu[1], Jafar Saniie[2], Mehdi Modares[3], and Jamshid Mohammadi[3]

[1,2]*Department of Electrical and Computer Engineering*
[3]*Department of Civil, Architectural, and Environmental Engineering*
Illinois Institute of Technology
Chicago, Illinois 60616, USA
E-mails: [1]erdal@ece.iit.edu, [2]sansonic@ece.iit.edu, [3]mmodares@iit.edu, [4]mohammadi@iit.edu

1. Introduction

Operational safety of infrastructures such as bridges, interstate highways and power grids is a significant issue with immediate public safety ramifications in addition to economic losses and road network disruption concerns. Currently, keeping roads and bridges in a safe operating condition is a major financial burden on state departments of transportation as well as many local agencies. State and local agencies have to rely on limited data (primarily from subjective ratings provided by inspectors) to prioritize structures for repair and retrofitting. In many instances, such information is not conclusive and may result in an oversight on the part of authorities. There have been instances when bridges have received favorable ratings, while in reality there were hidden problems that were missed by visual inspections. A significant example of this scenario is the I-35W Minnesota Bridge that failed due to design anomalies in its gusset plates (Holt and Hartmann, 2008). There also have been cases of corrosion of pre-stressing tendons in box-girder bridges leading to severe damage (e.g., Mid-Bay Bridge near Destin, Florida) (Hartt and Venugopalan, 2002)—conditions that may often be missed using visual inspections. It is evident that the current condition of roads and bridges imposes a critical national issue. The American Society of Civil Engineers (ASCE) gives a score of "D" in its 2009 report card of infrastructure conditions in the USA, and specifically lists bridges among critical infrastructure systems (ASCE 2009). According to this report, The five-year spending estimate for infrastructure needs reaches $2.2 trillion.

Wireless sensor networks have been proposed extensively over the past several years as a means of alleviating instrumentation costs associated with structural health monitoring of civil infrastructure. However, low data throughput, unacceptable packet yield rates, and limited system resources have generally plagued many deployments by limiting the number of sensors and their sampling rate. The sensor networks present challenges in three broad areas: energy consumption, network configuration and interaction with the physical world. Therefore, the development of sensor networks requires technologies from three different research areas: sensing, communication, and computing (including hardware, software, and algorithms). The next generation of structural health monitoring sensors need to be low-cost, low-power, self-healing, self-organized, and compact.

The key to achieving these objectives is the seamless integration of sensor clusters, processing engines for on-site signal processing, operational control, and wireless mesh network communications. Figure 1 shows the main components of a distributed sensor system. In this system all hardware (sensors, network, processing core) and software (signal processing, operational control and power management) components are designed to be application specific, thereby eliminating the ad-hoc approach prevalent today for structural sensing and monitoring.

The organization of the chapter is as follows: Section 2 presents the continuous, passive sensor operations using acoustic emission. An advanced sensing and signal processing methodology is described for automatic defect monitoring related to generation and propagation of cracks using smart processing nodes. Section 3 highlights the need for an improved network layer design for wireless health monitoring applications concerning the network lifetime and stability, reliability, self-formation, and distributed processing. Section 4 discusses the smart sensor system-on-chip (SoC) design based on reconfigurable hardware and shows the necessity of smart sensors for executing more advanced algorithms, reducing the frequency of RF transmissions and achieving better power management. Section 5 describes the power consumption for key components and overall energy requirements in the smart wireless sensor network. A redundant energy harvesting and power management system is shown for achieving autonomous and sustainable structural health monitoring operations. Section 6 provides references for further information in this topic and Section 7 concludes the chapter.

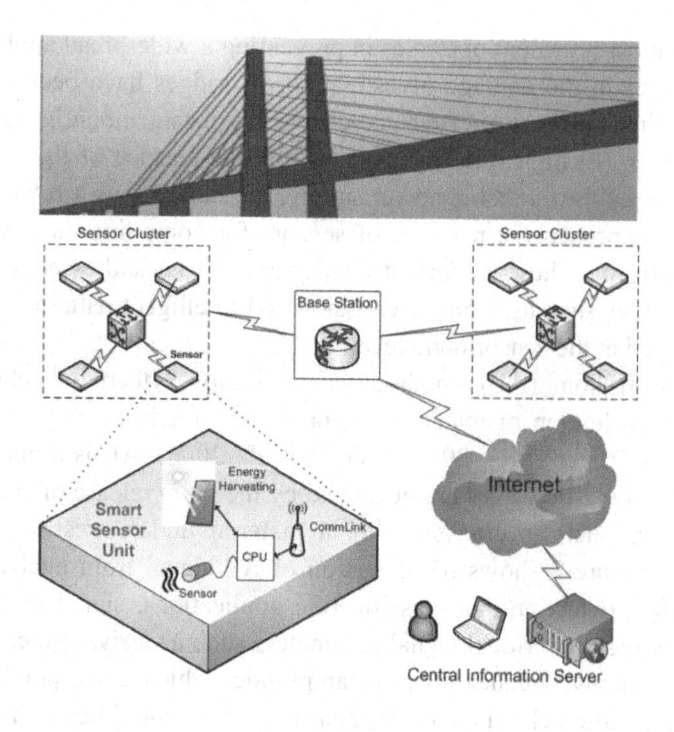

Fig. 1. Sensor networks for structural health monitoring.

2. Structural Health Monitoring Using Smart Acoustic Emission Sensors

The current methods of structural monitoring, for example in fatigue investigation of bridges, have in most part relied on data gathered using strain gages installed at critical locations (Mohammadi *et al.*, 2004) and a host of other types of sensors, such as accelerometers, in other applications. Specific to fatigue application, the data is then used to develop stress histories for use along with fatigue characteristics of critical components to (a) estimate the extent of damage and (b) predict the remaining life. Recent advances in sensor technologies and on-site data analysis have been instrumental in expediting the process and providing for a quick access to data (Howell and Shenton, 2006)) and an opportunity to expand the data collection process for many structural and environmental parameters. Although there have been many attempts to incorporate a diverse set of damage analyses within a comprehensive bridge management system (Messervey *et al.*, 2006) or a global bridge health monitoring system (Nassif *et al.*, 2006; Sumitro and Hodge, 2006), the underlying method still relies on using field data from sensors installed at critical locations. The most prevalent obstacles in preventing a widespread application of health monitoring in the damage investigation of bridges have been (a) sensor installations being cumbersome and expensive (requiring mounting sensors at critical locations), (b) high cost associated with maintenance of the monitoring system and its need for a reliable power supply, (c) lack of sensor reliability, and (d) lack of a comprehensive network of sensors for compiling data on a multi component platform. These important challenges can be addressed with smart sensor technologies such as acoustic emission and intelligent pattern recognition methods executed at the sensor node level.

Acoustic Emission (AE) method is a highly effective technique in nondestructive evaluation of materials, in particular, for inspecting steel bridge superstructures (Kruger *et al.*, 2007; Wilcox et al., 2006). AE is a phenomenon whereby transient elastic waves are generated by the rapid release of energy from localized sources such as cracks within a material under stress (Grosse and Ohtsu, 2008). Figure 2 shows the detection of AE signal from nucleation of a crack. For defect detection and classification applications, signal processing is necessary to extract the critical signal parameters such as arrival time, rise time, duration time, energy, frequency, peak amplitude, which corresponds to size, type and location of cracks (Ince *et al.*, 2009). For structural health monitoring, multiple array of transducers can be mounted at several points on the structure, with the aim of detecting the presence, location, and intensity of acoustic signals generated by cracks and fractures (see Figure 3). AE technique for structural health monitoring offers several advantages:

- AE provides passive and global monitoring of defects for nondestructive testing applications.
- AE signal parameters (arrival time, rise time, duration time, energy, frequency, peak amplitude) provide critical information about defects inside the material.
- With the recent advances in embedded SoC systems, on-going and unattended monitoring of structures using acoustic emission technique is feasible.
- Real-time monitoring of AE signal is highly practical, cost-effective and consumes minimal power using the proposed SoC hardware.
- AE sensors passively detect emissions from acoustic sources, unlike pulse-echo ultrasonic testing methods where ultrasonic waves are generated actively.
- AE signal reveals information about the defects, and the severity of the load and the strain impacting the structure.
- AE contains frequency signatures (ranging from kHz to MHz) that can be correlated to the characteristics of structural defects.

In order to achieve a comprehensive evaluation of source, location, size, severity, and type of structural defects; advanced time-frequency signal processing methods (such as wavelet transform (Oruklu and Saniie, 2004), split-spectrum processing (Oruklu and Saniie, 2009), Hilbert-Huang transform (Oruklu *et al.*, 2009), chirplet signal decomposition (Lu *et al.*, 2006), and neural networks (Yoon *et al.*, 2007) can be integrated in smart AE sensing systems. These methods require a significant computing power and often have been dismissed due to their complexity. Nevertheless, a smart SoC based sensor platform is capable of handling the computational demand of these techniques for improved defect detection. Therefore, AE sensor arrays can be used as the primary tool for on-going sensing operations in detecting structural defects, such as microcracks in structures.

2.1. *AE sensing methodology*

The interpretation of the AE signals based on parameters such as duration, peak amplitude and energy, event number, ring-down count and time-of-arrival enables waveform analysis directly related with the geometrical shape, size and frequency of the acoustic discharge source. Hence, it provides continuous, automatic defect monitoring related to generation and propagation of damage. Nevertheless, it is challenging to differentiate the signals associated with damage growth under stress from other noise sources. Another critical task is to identify and precisely locate the source of a structural defect, even though it may be in its early stages of development.

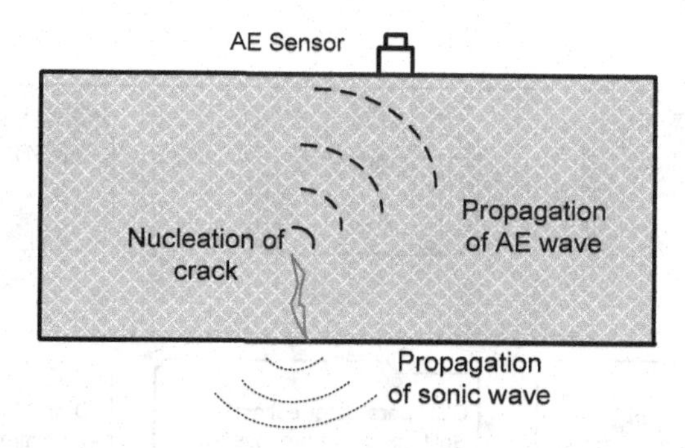

Fig. 2. Detection of AE signal from nucleation of crack.

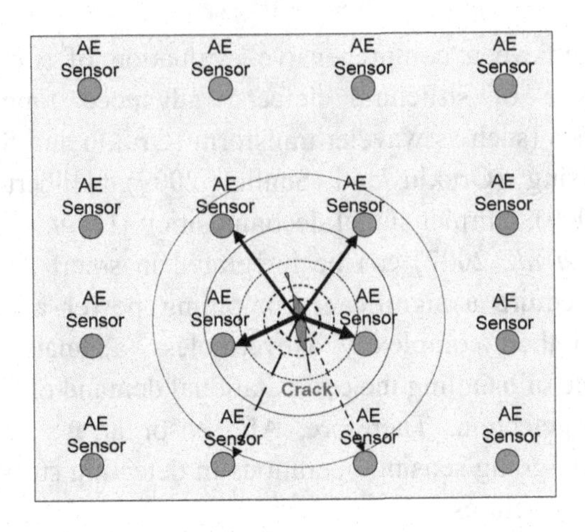

Fig. 3. Defect localization with AE sensor arrays.

In order to address these challenges specifically, a multi-stage signal processing methodology can be applied for analyzing acoustic emission signals in structural health monitoring applications (Grosse *et al.*, 2006). Figure 4 shows the algorithm stages and the host system hardware used in distributed processing nodes.

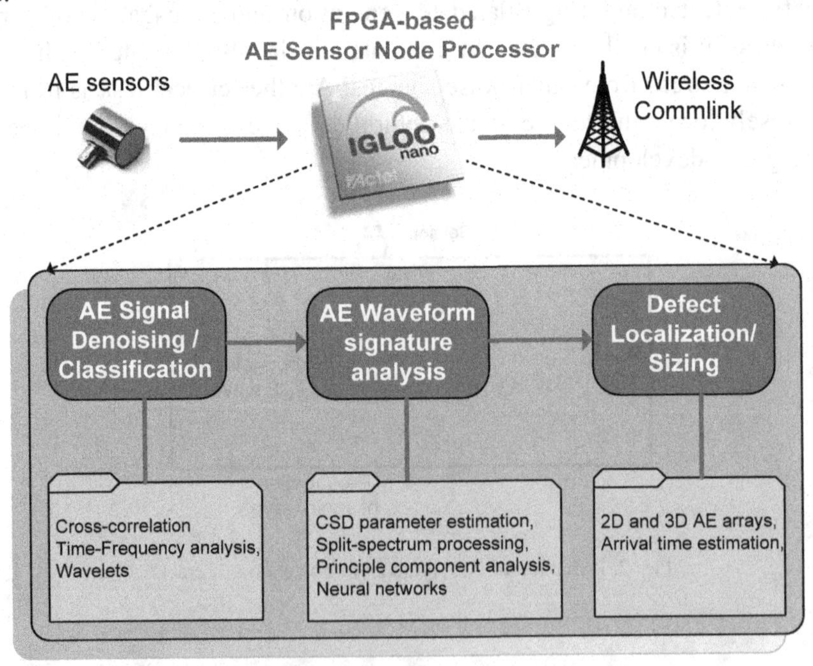

Fig. 4. Acoustic emission signal processing methodology for smart sensors.

The algorithm stages are explained below:

AE signal denoising and classification: In addition to external sources and environmental effects, acoustic emission signals are also degraded by the experimentation setup, electronic and other sensor related noise. Smart sensor nodes make it possible to use advanced algorithms (Grosse and Reinhardt, 2002) for real-time denoising and signal classification of the AE signal.

AE waveform signature analysis: After signal denoising, the next step for analyzing any specimen under test is to isolate the acoustic emission signal induced by structural deformation from environmental interfering signals. This could be a very challenging task due to different operational environments such as railroad bridges where vibration during train crossings could be overpowering other signal sources. Before any analysis can be done, acoustic emission signature need to be identified from the incoming signal. The implementation of this signature analysis (i.e., pattern recognition task) (Ziola and Gorman, 1991) requires a significant computation power due to correlation operations and necessary storage of signatures. Smart processing nodes based on FPGAs can handle these computations unlike most Mote-based designs that utilize simple microcontrollers.

Structural defect localization and sizing: A major advantage of using distributed AE sensors (transducers) is the capability to find and pinpoint the exact location of the anomaly within the structure (Ince *et al.*, 2009, Gross *et al.*, 1993). Furthermore, the size and the geometrical shape of the defect can be recognized. For enhanced defect localization and sizing, multiple AE sensors can be used in a planar area requiring synchronization and time-of-arrival signal analysis among sensors. This necessitates communication among the sensor nodes and situational awareness (i.e., distance/location of neighboring AE sensors, network topology). Using positional data from AE sensors significantly increases the accuracy and performance of defect detection and characterization (Grosse and Ohtsu, 2008). With the reconfigurable hardware coupled to each AE sensor array, smart arbitration and estimation of AE events can be implemented.

3. Wireless Sensor Networks for Structural Monitoring

Recently, wireless embedded sensor networks have emerged that can be characterized by local processing capabilities that minimize the amount of data transmitted in a single- or multi-hop strategy to extend the lifetime and robustness of the network. The multi-hop Wisden system (Xu *et al.*, 2004), which uses the small mica motes developed at the University of California at Berkeley (Horton et al., 2002), provides an example. In this system, by avoiding the transmission of lengthy time histories, the battery life of the wireless nodes can be extended, while the issues of strict time synchronization and loss intolerance are marginalized. The BriMon system (Chebrolu *et al*., 2008) provides an easy to deploy, long term and low maintenance system using battery-operated wireless sensor motes as an alternative for communication and data logging needs. While such developments in wireless sensor networks have demonstrated their potential to provide continuous structural response data to quantitatively assess structural health, many important issues including network lifetime and stability, reliability, time-synchronization, distributed processing and overall effectiveness when using low-cost sensors must be realistically addressed.

Owing to the specific requirements of the structural health monitoring systems, many additional challenges need to be solved at the network level itself, apart from the coordination needed with the application layer. Specific problems include:

- Self-formation: The network topology should be self-adjustable, i.e. addition of new sensor devices should be handled automatically in the network without manual intervention. Similarly, the sensor devices may drop out of the network if enough energy is not harvested. In this case, the rest of the network should be able to adjust and find alternate routes for transmitting the information and coordinating the sensing activities.
- Time Synchronization: The distributed sensors collect information for transmission to the local base station. The various sensors should be time synchronized such that the events causing the observation can be correlated and uniquely identified.
- Transceiver frequency: Selection of a suitable RF band for low-power, non-interfering, high-throughput operation is needed.
- Prioritization: The communication between the sensor nodes and the local base station could be synchronous (periodic update messages) or asynchronous (as a result of an anomalous event, which could create a trigger for the monitoring system). The messages sent asynchronously should be allocated higher priority in the network, because of their alert-like nature.
- Hierarchy: The sensor network needs to be organized in an adaptive hierarchy based on the application requirements. The hierarchy among the sensor nodes can ease the routing as well as provide the capability to make distributed decisions. Distributed decisions minimize the transmission of unnecessary raw data to the local base station.
- Information Storage and Retrieval: In the case of communication failure (with the remote central information server) due to inadequate power for

communication or interruption in communication link, the sensor network needs to be designed with limited storage capability. This storage capacity can be optimized with respect to the type and number of sensors.

- Protocol Design: Standard communication protocols needs to be customized in order to: (1) Minimize the communication overhead, and (2) Make the sensing system reliable and robust. This customization will provide application oriented network features unique to the continuous monitoring system.

The next generation of the wireless networks for sensor applications need to be designed accordingly to resolve these important challenges.

4. System-on-Chip Design for Smart Sensor Nodes

In order to provide decision-making capabilities and perform complex signal processing at the sensor level, dedicated reconfigurable System-on-Chip (SoC) devices are closely coupled with micro-electro-mechanical sensors (MEMS). This combined architecture forms the basis of smart sensor nodes.

Reconfigurable devices facilitate fast development time and adaptable architectures for signal processing applications in many domains, including ultrasonic testing and measurements (Rodriguez-Andina *et al.*, 2007). Until recently, FPGAs were not seriously considered for battery-powered sensor applications due to their relatively higher cost and power consumption. Today, various FPGA technologies have significantly different power profiles, and these differences can have a profound impact on the overall system design and power budget. Flash-based FPGAs such as Actel IGLOO (ACTEL, 2009) offer ultra-low power consumption with a selection of power management modes to drastically reduce power requirements while providing programmability and high computation power in small form-factor packages at a low cost. Due to the unparalleled adaptability and scalability of FPGAs they can be re-programmed and improved continuously and their designs can be modified, with no extra overhead cost.

Smart sensor nodes can be implemented using low-power FPGA devices as shown in Fig. 5. In order to meet all the design metrics optimizations are required at both algorithm level and architectural level. To address this issue, a hardware/software co-design scheme is necessary where an embedded processor core is used for pre-processing and synchronizing the streaming input data from multiple sensors. A point-to-point channel bus is used to perform fast communication between external hardware accelerator blocks and the processor(s) on the FPGA. These accelerator blocks implement the required datapath functions found in signal processing algorithms (i.e., filtering, wavelet transforming, and neural networks as described in AE sensing methodology) via specialized Processing Elements (PEs).

Recently, FPGA-based ultrasonic signal processing hardware have been successfully used in real-time flaw detection (Weber et al., 2008), ultrasonic data compression (Oruklu *et al.*, 2007) and parameter estimation applications (Lu *et al.*, 2008) demonstrating its versatility. In addition, power and area efficient implementations based on recursive filter structures for subband decomposition have been proposed (Oruklu *et al.*, 2008). These implementations are especially suitable for ultra-low power smart sensors used in structural monitoring applications.

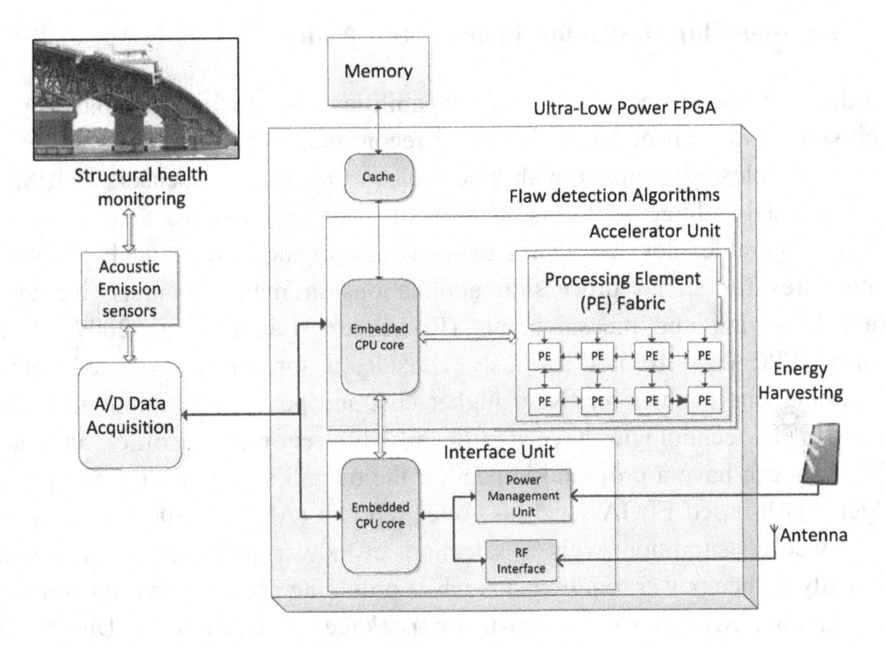

Fig. 5. Reconfigurable SoC for sensor nodes in health monitoring applications.

5. Sustainable Operation of the Wireless Sensor Network

Sensors, data acquisition systems, communication and processing units require sustainable power for truly autonomous operation. Sustainable operation of an intelligent sensor network platform is determined by the interrelation of three metrics: (1) Peak energy consumption of the sensor node components, (2) energy harvesting/generation capability, and (3) rechargeable battery capacity. If peak sensor energy consumption can eventually drain the battery, the system is deemed not sustainable. Therefore, designing intelligent hardware and software protocols is necessary for achieving energy and service equilibrium to enable the on-going sensing operations. In the following subsections, design decisions for achieving sustainability are highlighted with respect to these metrics.

5.1. *Power consumption in structural health monitoring applications*

There are three major components that consume power within the smart sensor nodes: (a) RF communication chip (b) smart processing core, and (c) sensor analog front-end. Smart sensors not only augment the capabilities for signal processing but also reduce data communication. This is possible since smart sensors only need to communicate when there is an anomaly or if an interrogation request arrives from the central server. In conventional wireless sensor networks applied to structural monitoring, the sensor nodes are programmed to transmit data periodically for data aggregation, increasing the communication needs significantly. The frequency and size of the transmission is extremely important since most of the energy consumption in wireless sensor networks comes from the RF front-end. Hence, although advanced processors used in smart sensor nodes may bring additional power requirements, the benefits of reducing RF transmissions outweigh this increase significantly.

Several standards exist for RF communications such as IEEE 802.11 (WLAN), Bluetooth and IEEE 802.15.4 (ZigBee). Among these standards, ZigBee has become a popular choice in WSN applications due to its low power requirements and adequate data rates (up to 250Kbps). For example, a common chip for ZigBee is Texas Instrument's CC2420 2.4 GHz RF transceiver (Texas Instruments, CC2420). It has been widely used in other smart sensors such as MicaZ and Mica2 (Lynch and Loh, 2006). The CC2420 transmission power output ranges from -25 dBm to 0 dBm; while the corresponding consumption ranges from 8.5mA to 17.4mA. In receiver mode, the typical current consumption is 19.7 mA. RF chip transmission power output is an important choice for the overall operation of the WSN. Transmission power determines (i) the communication distance, (ii) current consumption and (iii) battery output capacity. Therefore, sensor deployment (i.e. the proximity of the neighboring nodes) for infrastructure structural system should be done carefully by analyzing the power requirements and sensor node distances. Studies in structural health monitoring applications show that power level -10dBm is sufficient for a 20m transmission distance to other nodes while consuming 11.2mA (Linderman *et al.*, 2010) in most bridge structural health monitoring applications.

RF communication consumes significantly more power than sensor front-end or microcontroller units such as Texas Instruments is MSP430 which consumes only 330μA while running at 1 MHz. However, distributed computing strategies (associated with smart sensors) for health monitoring requires complex computation and processing (Gao, 2005). Several new sensor technologies follow this trend such as iMote2 (Kling *et al.*, 2005) which is based on Intel XScale processor running at up to 100 MHz. On the other hand, FPGAs provide reconfigurable logic, tremendous flexibility and dedicated data-path logic for custom data processing. This enables previously unattainable computation and control to be realized at the sensor node. New FPGA technologies target low-power sensor applications with ultra-low power standby and active mode selections. For example, an ACTEL IGLOO FPGA (which contains an ARM Cortex-M1 processor and 250,000 gates) uses a quiescent current of only 24μA (Actel, 2009).

The third major component of the sensor node is the acoustic emission sensor. Acoustic emission sensors are passive devices; they do not need power. However, additional circuitry is necessary to amplify, filter and convert AE signals to digital. All of these operations can be handled by a single analog front-end chip (Texas Instruments, AFE-5801) which has maximum 50mW power consumption at 30 MSPS and supports full power-down and standby modes.

5.2. *Energy harvesting*

Among all energy harvesting techniques, solar energy is the most convenient and suitable for structural health monitoring applications. Photovoltaic cells provide the highest power concentration (100 mW/cm2) (Roundy *et al.*, 2003). Structural systems such as bridges can utilize the ambient solar energy by coupling solar panels with sensor nodes. Another energy source, although limited in power generation, is piezoelectric material in which mechanical strains across a material layer generate a surface charge. Several companies such as Microstrain, Inc. produce piezoelectric energy harvesters (PVEH, 2011). These harvesters can produce up to 30mW at 3.2 VDC with 1.5 *g* input vibrations.

Solar and piezoelectric energy harvesters can be simultaneously deployed for redundant and fault-tolerant monitoring operation: The sensors located under direct sunlight are equipped with two sustainable sources: solar and piezoelectric. The system can utilize an intelligent controller to switch between available sources (solar or vibration). For instance, during the daytime—when sun irradiation is plentiful—solar provides the main power not only to power the smart sensors but also to charge up the backup battery. During the night or cloudy situations, the backup battery and the piezoelectric harvesters can act as the power source. A multiple-input power electronic converter needs to be implemented to add the energy of both these sources for energy diversification and increased reliability. The combination of various energy sources provides adequate power to energize the sensors and supply the power required for the SoC computation and RF communication.

Other energy harvesting systems, such as mini generators using bridge vibrations as the power source, or micro wind turbines, have also been considered for health monitoring applications, however to a much lesser extent than solar power systems.

5.3. *Power management*

For sustainability, not only is energy harvesting is critical, but efficient power management is also necessary. Power management and maintenance of a reliable operation is ensured by:

- Removing the load from energy harvesting source (putting the SoC system into sleep mode) if power reserves are less than a threshold.
- Minimizing the operations requiring high power consumption such as frequent and/or redundant raw data transmissions.
- Reserving minimum emergency power for critical instances of sensing such as when the traffic load is heavy and/or unexpected severe environmental changes occur.
- Utilizing ultra-low power components with minimal standby current.
- Transmitting only when there is an anomaly or a major change in the structural health to be reported.
- By integrating smart SoC processors, sensor nodes are capable of self-monitoring their power generation and power consumption continuously.

For continuous use, a power reserve must be provided in the form of battery to avoid power shortages. Using rechargeable energy storage such as high capacity (>5000mAh) Li-Ion/Polymer batteries, the power harvested by photovoltaic cells or piezoelectric energy sources can extend over a long period of time.

A sustainable sensor node system for structural health monitoring applications is shown in Fig. 6. Here, a power manager and ultra-low voltage step-up converter chip (Linear, LTC3108) is used for harvesting current from the photovoltaic cell or piezoelectric sensors and providing power to smart sensor node and wireless transmitter chips. In addition, the harvested current can be diverted to recharging the battery during stand-by mode, in order to power the system when the energy harvesting source is insufficient. For charging Li-Ion/Polymer batteries, a battery charger system (Linear, LTC4070) is used. This charger is optimized for intermittent or continuous charging sources, making it ideal for energy harvesting.

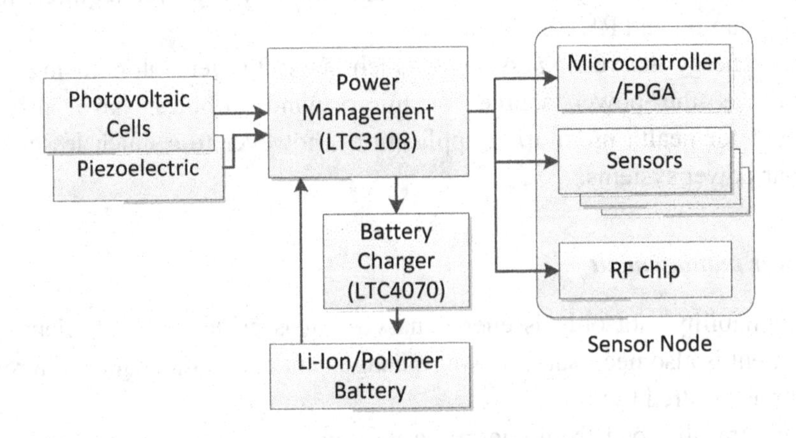

Fig. 6. Sustainable smart sensor nodes

6. Further Information

More information in this topic can be found in the journal publications: Structural Health Monitoring, An International Journal, NDT&E International, IEEE Sensors Journal, IEEE Transactions on Mobile Computing, IEEE Transactions on Ultrasonics, Ferroelectrics and Frequency Control, Journal of the Acoustical Society of America, and in the proceedings of IEEE Sensors: IEEE Conference on Sensor, Mesh and Ad Hoc Communications and Networks (SECON), International Conference on Information Processing in Sensor Networks (IPSN), and IEEE Ultrasonics Symposium.

7. Concluding Remarks

A major challenge facing our society is the rapidly aging infrastructure. The funds needed to reconstruct infrastructure systems and to maintain their service at a level that is considered satisfactory are prohibitive. Therefore, it is imperative that any planning for reconstruction should specifically look into a prioritization scheme. The availability of a versatile and smart monitoring system with the ability to provide information on structural health conditions on a routine basis, will substantially enhance the capabilities of various agencies when they plan for prioritizing their infrastructure systems for maintenance.

Many structural health monitoring systems that are available today are only applicable to specific structures and lack the versatility needed to cover a whole host of distress conditions. To address these shortcomings, implementation of structural health monitoring systems should include:

- Design and realization of application-oriented network for wireless communication.
- Design and realization of smart computing engines in the sensor nodes for on-going real-time monitoring.
- Design and realization of power harvesting and power usage optimization for self-sustainable operation.
- Design and synthesis of advanced signal processing algorithms for defect detection and characterization.
- Data archiving and analysis for damage assessment and maintenance scheduling.

References

ACTEL, (2009). Low-Power Flash FPGAs Handbook Igloo Handbook, *http://www.actel.com/documents/IGLOO_HB.pdf.*

ASCE, American Society of Civil Engineers, (2009). *Report Card for America's Infrastructure.* Available at *www.asce.org/reportcard*.

Chebrolu, K., Raman, B., Mishra, N., Valiveti, P. K. and Kumar R. (2008). BriMon: A sensor network system for railway bridge monitoring, *6th Annual International Conference on Mobile Systems, Applications and Services, MobiSys.*

Gao, Y. (2005). *Structural health monitoring strategies for smart sensor networks,* Doctoral dissertation, University of Illinois at Urbana-Champaign.

Gross, S. P., Fineberg, J. M., McCormick, W. D. and Swinney, H. (1993). Acoustic Emissions from Rapidly Moving Cracks, *Physical Review Letters*, vol.71, no. 19, pp. 3162- 3165.

Grosse, C. U. and Reinhardt, H. (2002). Signal conditioning in acoustic emission analysis using wavelets, *NDT.net*, vol. 7, no. 9.

Grosse, C. U., Kruger, M. and Glaser, S. D. (2006), Wireless Acoustic Emission Sensor Networks for Structural Health Monitoring in Civil Engineering, *Proceedings of ECNDT.*

Grosse, C. U. and Ohtsu, M. (2008). *Acoustic Emission Testing*, (Springer-Verlag), Berlin, Heidelberg.

Hartt, W. H. and Venugopalan, S. (2002). *Corrosion Evaluation of Post Tensioned Tendons on the Mid Bay Bridge in Destin, Florida.* (Department of Transportation Research Center Report).

Holt, R. and Hartmann, J. (2008). *Adequacy of the U10 & L11 Gusset Plate Designs for the Minnesota Bridge No. 9340 (I-35W over the Mississippi River)*, Turner-Fairbank Highway Research Center Report, (Federal Highway Administration).

Horton, M., Culler, D., Pister, K. S. J., Hill, J., Szewczyk, R. and Woo, A. (2002). MICA: The Commercialization of Microsensor Motes, *Sensor*.

Howell, D. A., and Shenton, H. (2006), System for in-service strain monitoring of ordinary bridges, *Journal of Bridge Engineering, ASCE*, 6, pp. 673-680.

Ince, N. F., Kao, C., Kaveh, M., Tewfik, A. and Labuz, J. F. (2009), Averaged Acoustic Emission Events for Accurate Damage Localization, *IEEE International Conference on Acoustics, Speech and Signal Processing, ICASSP*, pp. 2201-2204.

Kling, R., Adler, R., Huang, J., Hummel, V. and Nachman, L. (2005). Intel mote-based sensor networks, *Structural Control and Health Monitoring, 12, pp.* 469-479.

Krüger, M., Grosse, C. U., and Kurz, J. (2007), Sustainable Bridges: Report on Wireless Sensor Networks using MEMS for Acoustic Emission Analysis including other Monitoring Tasks *Report D5.5, Technical report, European Commission within the Sixth Framework Programme*.

Linderman, L. E., Rice, J. A., Barot, S., Spencer, B. F. and Bernhard, J. T. (2010). Characterization of Wireless Smart Sensor Performance, *NSEL Report Series, Report No. NSEL-021*.

LINEAR Technology, Ultralow Voltage and Power Manager, Data Sheet, *http://www.linear.com/product/LTC3108*

LINEAR Technology, Li-Ion/Polymer Battery Charger System, Data Sheet, *http://www.linear.com/product/LTC4070*

Lu, Y., Demirli, R., Cardoso, G. and Saniie, J. (2006). A successive parameter estimation algorithm for chirplet signal decomposition, *IEEE Transactions on Ultrasonics, Ferroelectrics and Frequency Control Symposium*, vol.53, pp. 2121-2131.

Lu, Y., Oruklu, E. and Saniie, J. (2008). Fast Chirplet Transform with FPGA-Based Implementation, *IEEE Signal Processing Letters*, vol.15, pp.577-580.

Lynch, J. P. and Loh, K. (2006). A summary review of wireless sensors and sensor networks for structural health monitoring", *Shock and Vibration Digest*.

Messervey, T., Frangopol, D., and Estes, A. (2006). Reliability-based life-cycle bridge management using structural health monitoring, *Proc., 3rd International Conference on Bridge Maintenance, Safety and Management*, Taylor and Francis/Balkema the Netherlands, pp. 545-546.

Mohammadi, J., Guralnick, S. A., and Polepeddi, R (2004). Use of stress range data in fatigue reliability assessment of highway bridges," in *NDT Methods in Fatigue Reliability Assessment of Structures* (Ed. J. Mohammadi), American Society of Civil Engineers, Reston, Virginia, pp. 56-71.

Nassif, H., Davis, J. C., and Suksawang, N. (2006). Fatigue performance of steel girder bridges based on data from structural monitoring, *Proc.,, 3rd International Conference on Bridge Maintenance, Safety and Management - Bridge Maintenance, Safety, and Management*, Taylor and Francis/Balkema Publishers, the Netherlands, pp. 719-720.

Oruklu, E. and Saniie, J. (2004). Ultrasonic flaw detection using discrete wavelet transform for NDE applications, *IEEE Ultrasonics Symposium*, vol. 2, pp. 1054-1057.

Oruklu, E., Maharishi S. and Saniie, J. (2007), Analysis of Ultrasonic 3-D Image Compression Using Non-Uniform, Separable Wavelet Transforms, *IEEE Ultrasonics Symposium*, pp. 154–157.

Oruklu, E., Weber, J. and Saniie, J. (2008). Recursive filters for subband decomposition algorithms in ultrasonic detection applications, *IEEE Ultrasonics Symposium*, pp.1881-1884.

Oruklu, E. and Saniie, J. (2009), Hardware-efficient realization of a real-time ultrasonic target detection system using IIR filters, *IEEE Transactions on Ultrasonics, Ferroelectrics and Frequency Control*, vol. 56, no. 6, pp. 1262-1269.

Oruklu, E., Lu, Y. and Saniie, J. (2009). Hilbert transform pitfalls and solutions for ultrasonic NDE applications, *IEEE Ultrasonics Symposium*, pp. 2004-2007.

PVEH, Piezoelectric Vibration Energy Harvester, (2011). Data Sheet, *Microstrain, Inc.*

Rodriguez-Andina, J. J., Moure, M. J. and Valdes, M. D. (2007). Features, design, tools, and application domains of FPGAs, *IEEE Transactions on Industrial Electronics*, vol. 54, pp. 1810–1823.

Roundy, S., Otis, B., Chee, Y. H., Rabaey, J. and Wright, P. (2003). A 1.9 GHz RF transmit beacon using environmentally scavenged energy, *IEEE Int. Symposium on Low Power Elec. and Devices.*

Sumitro, S., and Hodge, M. H. (2006). Global smart bridge monitoring system, *Proc., 3rd International Conference on Bridge Maintenance, Safety and Management - Bridge Maintenance, Safety, and Management*, Taylor and Francis/Balkema Publishers, the Netherlands, pp. 673-674.

Texas Instruments, AFE5801, Datasheet, 8-Channel Variable-Gain Amplifier (VGA) With Octal High-Speed ADC, available at: *http://focus.ti.com/docs/prod/folders/print/afe5801.html.*

Texas Instruments, CC2420, (2010). 2.4 GHz IEEE 802.15.4 / ZigBee-ready RF Transceiver, available at: *http://www.ti.com/lit/ds/swrs041b/swrs041b.pdf.*

Weber, J., Oruklu, E. and Saniie, J. (2008). Configurable hardware design for frequency-diverse target detection, *IEEE 51st Midwest Symposium on Circuits and Systems, MWSCAS 2008*, pp.890-893.

Wilcox, P. D., Lee, C. K., Scholey, J. J., Friswell, M. I., Wisnom, M. R. and Drinkwater, B. W. (2006). Quantitative structural health monitoring using acoustic emission, *Smart Structures and Integrated Systems, Proc. of SPIE*, vol. 6173.

Xu, N., Rangwala, S., Chintalapudi, K.K., Ganesan, D., Broad, A., Govindan, R. and D. Estrin, (2004), A wireless sensor network for structural monitoring, *Proceedings of the 2nd international conference on Embedded networked sensor systems.*

Yoon, S., Oruklu, E. and Saniie, J. (2007). Performance evaluation of Neural Network based ultrasonic flaw detection, *IEEE Ultrasonics Symposium*, pp. 1579-1582.

Ziola, S. M. and Gorman, M. R. (1991). Source location in thin plates using cross-correlation, *J. Acoust. Soc. Am.*, vol. 90, no.5, pp. 2551-2556.

Index